Advanced Research in Starch

Nirmal Mazumder • Md. Hafizur Rahman
Editors

Advanced Research in Starch

Springer

Editors
Nirmal Mazumder
Department of Biophysics, Manipal School of Life Sciences
Manipal Academy of Higher Education
Manipal, Karnataka, India

Md. Hafizur Rahman
Department of Quality Control and Safety Management, Faculty of Food Sciences and Safety
Khulna Agricultural University
Khulna, Bangladesh

ISBN 978-981-99-9526-4 ISBN 978-981-99-9527-1 (eBook)
https://doi.org/10.1007/978-981-99-9527-1

This Springer imprint is published by the registered company Springer Nature Singapore Pte Ltd.
The registered company address is: 152 Beach Road, #21-01/04 Gateway East, Singapore 189721, Singapore

Preface

In the dynamic landscape of modern scientific research, our understanding of fundamental elements often unfolds in unexpected and enlightening ways. The exploration of starch, a ubiquitous and complex carbohydrate, has been no exception. *Advanced Research in Starch* is a testament to the relentless pursuit of knowledge in this field, capturing the essence of breakthroughs and innovations that propel starch science into new frontiers. This book brings together a collection of chapters penned by esteemed researchers and scholars who have dedicated themselves to advancing our comprehension of starch, from its molecular intricacies to its wide-ranging applications. The multidisciplinary approach adopted in these chapters reflects the diverse facets of starch research, encompassing chemistry, biology, physics, engineering, and applied sciences. As we delve into the pages of this book, readers will embark on a journey through the latest discoveries, methodologies, and technologies that define the forefront of starch-related studies. From the elucidation of starch biosynthesis pathways to the exploration of novel applications in various industries, each chapter contributes a valuable piece to the intricate puzzle of starch science.

Starch, a crucial carbohydrate for plants and animals, is a primary energy source. Its diverse functional properties make it valuable in the food industry. However, native starches may not always meet specific functional requirements for certain foods. Thus, starch is modified to enhance its functional attributes for application in different food formulations. The first chapter by Navaf and Sunooj titled 'Physical Modifications of Starch' discusses the different physical modifications of starch and their techniques. The second chapter authored by Ali et al. elucidates the preparation, applications, and safety assessment of cross-linked, esterified, etherified, oxidized/bleached, and acid-thinned starches along with the impact of these modifications on the structural, functional, and digestibility characteristics of starches. After the physical and chemical modification of starches, the third chapter authored by Saeid et al. dives into the topic of enzymatic modification of starches. The main focus here is the starch modifying enzymes, which are discussed in detail, including their properties, mechanisms of action, and industrial applications. Chapter 4 by Hadi and Rahman discusses the modification of starches by solvent methods. The chapter also elucidates the application of supercritical carbon dioxide ($ScCO_2$) and ionic liquids (ILs) in the modification of starches, with a focus on sustainable starch modifications. The next few chapters discuss the characterization of starch through various methods. Chromatographic techniques such as gel permeation chromatography, high-performance size exclusion chromatography coupled with different detectors (multi-angle laser-light scattering, refractive index, intrinsic viscosity), and high-performance anion-exchange chromatography used for the characterization of starch are discussed in Chap. 5 by Luis et al. The chapter also discusses the existing gaps in the study of starch structures using chromatographic methods. The surface morphology of starch granules are observed with greater resolution through electron microscopes such as scanning electron microscopes than optical microscopes. Chapter 6 authored by Anusha and Mazumder elucidates the characterization of starch granules derived from various sources. The chapter also discusses the application of electron microscopes to study the effect of enzymatic degradation patterns on starch. Vibrational spectroscopy plays a very vital role in the characterization of materials. It is one of the first pillars of spectroscopic tools that enable researchers to investigate the primary configuration of the object under investigation. Rajib Biswas in Chap. 7 outlines the role of vibrational spectroscopy in the analysis and characterization of starch. In Chap. 8, Deka et al. elucidate the various tools used for microscopic assessment, thermal characterization, rheology measurement, and the digestibility analysis of starch. This chapter is most useful for those with basic knowledge on the principles, working, and applications of analytical tools. After studying the characterization of starches, the book delves into the applications of modified starches in various fields such as biomedical, industrial, and food technologies. Chapter 9 by Mateti et al. elucidates starch modification methods for its use in the food industry and as a biomaterial for biomedical applications such as wound healing and its prospects. Chapter 10 authored by Lobo et al. explores the Andean grains used in traditional cooking. The writers elucidate the characteristics, modifications performed for industrial and nutraceutical applications. Eventually, the book takes us through genetic modifications of starch and the advantages of resistant starch. The last Chap. 11 by Sarmah et al. sheds light onto resistant starch and its various applications in the biomedical and nutraceutical fields. The chapter also discusses the exciting possibilities of genetically modified resistant starches, and their great potential to improve human health.

As we delve into the pages of this book, readers will embark on a journey through the latest discoveries, methodologies, and technologies that define the forefront of starch-related studies. From the elucidation of starch biosynthesis pathways to the exploration of novel applications in various industries, each chapter contributes a valuable piece to the intricate puzzle of starch science. We express our deepest appreciation to the contributors, whose intellectual prowess has shaped this compilation into a comprehensive reference for scientists, academics, and industry professionals. We are grateful to the reviewers for devoting their time to providing insightful recommendations and comments on the chapters. We also acknowledge our colleagues who helped us with this volume's suggestions. It is our hope that *Advanced Research in Starch* will not only serve as a reservoir of knowledge but also inspire future research endeavours, fostering a community committed to unravelling the complexities of starch in all its forms. Machi Sugimoto and Swati Sharma

(Springer Nature Publishing) deserve special thanks for their contributions to the book. May this book be a source of inspiration, a catalyst for scientific inquiry, and a cornerstone for the ongoing advancement of starch research. More importantly, we wish our readers a pleasant and productive browsing experience.

Manipal, Karnataka, India Nirmal Mazumder
Khulna, Bangladesh Md. Hafizur Rahman

Contents

Editors and Contributors

About the Editors

Nirmal Mazumder is an Associate Professor at the Department of Biophysics, Manipal School of Life Sciences, Manipal Academy of Higher Education (MAHE), Manipal, India. He obtained his Ph.D. in 2013 from National Yang Ming Chiao Tung University, Taipei, Taiwan. From 2013 to 2016, he worked as a Postdoctoral Fellow at the University of Virginia, USA, and the Italian Institute of Technology, Genoa, Italy. He has been developing nonlinear optical microscopes, including two photon fluorescence, second harmonic generation, and coherent anti-Stokes Raman scattering for biomedical applications. His research interests include development of microscopy–spectroscopy techniques, biomedical devices, biophotonics, smart materials, and microfluidics and its biomedical applications. He has published more than 80 research articles in peer-reviewed international journals, 7 books, 25 book chapters, and several poster/oral/invited presentations in national and international conferences. He is a member of several national and international scientific societies and organizations including the Optical Society (OPTICA, Senior Member), SPIE—The International Society for Optical Engineering (Senior Member), Society of Biological Chemists (I) (Life Member), Environmental Mutagen Society of India (Life Member), Asian Polymer Association (Life Member), Indian Laser Association (ILA, Life Member), and Research Society for the Study of Diabetes in India (RSSDI, Life Member).

Md. Hafizur Rahman is currently working as a Lecturer at the Department of Quality Control and Safety Management, Faculty of Food Sciences and Safety, Khulna Agricultural University (KAU), Bangladesh. He acquired his Ph.D. from the Laboratory of Food Chemistry and Nutrition Science, Institute of Food Science and Technology, Chinese Academy of Agricultural Sciences (CAAS), China, in August 2020. Then, he worked as a Postdoctoral Fellow in the Department of Agro-Product Processing Technology, Jashore University of Science and Technology, Bangladesh, from June 2021 to June 2022. He has extensively worked for a long time on the comparative study of different starch properties after physical modification and its effects on the different gluten-free model dough. He has also investigated the effects of different sugars and different level of salt concentrations on the rheological, fermentation, water mobility, and microstructural properties of starch-based gluten-free model dough prepared from corn, potato, and sweet potato starches and he developed some novel starch-based gluten-free model dough. He has published several research articles in different peer-reviewed international journals.

Contributors

Farzana Akter Department of Nutrition and Food Engineering, Daffodil International University, DSC, Birulia, Savar, Dhaka, Bangladesh

Md. Aslam Ali Department of Agro-Processing, Faculty of Agriculture, Bangabandhu Sheikh Mujibur Rahman Agricultural University, Gazipur, Bangladesh

Tahira Mohsin Ali Department of Food Science and Technology, University of Karachi, Karachi, Pakistan

Anusha Department of Biophysics, Manipal School of Life Sciences, Manipal Academy of Higher Education, Manipal, Karnataka, India

Vishwa Jyoti Baruah School of Integrative Plant Science, Cornell University Ithaca, NY, USA

Luis A. Bello-Perez Instituto Politécnico Nacional, CEPROBI, Yautepec, Morelos, Mexico

Rajib Biswas Department of Physics, Tezpur University, Tezpur, Assam, India

Natasha Abbas Butt Department of Biomedical Engineering, Faculty of Engineering, Science, Technology and Management, Ziauddin University, Karachi, Pakistan

Sonia R. Calliope Centro Interdisciplinario de Investigaciones en Tecnologías y Desarrollo Social para el NOA (CIITED, UNJu-CONICET), Facultad de Ingeniería-Universidad Nacional de Jujuy, San Salvador de Jujuy, Argentina

Smarak Islam Chaudhury Materials Research Centre, Indian Institute of Science, Bangalore, Karnataka, India

Baishali Das Department of Biophysics, Manipal School of Life Sciences, Manipal Academy of Higher Education, Manipal, Karnataka, India

Gunjana Deka Department of Food Technology, Mizoram University, Aizawl, Mizoram, India

Aditi Duarah Department of Food Technology, Mizoram University, Aizawl, Mizoram, India

Himjyoti Dutta Department of Food Technology, Mizoram University, Aizawl, Mizoram, India

Gitanjali Gautam Department of Food Technology, Mizoram University, Aizawl, Mizoram, India

Nabilah Abdul Hadi Faculty of Fisheries and Food Science, University Malaysia Terengganu, Kuala Nerus, Terengganu, Malaysia

Salman Haider Department of Food Science and Technology, University of Karachi, Karachi, Pakistan

Malti Kumari Materials Research Centre, Indian Institute of Science, Bangalore, Karnataka, India

K. Likhith Department of Biomedical Engineering, Manipal Institute of Technology, Manipal Academy of Higher Education, Udupi, Karnataka, India

Manuel O. Lobo Centro Interdisciplinario de Investigaciones en Tecnologías y Desarrollo Social para el NOA (CIITED, UNJu-CONICET), Facultad de Ingeniería-Universidad Nacional de Jujuy, San Salvador de Jujuy, Argentina

Tarun Mateti Centre for BioSystems Science and Engineering, Indian Institute of Science, Bangalore, Karnataka, India

Materials Research Centre, Indian Institute of Science, Bangalore, Karnataka, India

Nirmal Mazumder Department of Biophysics, Manipal School of Life Sciences, Manipal Academy of Higher Education, Manipal, Karnataka, India

Rita M. Miranda Centro Interdisciplinario de Investigaciones en Tecnologías y Desarrollo Social para el NOA (CIITED, UNJu-CONICET), Facultad de Ingeniería-Universidad Nacional de Jujuy, San Salvador de Jujuy, Argentina

Josue Moreno-Zaragoza Instituto Politécnico Nacional, CEPROBI, Yautepec, Morelos, Mexico

Muhammed Navaf Department of Food Science and Technology, Pondicherry University, Puducherry, India

Md. Hafizur Rahman Department of Quality Control and Safety Management, Faculty of Food Sciences and Safety, Khulna Agricultural University, Khulna, Bangladesh

Arnab Roy Department of Food Technology, Mizoram University, Aizawl, Mizoram, India

Abu Saeid Department of Nutrition and Food Engineering, Daffodil International University, DSC, Birulia, Savar, Dhaka, Bangladesh

Norma C. Sammán Centro Interdisciplinario de Investigaciones en Tecnologías y Desarrollo Social para el NOA (CIITED, UNJu-CONICET), Facultad de Ingeniería-Universidad Nacional de Jujuy, San Salvador de Jujuy, Argentina

Bhaswati Sarmah Section of Plant Breeding and Genetics, Cornell University, Ithaca, NY, USA

Savita Materials Research Centre, Indian Institute of Science, Bangalore, Karnataka, India

Cristina N. Segundo Centro Interdisciplinario de Investigaciones en Tecnologías y Desarrollo Social para el NOA (CIITED, UNJu-CONICET), Facultad de Ingeniería-Universidad Nacional de Jujuy, San Salvador de Jujuy, Argentina

Marium Shaikh Department of Food Science and Technology, University of Karachi, Karachi, Pakistan

Kappat Valiyapeediyekkal Sunooj Department of Food Science and Technology, Pondicherry University, Puducherry, India

Goutam Thakur Department of Biomedical Engineering, Manipal Institute of Technology, Manipal Academy of Higher Education, Udupi, Karnataka, India

Nusrat Zehra Department of Biomedical Engineering, Faculty of Engineering, Science, Technology and Management, Ziauddin University, Karachi, Pakistan

Physical Modifications of Starch

1

Muhammed Navaf
and Kappat Valiyapeediyekkal Sunooj

Abstract

Starch is a carbohydrate that serves as a primary source of energy for plants and animals. Various functional properties of starch make it an ideal ingredient for the food industry. It is generally considered that native starches do not always meet the necessary functionalities required for specific foods. Starch is typically modified to enhance its functional attributes for application in different food formulations. Different techniques are employed to synthesize starch with various functional attributes and applications. Food processors prefer to use physically modified starches rather than chemically modified starches, so it is becoming increasingly important to use functional starches that have not been chemically modified. The physical modification of starches via physical treatment is investigated as a method to modify starch properties without chemically modifying them. Since they are free of chemicals or biological agents, physically modified starches are a less expensive option than chemically and enzymatically modified starches.

Keywords

Physical modifications · Heat moisture treatment · Annealing · Irradiation · Ultrasonication · High pressure treatment · Cold plasma treatment

---

M. Navaf · K. V. Sunooj (✉)
Department of Food Science and Technology, Pondicherry University, Puducherry, India

N. Mazumder, M. H. Rahman (eds.), *Advanced Research in Starch*,
https://doi.org/10.1007/978-981-99-9527-1_1

1.1 Introduction

Starch is a natural polymer originating from various plant sources, i.e., fruits, vegetables, cereals, etc., and it has great utility in different industries, e.g., paper, textile, pharmaceuticals, etc. Starch granules mainly consist of two polymeric chains, amylose and amylopectins. Starch in its native form cannot always meet the demand for the wide utility of starch in industrial sectors. The properties like low gelatinization temperature, higher swelling, etc. may affect the industrial demand for starches in their native form. It is important to consider the composition and structure of starches when determining their functional properties. For example, starch with high amylopectin facilitates film formation, while starches with high amylose form soft gels. Therefore, starches are modified in such a way that they provide specific functionalities.

Figure 1.1 illustrates four general methods of modifying starches: physical, chemical, enzymatic, and genetic engineering. Physical modification is widely accepted and has been practiced for decades. Physical modifications of starch have the advantages of being non-toxic, cost-effective, and green methods. It offers the advantage of the absence of any reagents in the process and is free from byproducts released into the environment.

There are different types of physical methods adopted to modify the starches. Hydrothermal treatment, i.e., heat moisture treatment, annealing, etc., is one of the early techniques in starch modification. During hydrothermal modification, water and heat break down starch intermolecular bonds (Adewale et al. 2022). Recent past, new nonthermal methods have been adopted to modify starches, e.g., pulsed electric field treatment, high-pressure treatment, plasma treatment, etc. (Wu et al. 2022). Physical modifications affect the structural and packing arrangement of starch granules, thus leading to significant changes in its various functionalities. In

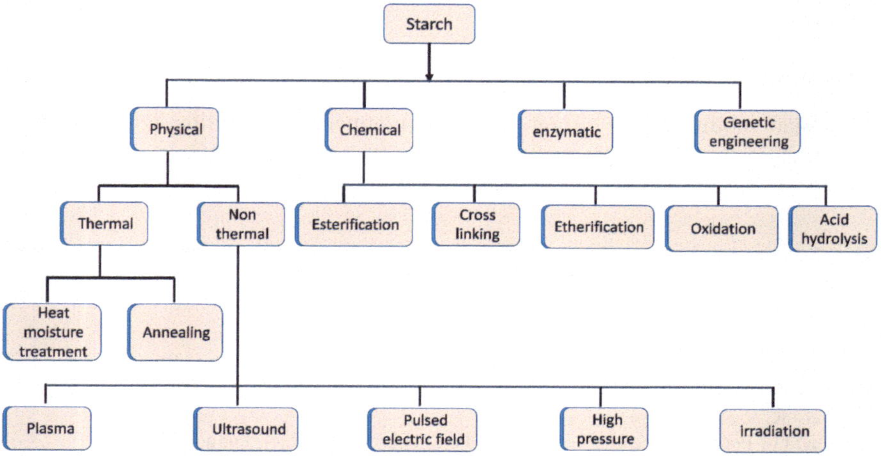

Fig. 1.1 An overview of starch modifications

addition, the type of starch and modification has a potential contribution to desired changes of physically modified starches.

1.2 Heat Moisture Treatment (HMT)

During HMT, starch with 10–30% moisture is exposed to a temperature of 90–120 °C for up to 16 h, viz., above the glass transition temperature (Tg) and below gelatinization. HMT alters starch properties and structure, varying with the starch source and composition. The mechanism of HMT includes facilitating the interactions of starch chains. In the beginning, HMT damages crystalline structures, dissociates double helix structures, and then reassembles them. HMT also alters the ordered arrangement of the starch crystallites (Hoover and Manuel 1997; Pratiwi et al. 2018). The changes in physicochemical and functional attributes of HMT-treated starches are discussed below.

1.2.1 Impact of HMT on Gelatinization Characteristics

Gelatinization and pasting are two terms used to explain the changes in starch when it is cooked in water. Gelatinization is the process of disordering the molecular structure of starch, resulting in permanent alteration in their functionalities. In its native form, starch can absorb roughly 30% (w/w) of moisture after being suspended in a high amount of water at certain conditions. The amorphous region of starch is easily accessible to water molecules. Granules of starch become amorphous and lose their crystalline structure when heated to a specific temperature with excess moisture content. This temperature-dependent transformation of starch structure is termed as gelatinization, and the temperature is termed gelatinization temperature (Mason 2009). Gelatinization results in changes in properties like swelling power (SP), starch solubilization, crystalline melting, etc. Differential scanning calorimetry is used to study the gelatinization characteristics of starch.

The impact of HMT on starch gelatinization characteristics varies with the starch source, treatment conditions, moisture content (mc) of starch, etc. HMT generally shifts the gelatinization temperature to a higher value and widens the temperature range. When mc and treatment temperature are increased, the gelatinization temperatures increase as well. An increase in gelatinization temperature has been noted in starches of wheat (Xie et al. 2018a), rice (Arns et al. 2015), pearl millet (Sharma et al. 2015a), kithul (Sudheesh et al. 2020a), grass pea (Piecyk et al. 2018), Indian water chestnut (Yadav et al. 2013a), etc. The changes in gelatinization characteristics of HMT starch are summarized in Table 1.1. A higher gelatinization attribute of HMT starch could be ascribed to its structural changes during HMT, and it induces interactions of starch chains within the granules. Additionally, it promotes starch diffusion and intermolecular links between amylose and amylopectin through inter- and intramolecular linkage (Schafranski et al. 2021). As the gelatinization process proceeds, crystallites and double helices melt, and starch granules become hydrated

Table 1.1 The gelatinization characteristics of starches

Starch	HMT conditions							
	Moisture (%)	Temperature	T_O (°C)	T_P (°C)	T_C (°C)	ΔT (°C)	ΔH (J/g)	References
Sorghum	20	110 °C, 8 h	66.46	70.29	76.41	9.95	10.11	Singh et al. (2011b)
	30		67.43	71.47	81.55	14.12	9.47	
	40		67.62	72.45	84.85	17.23	10.96	
Chinese yam	20	100 °C, 10 h	86.8	89.2	91.5	4.7	6.21	Yu et al. (2021)
Field bean	15	100 °C, 2 h	60.7	66.4	74.4	13.7	10.6	Piecyk and Domian (2021)
	30		65.6	77.7	83.0	17.4	10.3	
	15	120 °C, 2 h	61.1	68.2	77.8	16.7	12.0	
	30		65.2	83.8	89.5	24.3	13.7	
Corn	30	100 °C, 2 h	71.9	82.1	90.6	18.7	7.5	Chung et al. (2009a)
Pea			69.4	83.4	94.2	24.8	7.9	
Lentil			72.5	84.7	94.5	22	9.6	
Corn	30	120 °C, 2 h	67.8	82.2	88.8	21.0	6.1	
Pea			64.3	88.6	97.5	33.2	7.1	
Lentil			65.9	86.4	96.3	30.4	8.0	
Potato	20	120 °C, 4 h	63.65	69.52	76.15	12.5	6.96	Yang et al. (2022)
Water chestnut	20	110 °C, 16 h	75.53	78.85	89.01	13.48	19.18	Yadav et al. (2013b)
	25		81.09	85.48	90.25	9.16	12.65	
	30		82.98	87.35	91.87	8.89	12.84	

and swell. In HMT-treated starch, the interactions result in increased swelling temperature, which destabilizes the starch crystalline domain, triggering a higher transition temperature (Zavareze and Dias 2011). On the contrary, Zhang et al. (Zhang et al. 2020) noticed a reduction in the gelatinization temperature of proso millet starch after HMT. Upon heating, the glycosidic linkages are broken resulting in molecular degradation, and the unwinding or breakdown of the double helix leads to disruption of starch chains.

The enthalpy of gelatinization (ΔH) implies the amount of double helix and crystalline order. In general, HMT decreases the ΔH value. A decrease in ΔH values by HMT has been reported in potato, brown rice starches (Chakraborty et al. 2021), peas, lentils, navy bean starches (Chung et al. 2010), etc. Contrary to these, the ΔH value increased in field bean starches treated at 15% and 30% mc at 120 °C (Piecyk and Domian 2021). The disordering of the double helix structure could account for the reduction in ΔH values after HMT. In addition, the researchers also added that the partial gelatinization of starch during HMT also ends in a decreased ΔH value, indicating reduced stability (Zavareze and Dias 2011).

1.2.2 Effect of HMT on Pasting and Rheological Attributes

In pasting, the viscosity of the sample is measured during and after gelatinization. When the starch is continuously heated, starch granules will swell, disrupt, and exudate molecular components, i.e., leaching of amylose, which gives viscosity to the aqueous medium. The process of development of viscosity in the above process is termed the pasting. Rapid Visco Analyzer (RVA) is used to analyze the pasting behavior of starch gels.

HMT significantly affects starch's pasting profile. It was reported that irrespective of HMT conditions, HMT-treated starch had lower peak viscosity (PV), final viscosity (FV), breakdown viscosity (BV), and setback viscosity (SV). However, it had a high pasting temperature (PT) and thermal shear stability. HMT promotes intermolecular interaction, and thus high energy is needed for the disruption of HMT starches. The restricted swelling and solubility in HMT starches may decrease the pasting profile (Hoover 2010). The PT of HMT-treated starch increased with the mc and heating temperature.

The HMT of waxy and normal potato starches decreased the PV and BV. However, the normal starch has reduced SV, whereas waxy starch exhibited higher SV (Varatharajan et al. 2010). The authors also added that the increased SV of waxy starch could be due to its higher SP. The PV, trough viscosity (TV), SV, and FV of the HMT potato starches were decreased with a prolonged processing time from 2 to 12 h. at 120 °C (Zhang et al. 2020). Similarly, the pasting profile of pearl millet starch was reduced at an exposure of 110 °C for 4 h. (Sharma et al. 2015a). Brahma and Sit (2020) studied the effect of HMT at different conditions (30 and 35% moisture and 100 and 120 °C) on potato starch. They observed that the starch treated at 100 °C had a higher PT than starch treated at 120 °C. However, the PT increased with a change in mc from 30% to 35%. In addition, the PV, TV, BV, and FV were lower for HMT treated at 100 °C as compared to the potato starch sample treated at 120 °C.

HMT significantly changes the rheological attributes, i.e., storage modulus (G') and loss modulus (G") of starches. The impact of HMT on starches varies with starch and treatment conditions (Sharma et al. 2015b). The molecular rearrangement caused by HMT hinders the swelling and amylose leaching, which results in increased G' and G" values. Starches, such as sohphlang starch (Marboh and Mahanta 2021), potato starch (Yang et al. 2022), etc., exhibited a higher G' and G" values. Contrary to the above, HMT of black rice starch decreased the G' and G" values (Dhull et al. 2021), as a result of molecular disintegration and degradation of starch structure (Ji et al. 2022) during HMT.

1.2.3 Impact of HMT on Structural Properties

HMT potentially impacts the starch structure, and the extent of structural alteration varies with the starch type and HMT parameters. It alters the polymorphic structure or the crystallinity of starch. A transformation of the B to a combination of A and B-type patterns was observed in HMT normal and waxy potato starches (Varatharajan et al. 2010). Studies revealed that crystalline structure transformation could result from water molecules' dehydration in the central cavity of B-type starches and the motion of double-helix molecules. Similarly, a change from A-type to a blend of A and V-type crystalline patterns was observed in maize starch (Wang et al. 2021). This kind of transformation to V type was prominent at HMT at high moisture levels. The HMT of lycoris starch at 120 °C for 6 h changed the crystalline pattern from type A to type B (Zhang et al. 2018).

HMT enhanced the relative crystallinity (RC) of Chinese yam starch treated at 100 °C for 10 h (Yu et al. 2021), amaranth starch treated at 85 °C (Sindhu et al. 2019), and sweet potato starch treated at 100 °C for 6 h. The improved crystallinity of HMT starches implies that the higher chain mobility during HMT promoted further interaction between starch chains.

The HMT of starches, like corn starch (Chung et al. 2009b), mungbean starch (Zhao et al. 2020), amaranth starch, etc., decreased the RC (Sindhu et al. 2019). The disruption of the amylopectin chain and the conversion of the semicrystalline fraction to an amorphous fraction during HMT treatment resulted in decreased RC. In addition, the partial gelatinization of starch during HMT may also contribute to a decreased RC (Brahma and Sit 2020).

1.2.4 Impact of HMT on Starch Digestibility

Starch is categorized into three categories: rapidly digestible starch (RDS), slowly digestible starch (SDS), and resistant starch (RS), depending on how much glucose is produced during enzymatic hydrolysis. After consuming RDS, blood glucose levels rise rapidly. However, RS is the fraction resistant to digestion. The SDS is the portion of starch that converts to glucose after 20–120 min of hydrolysis.

The HMT process is widely reported to increase SDS and RS content in starch-based products, improving their nutritional value. The digestibility of HMT starch varies with the source, crystallinity, granule size, amylose, and amylopectin ratio, and the HMT conditions, which are connected to the alteration in crystallinity and morphology of treated starches (Zavareze and Dias 2011). In general, HMT improves the crystalline perfection of starches and thus decreases their digestibility. The HMT of rice starch reduced the digestibility (Wang et al. 2018), whereas SDS content increased in waxy rice starch (Zeng et al. 2015). Similarly, the HMT of maize starch increased the SDS and RS content (Park et al. 2018). At the same time, waxy maize starch increased the RDS and SDS content (Park et al. 2018).

1.2.5 Impact of HMT on Granule Morphology

The surface attributes of starch have a potential role in its various applications. It is mainly studied using a scanning electron microscope, polarized microscope, confocal microscope, etc. In HMT starches, the extent of morphological changes may vary, and it depends on treatment conditions like moisture, time and temperature, and source of starch. A higher mc promotes their expansion and provides morphological changes during heating (Schafranski et al. 2021). The HMT of starches like potato, sweet potato, yam, etc., at lower mc and temperature, did not cause any visible changes in granule morphology (Hoover 2010).

Piecyk and Domian (Piecyk and Domian 2021) reported that the HMT of field pea starch with 15% mc at 120 °C did not cause any noticeable effect on the granule morphology. However, indents in the center of particular granules are evidenced in 30% mc. Similarly, the HMT of rice starch with 10% mc. At 20 and 30% mc, the surface became rough and showed evidence of agglomeration (Wang et al. 2018). Similarly, the HMT-treated potato starch also exhibited agglomeration (Brahma and Sit 2020). The morphology of pearl millet starch was evidenced by dents or holes (Sharma et al. 2015b). A decrease in particle size was noticed in HMT-treated sorghum starch treated at 40% mc (Singh et al. 2011b) (Fig. 1.2).

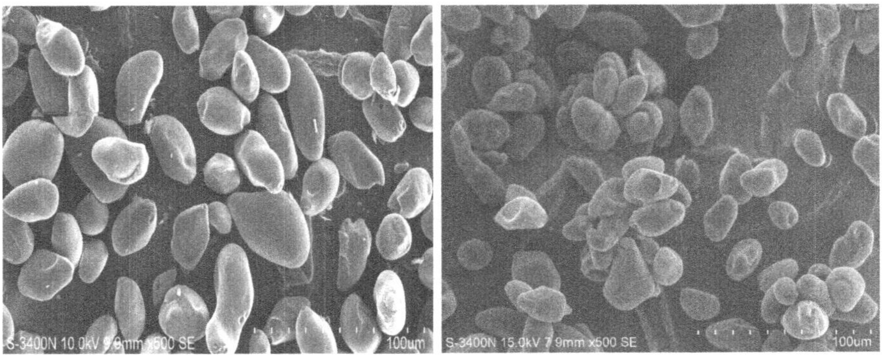

Fig. 1.2 Micrograph of native and heat moisture-treated talipot starches

1.2.6 Impact of HMT on Starch Functional Attributes

The HMT-induced alteration in the structural characteristics of starches leads to modification in their functional properties. The improved crystalline perfection by HMT may diminish the SP of starches. A decrease in SP after HMT was noticed in many starches, such as maize (Liu et al. 2016a; Xie et al. 2018b), pearl millet (Sharma et al. 2015a), grass pea (Piecyk et al. 2018), rice bean (Thakur et al. 2021), normal and waxy potato (Varatharajan et al. 2010), etc. It was observed that the impact of HMT on SP depends on its mc, and it could be due to the rearrangement of starch chains, thus limiting the granules' hydration capacity.

The solubility of starch is primarily due to the leaching of amylose molecules during heating and swelling. Solubility indicates the change in molecular structure from a stable to a destabilized structure during gelatinization. The restricted swelling and increased rigidity of HMT starches reduce the amount of leached amylose and its solubility. The solubility of potato starch was decreased when it was treated at different mc and temperatures (Brahma and Sit 2020). Similarly, a decrease in solubility was reported in starches, such as oat (Kaur and Singh 2019), sorghum (Singh et al. 2011b), proso millet starch (Zhu et al. 2020), etc.

The HMT significantly alters the gel textural properties of starch. The HMT of talipot starch (Navaf et al. 2021), potato starch (Yang et al. 2022), water chestnut starch (Yadav et al. 2013b), sorghum starch (Singh et al. 2011b), potato starch (Brahma and Sit 2020), etc. significantly increased the gel hardness. The higher chain interaction, restricted swelling, and the development of more junction zones in starch suspension, etc. are responsible for HMT starches' improved gel hardness. It also depends on mc and temperature. The HMT with high moisture and temperature shows higher gel hardness (Brahma and Sit 2020). On the contrary, hardness was reduced in HMT amaranth starch, possibly due to its decreased solubility and increased SP (Sindhu et al. 2019).

The ability of starch to transmit light is an important function that has practical applications in various industries. The gel's light transmittance (LT) measures its paste clarity, mainly influenced by the remnants of swollen granules, leached amylose, etc. (Jyothi et al. 2010). HMT significantly decreases starch's LT and paste clarity. Researchers observed a reduction in LT of mango kernel starch (Bharti et al. 2018), cassava, potato, arrowroot (Jyothi et al. 2010), oat (Kaur and Singh 2019), water chestnut (Yadav et al. 2013b) starch, etc., after HMT treatment. The decrease in LT of HMT starch could be due to the enhanced flexibility of its chains in the amorphous area, amylose leaching, and granule remnants in the starch gel.

The percentage syneresis value was increased in HMT potato (Brahma and Sit 2020) and amaranth (Sindhu et al. 2019) starches. However, it was decreased in water chestnut starch (Yadav et al. 2013b). The higher reassociation between starch chains and exudation of water molecules increases the syneresis in HMT-treated starch. HMT of amaranth starch increases its syneresis value and depends on the treatment intensity (Sindhu et al. 2019).

1.3 Annealing

Annealing (ANN) involves the holding of starch in excess (>60% [w/w]) or at intermediate (40% [w/w]) mc at a temperature above the Tg and below the To of gelatinization for a defined period of time (Jayakody and Hoover 2008). The annealing of starch alters the physicochemical properties, such as crystallinity, by enhancing interactions between its chains. Amylopectin and starch molecules are reorganized by ANN so that their structures acquire a more organized state. The pictorial representation of annealing on starch granules is given in Fig. 1.3. Depending on the origin and kind of starch, the degree of starch chain interaction and the realignment of double helices may differ. The annealing of starch brings the following changes: (1) reorganization of the granular structure, (2) improved granular stability, (3) enhanced crystallinity, (4) enhanced interaction of starch chains, and (5) development of double helices. The effect of annealing on the physicochemical and functional attributes of starch is discussed below.

1.3.1 Impact of Annealing on Gelatinization Properties

Annealing potentially impacts the gelatinization characteristics of starch. Annealing has been extensively studied for its impact on starch gelatinization properties. Annealing increases the gelatinization temperature. However, it decreases the range of gelatinization temperature (ΔT). Annealing treatment will also alter the ΔH, and it varies among the different sources of starches. The gelatinization characteristics of some annealed starches are summarized in Table. 1.2. The gelatinization temperature increases with an increase in annealing time and is more visible in the onset gelatinization temperature, least at the conclusion temperature. The rise

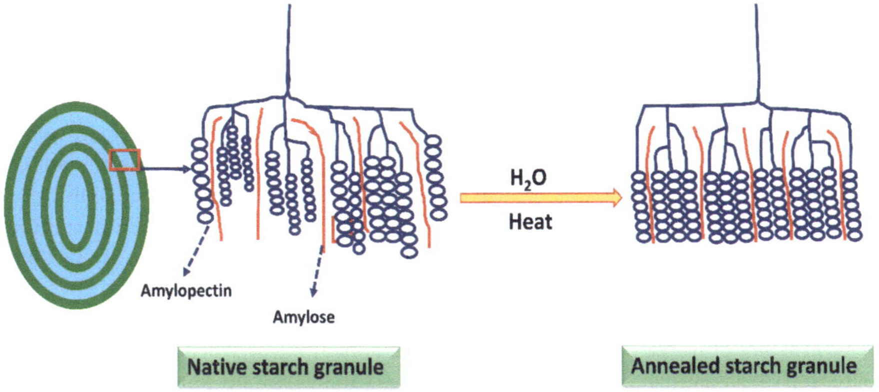

Fig. 1.3 The pictorial representation of annealing on starch granule

Table 1.2 The gelatinization characteristics of annealed starches

Starch	Annealing conditions	T_O (°C)	T_P (°C)	T_C (°C)	ΔT (°C)	ΔH (J/g)	References
Corn	55 °C, 24 h	70.9	75.4	82.4	11.5	8.2	Ariyantoro et al. (2018b)
Pea	70% moisture, 24 h	67.5	71.5	84.2	14.7	14	Chung et al. (2009a)
Lentil		71.8	75.0	86.8	15	13	
Kithul	1:4 (starch/water), 60 °C, 24 h	77.75	82.13	90.11	12.36	5.65	Sudheesh et al. (2020b)
Chestnut	1:2 (starch/water) 65 °C, 24 h	81.58	84.96	88.12	6.54	12.59	Yadav et al. (2013a)
Sorghum	1:4 (starch/water), 50 °C, 24 h	67.83	70.42	74.59	6.76	12.71	Singh et al. (2011a)
Sweet potato	90% water, 55 °C, 72 h	72.1	76.4	81.0	8.9	8.9	Song et al. (2014)

in gelatinization temperature of annealed starch could be due to the perfection of pre-existing crystallites (Ariyantoro et al. 2018a), and it can also be associated with decreased swelling of annealed starches (Zavareze and Dias 2011). The development of more homogeneity and cooperative melting of crystallites results in narrowing the gelatinization range of annealed starches (Jayakody and Hoover 2008). A higher gelatinization temperature was reported in normal and waxy corn starches after ANN (Rocha et al. 2012). However, the ΔH was increased only in waxy starch. Similarly, the ΔH value was increased in maize starch annealed with 80% moisture at 50 °C for 24 h. (Chen et al. 2021). Contrary, the annealing (1:4 starch/water) of maize and potato starch at 50 °C for 24 h. significantly decreased the ΔH values (Chi et al. 2019).

1.3.2 Impact of Annealing on Pasting and Rheological Properties

Several factors, like amylose leaching, chain length distribution, amylopectin, RC, etc., influence annealed starches' pasting properties. The annealing-induced variations in the pasting profile of starch are complex and vary among starch sources (Zavareze and Dias 2011). According to Yu et al. (Yu et al. 2021), annealing of Chinese yam starch at 50 °C for 24 h. (70% moisture) significantly decreased the PV, TV, BV, and SV. Similarly, the PV, BV, and FV values decreased in talipot starch (Navaf et al. 2021). Contrary to these observations, a higher pasting profile was observed in annealed kithul starch (Sudheesh et al. 2020a) and wheat starch (30 and 40 °C) (Wang et al. 2017a). At the same time, annealing of wheat starch at 50 °C decreased the pasting viscosities and increased the PT. The annealing of waxy maize starch for 24 h. at 45 °C did not alter its pasting attributes. However, the extended period decreased the PV, TV, and FV (Wang et al. 2014). The annealing of sweet potato starch at 45 °C for 24 h. did not show any significant changes in PT, whereas it decreased the pasting viscosities (Trung et al. 2017).

The higher PT of annealed starches can be ascribed to their higher crystalline perfection and decreased swelling (Sudheesh et al. 2020a). The reorganization within modified starch is believed to reduce the pasting profile. Consequently, SP was limited, and a minimal amount of amylose could leach into the water to increase viscosity (Yadav et al. 2013a). Contrary to this, the lower degree of deformation and increased rigidity of the granules could cause the increased pasting profile of annealed starch (Sudheesh et al. 2020a).

The dynamic rheological characteristics, i.e., G' and G" of annealed starches, were found to be increased. Since annealing results in a more ordered structure, the annealed starches show improved rheological characteristics. Yan et al. (Yan et al. 2022) reported the annealing of potato and pea starch with normal and plasma-activated water considerably increased the G' and G" values. Similarly, a rise in rheological attributes by annealing is reported in starches like Andean oca (Puelles-Román et al. 2021), talipot (Navaf et al. 2021), kithul (Sudheesh et al. 2020a), etc. It implies that the annealing of starch enhances the stiffness of macromolecular chains, and it forms a strong gel.

1.3.3 Impact of Annealing on Structural Attributes

Annealing potentially impacts the structural characteristics of starch, and it varies with the type of starches and annealing conditions. It was noted that annealing has a more dominant effect on type B starches than on type A starches (Jayakody and Hoover 2008). In a few starches, the annealing changes the weak polymorphic structure to a more stable one. Genkina et al. (Genkina et al. 2004) reported that annealing of sweet potato starches for up to 8–10 h resulted in a gradual change in crystalline pattern from B to A type. Similarly, changes from B-to A-type crystalline patterns were observed in cassava starch annealed at 50 °C up to 192 h (Gomes et al. 2004). Likewise, upon annealing, a shift to A-type polymorphic structure was also observed in barley starches (Waduge et al. 2006).

On the other hand, the annealing did not impact the crystalline pattern for many starches but altered the RC. The annealing of yam starch with 70% moisture at 50 °C didn't affect its crystalline pattern. However, it increased the crystallinity (Yu et al. 2021). Similar observations were made in talipot (Navaf et al. 2021) and kithul (Sudheesh et al. 2020a) starches. Yeum et al. (Yeum et al. 2021) stated that the mc has a potential role in the crystalline-type formation of annealed starches, and higher mc favors the reordering of the crystalline area.

The increase in RC of annealed starches is due to alteration in the orientation of starch crystallites, crystalline perfection, amylopectin content, etc. (Yu et al. 2021). In ANN, starch chains are realigned both amorphous and crystalline areas, enhancing the granule's stability and increase the crystallinity (Zavareze and Dias 2011). Additionally, the association of smaller crystallites during annealing also facilitates increased crystallinity (Liu et al. 2016a).

Chung, Hoover, and Liu (Chung et al. 2009b) noticed that annealing of corn starch at 50 °C for 24 h did not affect the polymorphic structure and relative crystallinity. However, prolonged annealing reduced the RC. Similarly, 48 h of annealing reduced the RC of maize and pea starches (Zheng et al. 2023). Rice starch with different amylose content also exhibited a reduction in RC after annealing treatment (Dias et al. 2010).

1.3.4 Impact of Annealing on Starch Digestibility

Annealing is a hydrothermal treatment used to alter the digestibility of the starches, and its digestibility varies with the starch origin and annealing parameters. Annealing may enhance the starch chain interactions. It increases crystalline perfection and reduces the granules' enzyme accessibility, resulting in decreased digestibility. Contrary, an increased digestibility was also observed in annealed starches. The development of pores on granules during annealing may facilitate more accessibility of the enzymes and thus increase the RDS content in annealed starches.

The annealing of maize starch with different amylose content for 24 h and 74 h did not impact their digestibility (Wang et al. 2014). In contrast, it increased the digestibility of sweet potato starch (Song et al. 2014). Likewise, the annealed waxy rice starch exhibited enhanced digestibility (Zeng et al. 2015). Contrary to the above, annealed yam (Yu et al. 2021) and potato starches (Chi et al. 2019) exhibited a decreased digestibility.

1.3.5 Impact of Annealing on Granule Morphology

Many researchers observed that annealing did not have a profound effect on granule morphology. The starches like banana (Cahyana et al. 2017), maize, barley, sorghum wheat (Ariyantoro et al. 2018a), etc. did not cause any visible change in granule morphology. However, there are some contrasting observations. The annealed kithul exhibited large holes and fissures in the granule surface (Sudheesh et al. 2020a). Likewise, dents and cracks were formed in annealed pea and potato starches (Yan et al. 2022) and pores in high-amylose rice starches (Dias et al. 2010) (Fig. 1.4).

1.3.6 Impact of Annealing on Functional Properties

Annealing promotes the interplay between starch chains, such as it promotes the interaction between starch chains, increased crystalline perfection, etc. Annealing restricts the hydration ability of starch, thereby reducing its SP (Zavareze and Dias 2011). The ANN has decreased the SP of Bambara ground nut starch (Oyeyinka et al. 2018), rice starch (Wang et al. 2017b), kithul starch (Aaliya et al. 2022), yam starch (Yu et al. 2021), potato starches (Xu et al. 2018), etc. It was noticed that as the annealing duration increased, the SP reduction intensity

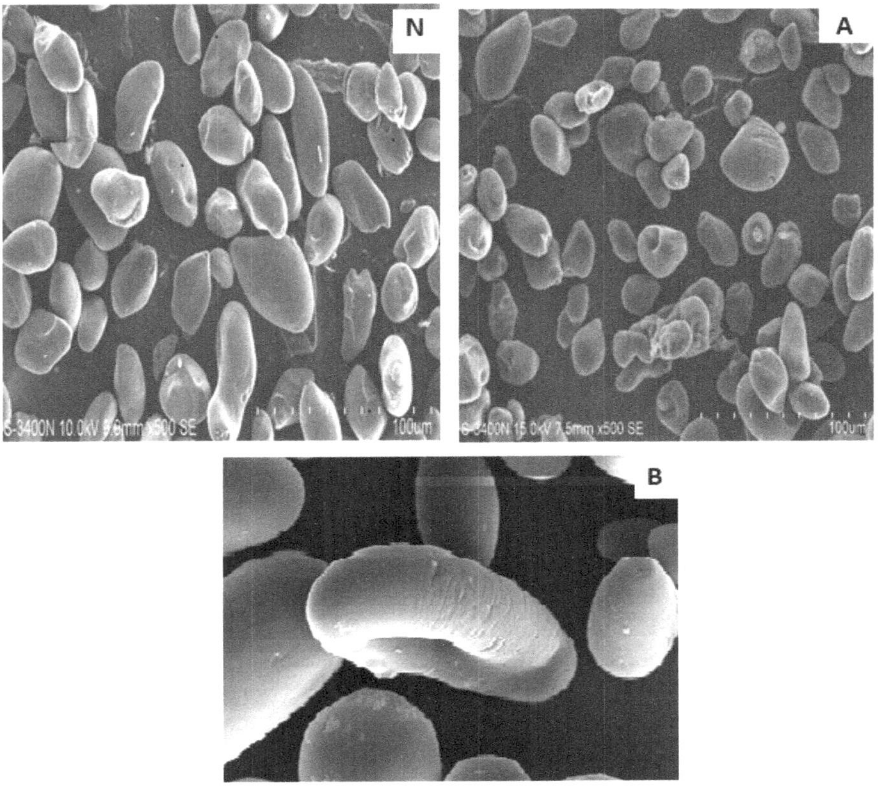

Fig. 1.4 SEM images of native (N) and annealed talipot starch (**a**), kithul starch (**b**)

increased, since it enhances the interaction between the starch chain and the crystallinity of starches.

Similarly, the ANN of starches significantly decreases the solubility. A decrease in solubility was reported in ANN kithul starch (Sudheesh et al. 2020a), water chestnut starch (Yadav et al. 2013b), sorghum starch (Ogundiran and Adeniyi Afolabi 2018), etc. The enhanced starch chain interaction by ANN improves the crystalline perfection, restraining the amylose leaching and resulting in a predominant reduction in solubility.

The impact of ANN on gel hardness may vary with the origin of starch and treatment parameters. The annealing of sorghum starch increases the gel hardness (Singh et al. 2011a), whereas it was decreased in maize starch (Liu et al. 2016a). The enhanced molecular arrangement and decreased swelling of annealed starches contributed to its higher gel hardness. ANN of starch significantly increases its LT. The ANN of corn starch (Ariyantoro et al. 2018b), banana starch (de Almeida et al. 2020), etc. pronouncedly increased the LT. The researchers stated that the increased interaction of starch chains restricts the leaching of amylose. The reduction in amylose leaching minimizes the interaction between amylose and other components of starch granules, resulting in increased LT (de Almeida et al. 2020). Nevertheless, a decrease in LT was reported in ANN kithul starch (Sudheesh et al. 2020b). It was speculated that a higher rate of retrogradation of annealed kithul starch decreases the LT and paste clarity.

The ANN treatment had a potential impact on the syneresis values of starches. The annealing of Bambara groundnut starch significantly decreased the syneresis (Nwaogazie et al. 2022). The syneresis value of maize, potato, and pea starches increased with an increase in annealing time (Zheng et al. 2023). The leached water molecules can be easily expelled from the granules of annealed starches, contributing to increased syneresis (Zheng et al. 2023).

1.4 Irradiation

Gamma (γ) rays are electromagnetic radiation. Food processing uses γ-rays that can penetrate up to several feet into the food product and kill microorganisms or insects or break bonds. γ-rays are created by decaying radioactive isotopes (cobalt-60 or cesium-137). The irradiation of starch with γ-rays results in the breakdown of H_2O into free radicals (OH* and H*) and electrons. They induce molecular changes in the starch by fragmentation or crosslinking, thus modifying it. The irradiation of starch and starch-containing foods undergoes significant changes in their physio-chemical and functional properties, which are discussed in the following sections. The pictorial representation of gamma irradiation on starch granules is given in Fig. 1.5.

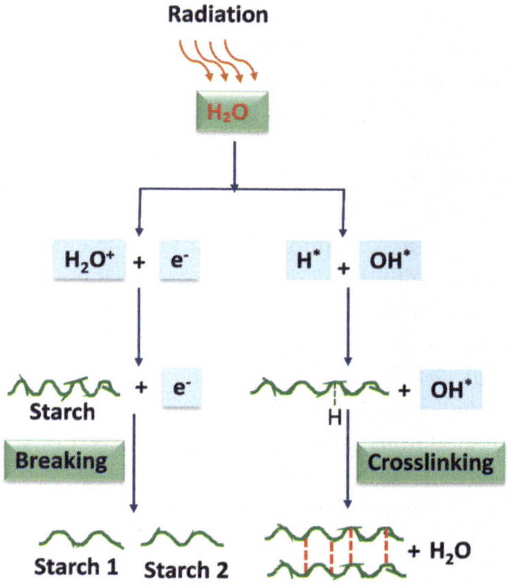

Fig. 1.5 The pictorial representation of γ-irradiation on starch granule

1.4.1 Impact of Irradiation on Gelatinization Properties

The impact of irradiation on the gelatinization attributes of starch is influenced by the starch type, its mc, and the irradiation parameters, i.e., radiation dose and dose rate (Zhu 2016). The gelatinization characteristics of some irradiated starches are summarized in Table 1.3. A 2–10 KGy irradiation of amaranth starch did not affect the gelatinization temperatures. Besides, the ΔH shows a slight fluctuation, but it is not obvious (Kong et al. 2009). The irradiation of potato starch with a low dose (0. 01, 0.05, 0.1, and 0.5) of radiation significantly increased the gelatinization temperature and ΔH values (Singh et al. 2011c). The increased gelatinization properties can be explained by the degradation of a weak crystalline structure, leaving crystallite with high stability. It can also be explained due to the increase in the metastable phases (Zhu 2016). Othman et al. (Othman et al. 2015) stated that the degradation

Table 1.3 The gelatinization characteristics of irradiated starches

Starch	Radiation dose (kGy)	T_O (°C)	T_P (°C)	T_C (°C)	ΔT (°C)	ΔH (J/g)	Reference
Talipot	20	78.79	82.24	85.52	6.73	8.40	Navaf et al. (2022a)
	50	77.47	82.43	86.63	9.16	8.13	
	80	72.53	76.97	81.57	9.04	7.48	
	100	72.04	76.72	81.45	9.41	7.41	
Amaranth	2	69.51	73.40	80.06	10.55	16.10	Kong et al. (2009)
	4	69.83	73.69	79.85	10.02	15.77	
	6	69.78	73.51	79.62	9.84	15.49	
	8	69.39	73.35	80.03	10.64	15.34	
	10	69.47	73.21	79.38	9.91	5.38	
Brown rice	5	63.05	65.89	70.11	7.06	7.11	Kumar et al. (2017)
	10	62.21	65.14	68.65	6.44	6.05	
	15	59.14	61.55	65.48	6.34	4.19	
	20	58.58	60.10	65.11	6.53	3.01	
Chickpea	0.5	63.63	69.17	76.00	12.37	8.17	Bashir and Aggarwal (2017)
	2.5	63.15	68.04	75.22	12.07	6.78	
	10	62.29	66.66	74.08	11.79	6.26	
Arrowroot	5	66.55	74.0	84.7	18.15	22.0	Barroso and del Mastro (2019)
	10	66.7	74.0	83.9	17.2	23.4	
	15	67.9	74.1	83.9	16	22.3	
Wheat	1	56.3	63.0	71.4	15.1	12.7	Kong et al. (2016)
	5	56.4	62.8	71.2	14.8	12.5	
	9	56.5	62.8	71.3	14.8	12.5	
Quinoa	1	44.3	55.8	72.3	28	9.5	
	2	44.7	55.9	72.0	27.3	9.2	
	4	44.7	55.5	72.0	27.3	9.0	
	8	45.0	55.0	71.7	26.7	9.1	
Rice (PR121)	2	60.1	64.2	71	10.9	10.2	Gul et al. (2016)
	5	60.0	63.7	71.4	11.4	10.9	
	10	58.6	63.0	70.7	12.1	11.3	

of molecules by irradiation leads to the formation of monosaccharides, thus increasing the gelatinization temperature.

The irradiation (3–50 KGy) of wheat starch did not cause any changes in gelatinization temperatures. At the same time, the ΔH value increased at a lower dose (3 and 5KGy) and did not show a significant difference at the higher doses (Atrous et al. 2015). The work concluded that the depolymerization of the starch chain into fragments did not influence the melting of starch crystallites; hence ΔH value did not change significantly. Similarly, the irradiation of kithul starch at up to 2.5 KGy significantly decreased the gelatinization properties. However, 5 and 10 KGy doses significantly increased the gelatinization properties (Sudheesh et al. 2019a). Contrary to the above observations, a decrease in gelatinization temperatures and ΔH values was reported in rice starch (1, 5, and 10 kGy) (Gul et al. 2016) and talipot starch (20, 50, 80, and 100 KGy) (Navaf et al. 2022a). The structural degradation and depolymerization of granules during irradiation lead to the loss of ordered structure, and it negatively impacts its gelatinization characteristics.

1.4.2 Impact of Irradiation on Pasting and Rheological Attributes

Irradiation of starch significantly changes their pasting profile. The impact of irradiation pasting properties varies with the irradiation dose, types of starch, mc, etc. An irradiation above 20 kGy significantly decreased the PT, and a dose of 500 kGy did not show any PT (Dikkala and Shirisha 2018). Irradiation of potato starch at 0.1 and 0.5 KGy radiation significantly increased the PT, PV, TV, FV, and SV (Ezekiel et al. 2007). Contrary to this observation, a progressive reduction in the PV of potato starch was noted at a dose of 0.1–0.5 kGy (Singh et al. 2011c). A considerable reduction in pasting profile was noted in maize starch with irradiation of 1–500 kGy (Liu et al. 2012). Similarly, irradiation of bean starch with 10 and 50 kGy decreased the PV, FV, and SV (Chung and Liu 2010). When starch is γ-irradiated, it destroys amylopectin and amylose, leading to a decline in water retention ability. As a result, PV is significantly reduced since starch granules are less prone to swelling. The PV of corn starch was decreased by 49, 83, and 95%, respectively, when irradiated at a dose of 3, 5, and 10 kGy, and a dose of 20 kGy did not show any pasting profile (Polesi et al. 2016). The limited SP and disintegration of intra-and intermolecular glycosidic linkage that damages the starch granules structure may cause the reduced pasting profile of irradiated starches.

γ-irradiation of starch causes molecular depolymerization, forming a gel with weak consistency. The rheological studies of irradiated starch show that irradiation reduces both G' and G''. The irradiation of kithul starch by 10 kGy (Sudheesh et al. 2019a), potato starch by 0.01–0.5 kGy (Singh et al. 2011c), and amaranth starch by 2–10 kGy reduced the G' and G'' values (Kong et al. 2009). Amaranth starch exhibited a crossover of G' and G'' at a higher dose of 6, 8, and 10 kGy. It indicates the liquid-like characteristics of irradiated amaranth starch. Likewise, higher-dose irradiation (20–100 kGy) of talipot starch considerably decreased its rheological attributes (Navaf et al. 2022a). Irradiation of starch causes the depolymerization and degradation of its macromolecular structure. This leads to a low gelation and retrogradation ability and results in weak gel formation (Navaf et al. 2022a), which reduces the G' and G" values.

1.4.3 Impact of Irradiation on Structural Characteristics

Depending on the irradiation dose and site of action, irradiation may increase or decrease the RC of starches. In general, RC will decrease with an increase in irradiation dose, and it is due to the partial degradation of the crystalline area of granules by irradiation (Gul et al. 2016). The A-type starches possess a more compact double helix structure than B-and C-type starches. Hence, it is less vulnerable to radiation than B-and C-type starches (Navaf et al. 2022a). A 20 KGy irradiation does not impact the crystalline pattern of buckwheat and oat starches. However, it decreased the RC (Dar et al. 2018). Similarly, the RC of starches, such as potato starch (0.5 kGy) (Singh et al. 2011c), kithul starch (10 kGy), talipot starch (100 kGy), etc., were decreased by irradiation. The deterioration of the original structure and double helix of the starch granules may be the cause of the reduced RC (Dar et al. 2018).

Atrous et al. (2015) showed that irradiation of wheat starch did not change the crystallinity until 50 kGy. On the contrary, an increased RC was reported in rice starch (white rice and IRGA 417) with radiation of 0.5 kGy (Bao et al. 2005) and 5 kGy (Polesi et al. 2016).

1.4.4 Impact of Irradiation on Starch Digestibility

The irradiation of starch has a pronounced effect on its digestibility, and it varies with the type of starch, mc, and irradiation condition (Sudheesh et al. 2019a). A low dose of irradiation of talipot starch (1 and 10 kGy) (Aaliyaa et al. 2021) and kithul starch (0.5–10 kGy) (Sudheesh et al. 2019a) increased the digestibility. The free radical-induced molecular depolymerization and cleavage of anhydrous glycosidic linkages lead to low molecular weight chains and decrease the RC of starch, which in turn facilitates easy enzyme hydrolysis, thereby increasing the digestibility. In addition, the morphological changes, i.e., cracks and fissures, provide higher enzyme accessibility to granules. All these collectively contribute to an increased digestibility of irradiated starches.

Contrary to the above, certain starches such as potato and bean (10 and 50 kGy) (Chung and Liu 2010), waxy and normal corn starch (5–20 kGy) (Yoon et al. 2010), etc. shows an increase in RDS and RS content. Irradiation-induced cleavage of glycosidic linkages may enhance the mobility of chains and structural changes in the granules to a packed starch matrix (Yoon et al. 2010); it may reduce starch digestibility. Further, in the formation of β-bonded starches, the higher level of the carboxyl group also reduces the enzymatic hydrolysis of starch granules (Chung and Liu 2010).

1.4.5 Impact of Irradiation on Granule Morphology

The treatment of starches with a low dose of radiation did not show an obvious change in granule morphology, but a higher dose of radiation resulted in obvious changes, such as fissures on granule surfaces. The exposure of up to 10 kGy radiation on rice starch did not produce any noticeable changes, except for the deformation of a few granules at 10 kGy (Gul et al. 2016). Similarly, the irradiation of starches like brown rice (5–20 KGy) (Kumar et al. 2017), horse chestnut (5–15 KGy) (Wani et al. 2014), etc. didn't have any impact on granule surfaces. However, fissures were observed in irradiated (5 and 10 kGy) chickpea starches (Bashir and Aggarwal 2017). Low-dose irradiation of kithul (Sudheesh et al. 2019a) and talipot (Navaf et al. 2022a) starch resulted in granule breakdown and the loss of surface smoothness, respectively. Gani et al. (2012) noticed surface fracturing of granules in bean starches. However, a higher dose (500 kGy) resulted in the agglomeration of granules (Liu et al. 2012). The evidence of cracks and fissures on irradiated starch is attributed to the disintegration of granules (Fig. 1.6).

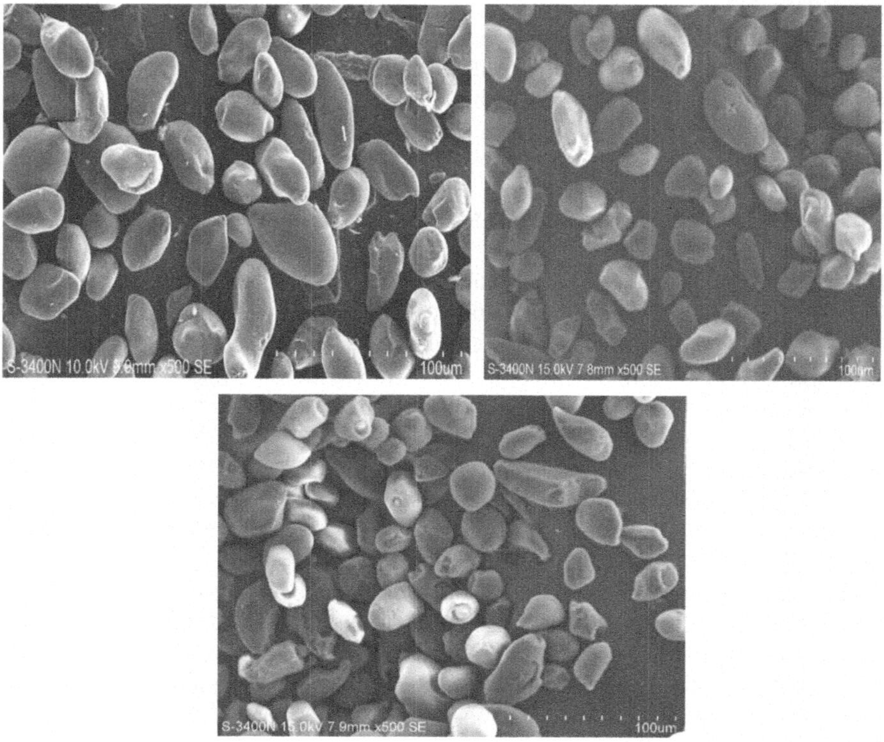

Fig. 1.6 SEM images of untreated and irradiated starches at 20 and 50 kGy

1.4.6 Impact of Irradiation on Starch Functional Attributes

The irradiation of native starch from diverse botanical sources with various doses and dose rates diminishes the SP but increases water solubility and amylose leaching. The irradiation of starches like buckwheat and oat starch (Dar et al. 2018), potato and bean (Chung and Liu 2010), cowpea starch (Verma et al. 2019), etc. decreased their SP. The amylopectin chains are primarily involved in starch swelling. Hence, the reduced SP of starch after irradiation results from the breakdown or destruction of amylopectin. Irradiation of starches induces the depolymerization of starch granules and forms simple sugars and starch fragments with low molecular weight. This increases the amount of hydrogen bonds with water molecules and increases solubility (Navaf et al. 2022a). Several studies reported a dose-dependent increase in the solubility of irradiated starches, such as wheat (Bashir et al. 2017), potato (Verma et al. 2017), kithul (Sudheesh et al. 2019a), etc.

The gamma irradiation of starch has a considerable effect on gel textural properties. In general, irradiation decreases the gel's hardness and forms a weak gel. The irradiation of wheat starch (1–9 kGy) (Kong et al. 2016), talipot starch (20 and 50 kGy) (Navaf et al. 2022a), and potato starch (0.01–0.5 kGy) (Singh et al. 2011c) decreased the gel hardness.

Irradiation significantly increases the LT and paste clarity of starch gels. The LT of starches such as broad bean (Sofi et al. 2013), horse chestnut (Wani et al. 2014), chickpea (Bashir and Aggarwal 2017), mung bean (Castanha et al. 2019), etc. increased by radiation. The lower swelling index and granule disintegration of irradiated starches may account for their better LT and paste clarity. In addition, the formation of carboxyl groups retains the H_2O through hydrogen bonding, giving higher LT (Navaf et al. 2022a).

Studies observed that γ-irradiation of starches had reduced syneresis or retrogradation tendency. The syneresis values of starches like buckwheat (Verma et al. 2017), kithul starch (Sudheesh et al. 2019a), horse chestnut starch (Wani et al. 2014), etc. were decreased with irradiation. The irradiation-induced degradation of starch granules forms simple sugars, i.e., mono-and disaccharides, which have high water retention capacity, leading to decreases in the syneresis value.

1.5 High-Pressure Treatment

High pressure (HP) is a non-thermal processing technique that has little impact on food's organoleptic and nutritional qualities while inactivating bacteria that cause foodborne illness. With little impact on covalent bonds, HP is a green and environmentally friendly technique that can change non-covalent bonds. Typically, the pressurization process takes place at room temperature and uses an applied pressure range of 100–600 MPa. The starch modification by HP involves three steps. Initially, the sample preparation involves making a starch suspension containing 3–30%

starch and vacuum sealing it. Then, the sample is loaded into a HP chamber containing a pressure-transmitting medium (i.e., water). After attaining desired pressure, the sample is treated for a pre-decided time. After treatment, the pressure is immediately reduced to atmospheric pressure. It induces structural changes in starch, such as granular disruption, changes in crystallinity, and starch chain interactions, which change their functional properties (Castro et al. 2020). The impact of high-pressure treatment on starch's physicochemical and functional attributes is discussed below.

1.5.1 Impact of HP Treatment on Gelatinization Properties

The impact of HP on the gelatinization characteristics of starch varies with treatment conditions and starch sources. The gelatinization attributes of HP-treated starch are summarized in Table 1.4. A 500 MPa and 600 MPa pressure-treated pea starch did not show any gelatinization peaks (Leite et al. 2017). However, 300 MPa and 400 MPa pressure-treated starches showed endothermic peaks, and 300 MPa did not cause any difference in gelatinization properties as compared to native starch. It indicates that 300 MPa is unable to promote any changes in the gelatinization characteristics of pea starch, whereas a 400 MPa treatment caused a significant decrease in the ΔH. Likewise, high hydrostatic treatment of pea starch at a pressure

Table 1.4 The gelatinization characteristics of HP-treated starches

Sample	Pressure (MPa)	T_O (°C)	T_P(°C)	T_C(°C)	ΔT (°C)	ΔH(J/g)	Reference
Pea	150	62.9	68.6	73.7	10.8	5.7	Liu et al. (2018)
	300	62.9	68.4	72.7	9.8	4.3	
	450	62.4	68.2	71.2	8.8	3.8	
Waxy rice	100	71.8	75.5	81.3	9.5	9.5	Zeng et al. (2018)
	200	72.5	76.3	81.8	9.3	9.3	
	300	75.6	79.3	83.2	7.6	7.7	
	400	77.7	81.5	85.6	7.9	7.9	
	500	75.3	79.2	82.9	7.6	7.6	
Rice	120	55.6	63.1	75.7	20.1	10.9	Li et al. (2012)
	240	55.1	62.8	75.1	20	10.3	
	360	55.2	62.6	73.5	18.3	9.2	
	480	54.8	60.1	72.3	17.5	7.2	
Red adzuki bean	150	66.97	70.75	76.2	9.23	6.30	Li et al. (2015)
	300	69.95	73.15	77.67	7.72	6.12	
	450	67.43	71.25	75.89	8.46	5.83	
Buckwheat	120	69.0	76.1	81.6	12.6	17.8	Liu et al. (2016b)
	240	66.7	73.9	78.8	12.1	15.9	
	360	65.2	71.5	76.7	11.5	12.3	
	480	64.2	70.8	74.9	10.7	8.4	
	600	62.1	68.5	71.6	9.5	6.6	

of 150–450 MPa decreased the gelatinization properties, and a high pressure of 600 MPa did not show any endothermic peaks corresponding to gelatinization (Liu et al. 2018). Similarly, 120–480 MPa pressure-treated rice starch exhibited decreased gelatinization characteristics. In addition, the 600 Mpa pressure-treated starch did not have any gelatinization peaks (Li et al. 2012). The absence of gelatinization peaks in high-pressure-treated starch shows that the starch is completely gelatinized during treatment.

During HP treatment, the crystalline regions are partially or completely degraded. The decrease in gelatinization parameters of HP-treated starches indicates its partial gelatinization. The pressure dependence of different starches varies. Tapioca and potato starch are less pressure sensitive than wheat starch. A complete gelatinization of wheat and tapioca starches was noticed at 400 MPa and 600 MPa, respectively. However, potato starch exhibited partial gelatinization even at 700 MPa (Bauer and Knorr 2005).

1.5.2 Impact of HP Treatment on Pasting and Rheological Properties

The pressure treatment can modify the pasting profile of starch. HP-treated (120 MPa to 480 MPa) rice starch exhibited a higher PT, PV, TV, and FV and lower BV than untreated (Li et al. 2012). A pressure of 600 MPa decreased the pasting profile except for the PT. Similar observations were also reported in HP-treated pea starch at a pressure of 150 MPa to 600 MPa (Liu et al. 2018). A 100–500 MPa pressure treatment of lotus seed starch increased the PT, PV, TV, and FV, whereas it decreased the BV and SV. Contrary to this 600 MPa treated lotus seed starch exhibited a decreased PV, TV, and FV and did not show any PT (Guo et al. 2015). These variations in the pasting profile of starch under various pressures may be the result of treatment-related changes to the crystalline structure. When treated at lower pressure, the higher pasting profile of starch is due to the increased swelling power brought about by the starch granule degradation. Contrary to the above observation, pea starch at 300 MPa and 400 MPa exhibited a reduction in the PV, and at 500 MPa and 600 MPa, it increased PV (Leite et al. 2017). A 600 MPa treatment of potato starch didn't show any changes in its pasting profile (Colussi et al. 2020). Liu et al. (Liu et al. 2016b) noticed a decreased pasting profile, except for an increased PT of buckwheat starch at a pressure of 120 MPa–600 MPa. The HP treatment causes structural alterations, i.e., crystalline structure, loss of molecular order, amylopectin entanglement, etc., increases the PT, and reduces the viscosity.

The rheological studies show that with an increase in the magnitude of pressure treatment in starch, loss modulus and storage modulus also increase. However, in certain starches, a higher pressure may cause a negative impact on rheological properties. The lentil starch exhibited higher G' and G" values after 400–600 MPa pressure treatment (Ahmed et al. 2016). Similar observations were reported in mung bean (Jiang et al. 2015) and quinoa starches at a pressure of 600 MPa (Ahmed et al. 2018). Starches like maize and quinoa showed an increased G' and decreased G" value with an increase in pressure (Li and Zhu 2018). The production of weak gels may result from the treatment at high pressure. A pressure of 600 MPa decreased the G' and G" values of lotus seed starch (Guo et al. 2015).

1.5.3 Impact of HP Treatment on Structural Characteristics

The effect of HP treatment on starch structure varies with their crystalline packing arrangement. The B-type starches (tuber starches) are more resistant to pressure than the A-and C-type starches. The crystalline structure of starches will not alter significantly by HP treatment. However, the intensity of crystalline peaks may be altered. Guo et al. (2015) reported a pressure ≤ 500 MPa didn't have any impact on the crystalline pattern of lotus seed starch. However, a pressure of 600 MPa caused a transformation from a C-to a B-type crystalline structure, and it also decreased crystallinity. A similar shift from C to B type was noticed in pea starch at a pressure of 600 MPa (Liu et al. 2018). The peak intensity of rice starch also diminished with an increase in pressure, and a pressure of 600 MPa resulted in a crystalline change from A type to a more stable B type. The A-type starch's crystalline structure is altered due to the double helix's ability to be rearranged by the fragmented, flexible, and fragile amylopectin structure (Li et al. 2012).

The high-pressure treatment of starch significantly alters the relative crystallinity. The RC of buckwheat starch was reduced from 37.6% to 26.2% by the application pressure from 120 MPa to 600 MPa (Liu et al. 2016b). Likewise, the RC of potato starch was decreased from 19 to 14% by exposure to a pressure of 600 MPa (Colussi et al. 2020). The B-type crystalline starch has less compressibility than the A-type starch, where water molecules are randomly distributed among amylopectin molecules, which may cause an apparent difference in pressure resistance between them (Kim et al. 2012).

1.5.4 Impact of HP Treatment on Starch Digestibility

The HP treatment of starch potentially impacts its digestibility and varies with starch origin and the magnitude of pressure used. In general, the digestibility of starch is limited with an increase in pressure. Bajaj et al. (Bajaj et al. 2022) observed that HP (300 and 600 MPa) treatment of starch with different crystalline types decreased their digestibility. Similarly, HP treatment of starches like buckwheat and sorghum starch (120–600 MPa) (Deng et al. 2014; Zhang et al. 2022a) and rice starch (200–600 MPa) (Deng et al. 2014) also decreased their digestibility. A 500 MPa pressure treatment of oat starch for 5–10 min also decreased its digestibility (Zhang et al. 2022a). However, further treatment time increased the digestibility. A low-pressure (100–200 MPa) treatment of waxy rice starch didn't impact its digestibility. However, a high pressure (300–500 MPa) increased the digestibility (Zeng et al. 2018).

The extent of digestibility of HP-treated starch is determined by its crystalline structure. HP treatment of starch may induce alterations in its polymorphic structure to a more stable form, thus reducing digestibility (Liu et al. 2018). In addition, the improved interaction between starch chains by high-pressure treatment also facilitates decreased digestibility. High-pressure treatment also promotes starch lipid complex formation, increasing SDS content (Bajaj et al. 2022). High-pressure treatment may cause structural alterations and starch granule rupture, leading to enhanced SDS content in high-pressure treated starch (Zeng et al. 2018).

1.5.5 Impact of HP Treatment on Granule Morphology

The HP treatment of starch may lead to partial gelatinization of granules that alters their integrity and morphology. A 500 MPa treatment of oat starch for 5 and 10 min led to the aggregation of the part of granules. However, a 15 min treatment resulted in swelling and deformation of granules (Zhang et al. 2022a). The 600 MPa treatment of potato starch did not alter its granule morphology (Colussi et al. 2020). However, exposure of lentil starch to 600 MPa for 10 min increased the average diameter of the granule from 20.22 to 23.77 μm (Ahmed et al. 2016). The morphological changes of HP-treated rice starch at 120–480 MPa pressure are not obvious, whereas, at 600 MPa, it caused irreversible changes in granular structure and formed a gel-like appearance (Li et al. 2012). The exposure of pea starch at 150–450 MPa resulted in broken starch granules with scraggy and plicated surfaces, and 600 MPa resulted in granule collapse and the formation of irregular nubby-spongy shapes (Liu et al. 2018).

1.5.6 Impact of HP Treatment on Starch Functional Attributes

The high-pressure treatment of starch increases its SP and solubility at a lower temperature; however, a higher temperature decreases SP and solubility. The HP treatment of rice and buckwheat starches at 600 MPa increased SP and solubility at a temperature of 50–60 °C, and a temperature above 60 °C reduced SP and solubility as compared to the untreated one (Li et al. 2012). Similarly, HP-treated lotus seed starch exhibited higher SP and solubility at a temperature of 55 and 65 °C, whereas it decreased at 85 °C and 95 °C (Guo et al. 2015). At lower temperatures (50–60 °C), amylose forms linkages with lipid molecules, which resists the granule's swelling. However, the higher SP and solubility of high-pressure treated starches at low temperatures could be due to the presumed aggregation of amylose molecules under pressure, which interferes with the lipid-starch linkage, leading to higher water retention (Liu et al. 2016c). The HP treatment enhances the ordered arrangement of starch granules, thus limiting the hydration capacity and decreasing SP and solubility at high temperatures (Li et al. 2012).

The HP treatment of starch has a varying impact on the textural characteristics of the starch gel. A 200–600 MPa pressure treatment increased the quinoa starch gel hardness, whereas it decreased for wheat starch gel (Li and Zhu 2018). The gel hardness of corn, waxy corn, and potato starches increased upon pressure treatment, whereas it decreased in wheat and sweet potato starches. Nevertheless, kidney bean starch increases the gel hardness at a pressure of 300 MPa and decreases at 600 (Bajaj et al. 2022).

The HP-treated starch exhibits a lower paste clarity and LT. A 120–600 MPa pressure treatment of red adzuki bean (Li et al. 2015) and mung bean starches (Li et al. 2011) decreased their LT. The LT values would differ depending on the swollen granule remnants in the starches. It was observed that starch pastes with swollen starch granules had less LT. As compared to high-pressure treated starches, the amount of swollen granules will be less in native starches, and hence it shows higher LT (Li et al. 2015). Additionally, during retrogradation, the leached amylose and amylopectins of high-pressure treated starches are reassociated or retrograde, which reduces the LT (Li et al. 2011).

1.6 Pulsed Electric Field

Pulsed electric field (PEF) technology applies short pulses of high electric fields with a very short duration of micro-to-milli-seconds and a field strength of ~1–80 kV/cm. The short duration helps to avoid excessive heat and unwanted electrolytic reactions. The samples are kept between a pair of electrodes. The electric field can be in the form of a bipolar square wave, exponential decaying, or oscillatory pulses. The effect of pulsed electric field treatment on the physicochemical and functional attributes of starch is discussed below.

1.6.1 Impact of PEF Treatment on Gelatinization Properties

Studies observed that the PEF-treated starches have lower gelatinization attributes. The gelatinization attributes of a few PEF-treated starches are summarized in Table 1.5. The treatment of waxy rice starch with PEF at 50 kV cm^{-1} reduced the ΔT from 11.8 °C to 9.8 °C and the ΔH value from 12.4 to 6.8 Jg^{-1} (Zeng et al. 2016). Similarly, a decrease in gelatinization temperatures and ΔH value was noticed in japonica rice starch at a field intensity of 8.57kVcm^{-1} (Wu et al. 2019). The PEF treatment of corn (Han et al. 2009a), potato (Han et al. 2009b), and tapioca (Han et al. 2009b) starches significantly decreased the gelatinization temperatures with a change in field strength from 30 to 50 kV cm^{-1}. The authors also observed that at a field strength of 50 kVcm^{-1}, the ΔH value decreased from 1.61 to 0.95 Jg^{-1}, from 11.61 to 9. 52 Jg^{-1}, and from 16.06 to 3.80 Jg^{-1}, respectively, in corn, potato, and tapioca starches. The reduced ΔT of PEF-treated starches is due to the disruption of the crystalline perfection with similar crystal structures and selectively destroys those with differing structures (Zhu 2018). In addition, PEF makes water molecules accessible to crystalline regions, leading to decreased ΔH value (Wu et al. 2019).

Table 1.5 The gelatinization characteristics of pulsed electric field-treated starches

Sample	PEF treatment conditions (kV cm^{-1})	T_O (°C)	T_P (°C)	T_C (°C)	ΔT (°C)	ΔH (J/g)	References
Waxy rice	30	69.3	73.4	81.1	11.8	9.9	Zeng et al. (2016)
	40	66.9	72.5	77.4	10.5	8.1	
	50	65.4	69.6	75.2	9.8	6.8	
Tapioca	30	62.73	68.12	75.15	12.42	8.87	Han et al. (2012)
	40	62.21	66.63	72.31	10.1	4.82	
	50	61.76	65.94	72.61	10.85	3.80	
Corn	30	68.60	73.38	78.49	9.89	1.22	Han et al. (2009a)
	40	67.46	72.21	77.68	10.22	1.14	
	50	67.34	71.51	77.40	10.06	0.95	
Potato	30	61.01	64.99	73.17	12.16	10.04	Han et al. (2009b)
	40	60.77	64.87	75.13	14.36	10.03	
	50	60.35	64.61	78.47	18.12	9.52	

1.6.2 Impact of PEF Treatment on Pasting and Rheological Attributes

The PEF treatment of japonica rice with a field intensity of 2.86–8.57 kV cm^{-1} does not affect the pasting profile, except for a slight decrease in SV (Wu et al. 2019). However, the pasting profile of starch treated with high field strength decreases with an increase in field strength. A field strength from 30 to 50 kV cm^{-1} reduced the PV and BV of potato starch (Han et al. 2009b), corn (Han et al. 2009a), and tapioca (Han et al. 2012) starches. Likewise, the PEF treatment of red rice starch at a field strength of 10 and 30 kV cm^{-1} significantly decreased the pasting profile (Almeida et al. 2022). The authors justified that the PEF treatment modifies the starch SP and thus decreases the pasting properties.

1.6.3 Impact of PEF Treatment on Structural Characteristics

PEF did not significantly alter the starch crystalline pattern, but its RC was significantly reduced. The extent of the decrease in RC varies with the starch type and PEF conditions. An increase in PEF intensity will impact the starch granule and decrease its RC. The RC of waxy rice starch was decreased from 38.1% to 28.7% by a field intensity of 50 kV cm^{-1} (Zeng et al. 2016). The PEF treatment of tapioca starch with 30, 40, and 50 kV cm^{-1} decreased its RC to 18.1%, 11.4%, and 7.2%, respectively (Han et al. 2012). Similarly, a 50 kV cm^{-1} PEF treatment of potato starch decreased the RC by 3.3% (Han et al. 2009b). The PEF treatment of wheat, potato, and pea starch did not impact their crystalline pattern. However, the RC was decreased in the order of potato starch, wheat, and pea starches. It is suggested that the phosphate ions in potato starch could be the transmitter for an electric pulse into deeper crystallites. It also speculates that each monoclinic crystal unit of type-A starch crystallite contains 8 water molecules, whereas hexagonal crystal units of type-B crystallites contain 36 water molecules with more open packing of helices (Li et al. 2019).

1.6.4 Impact of PEF on Starch Digestibility

The PEF treatment of starch increases their susceptibility to digestive enzymes. PEF causes damage to the granular surfaces and crystalline perfection and facilitates the passage of enzymes into the granules. The PEF treatment of japonica rice (Wu et al. 2019) and waxy rice starch (Zeng et al. 2016) increased their digestibility with increased field intensity. The PEF wheat starch exhibited a higher RDS content and a lower SDS content when the field strength was increased. However, the RS content was increased at lower field intensity (2.86 and 4.29 kV cm^{-1}) and did not show any changes at a higher intensity (7.14 and 8.57 kV cm^{-1}). A similar observation was made in potato starch. However, in pea starch, the RS did not show any significant change from its native counterpart (Li et al. 2019).

1.6.5 Impact of PEF on Granule Morphology

The changes in surface morphology of PEF-treated starches vary with treatment conditions. The tapioca (Han et al. 2012), potato (Han et al. 2009b), waxy rise (Zeng et al. 2016), and corn (Han et al. 2009a) starch granular surface became rough after a PEF treatment of 30 $kVcm^{-1}$. However, treatment with 40 $kVcm^{-1}$ resulted in the formation of pits and aggregation of smaller particles. The further increase to 50 $kVcm^{-1}$ altered granular shape and led to granular aggregation.

1.6.6 Impact of PEF on Starch Functional Properties

The structural changes that occurred during PEF treatment brought changes to their functional properties. A rise in pulse intensity from 10 to 30 kV cm^{-1} considerably increased red rice starches' water-holding capacity and solubility (Almeida et al. 2022). However, it decreased the paste clarity and syneresis index. Similarly, the PEF-treated starch gel showed significantly reduced gel hardness.

1.7 Ultrasonication

Ultrasounds are mechanical waves with a frequency above the threshold level of human hearing (>16 kHz). Ultrasound technique is regarded as one of the most promising utilities in the food industry (Schafranski et al. 2021). It is a cutting-edge processing method that has the potential for the physical depolymerization of starch. Studies proposed that ultrasonication especially destroys amorphous regions, deeply degrades the crystalline structure and double helix of starch granules, and more readily invades amylose areas than amylopectin areas (Yang et al. 2019).

Ultrasonication degrades the starch chains and produces dextrins with smaller molecular sizes and narrower molecule weights. The pictorial representation of ultrasonication on starch granules is given in Fig. 1.7. It could be explained that the

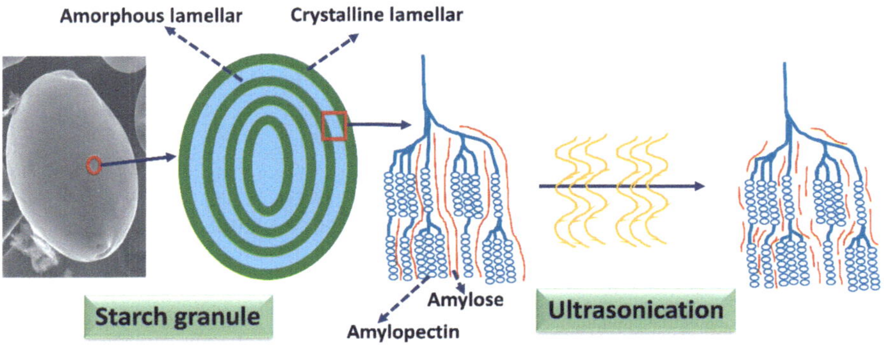

Fig. 1.7 The pictorial representation of ultrasonication on starch granule

cavitation process produces high pressure, temperature, and shear force in particular areas that can degrade the starch granules (Zhu 2015). This method can be used to modify material's functionalities and has the advantages of higher selectivity, quality, lower chemical usage, and shorter processing time. The effect of ultrasonication on starch's physicochemical and functional attributes is discussed below.

1.7.1 Impact of Ultrasonication on Gelatinization Characteristics

Depending on the starch origin and the treatment conditions, starches vary their gelatinization characteristics. The degradation of amorphous and crystalline areas of starch during sonication can alter their thermal attributes. The gelatinization characteristics of ultrasonicated starches are summarized in Table 1.6. The ultrasonication of sweet potato starch for different periods did not significantly change the onset and peak temperature. However, the ΔH was significantly decreased (Wang et al. 2019). A 50 kHz for 10 and 20 min of cassava starch and corn starches barely affected the gelatinization temperature; however, the ΔH was decreased by ultrasonication (Rahaman et al. 2021). Ultrasound treatment of maize starch with 20 kHz for 60 min reduced T_c and ΔH values. However, in similar treatment conditions, the T_c value of potato starch was increased (Wang et al. 2022). The decreasing trend of the ΔH value in ultrasonicated starch could be due to the gradual degradation of the crystalline perfection. S. Yu et al. (Yu et al. 2013) noted that the mc of the medium primarily influences the gelatinization properties of ultrasonicated starch.

1.7.2 Impact of Ultrasonication on Pasting and Rheological Properties

In general, ultrasonication of starch will alter the pasting profile and varies with treatment conditions and source of starch. According to the studies on ultrasonicated sweet potato (Wang et al. 2019), and waxy corn starch (Yang et al. 2019), a

Table 1.6 The gelatinization characteristics of ultrasonicated starches

Starch	Ultrasonication conditions	T_O (°C)	T_P (°C)	T_C (°C)	ΔT (°C)	ΔH (J/g)	References
Maize	20 kHz, 60 min	67.7	72.1	76.5	8.8	1.54	Wang et al. (2022)
Potato		56.1	63.5	73.2	17.1	1.65	
Plantain	25 kHz, 20 min	67.8	71.45	76.71	8.91	8.28	Carmona-García et al. (2016)
	25 kHz, 50 min	67.35	71.70	76.84	9.49	8.83	
Taro	25 kHz, 20 min	74.40	79.91	85.16	10.76	10.36	
	25 kHz, 50 min	74.02	79.35	84.99	10.97	10.63	
Corn	40 kHz, 10 min	71.5	76.66	85.13	13.63	8.13	Rahaman et al. (2021)
	40 kHz, 20 min	71.98	77.72	87.99	16.01	11.5	
Cassava	40 kHz, 10 min	66.15	73.50	85.6	19.45	8.86	
	40 kHz, 20 min	66.35	74.12	84.55	18.2	10.32	
Sweet potato	20 kHz, 15 min	64.31	69.36	81.51	17.2	15.62	Wang et al. (2019)
	20 kHz, 20 min	64.38	69.57	80.33	15.95	15.74	
	20 kHz, 25 min	64.15	69.22	79.70	15.55	14.90	
	20 kHz, 30 min	64.26	69.26	80.15	15.89	14.93	

reduction in viscosity profiles was observed. Contrary to these, PV and FV increased in taro starch (Martins et al. 2020) and plantain starch (Carmona-García et al. 2016). Certain starch shows contrasting behavior on pasting profiles at different treatment conditions. For example, the ultrasonication of rice starch at 16 kHz with 2000 W power increased the PV and FV (Cui et al. 2010). It is due to the fact that starch granules are softened by sonication, thereby improving the water absorption ability of starch (Bonto et al. 2021). However, the pasting profile of rice starch decreased at 211 kHz (2.5 and 4.1 W) (Zuo et al. 2009). The disruption of starch granules led to lower SP, which may cause the poor pasting profile of ultrasonicated starches. The granular disruption by sonication depends on the treatment conditions. A mild treatment condition may not cause molecular degradation while showing reduced viscosity. It could be due to the physical damage of the starch granule by sonication.

Many researchers studied the flow behavior of ultrasonicated starch. Most of the sonicated starches generally exhibit a higher G' than G", denoting its solid-like or elastic behavior. The sonication of potato starch with different duration increased its rheological attributes, and it depends on the treatment time and angular frequency (Wang et al. 2019). Similar observations were noted in G' value of cassava starch. However, it did not alter G" values significantly (Monroy et al. 2017). Contrary to the above, the sonication of cassava starch decreased the G" values (Krishnakumar and Sajeev 2018). The morphological alteration, increased amylose content, and disruption of the crystalline lamella are possible causes of a rise in G' after ultrasonic treatment. In this way, amylose and amylopectin were entangled and interacted, forming a strong gel network.

Nevertheless, the ultrasonication (80 W) of starch with different granular sizes decreased the G'and G" values. Starch with a larger particle size had a greater impact on its rheological properties (Carmona-García et al. 2016). Similar observations were made in talipot starch treated at 24 kHz for 30 min (Aaliyaa et al. 2021).

It was suggested that ultrasonication greatly affects the branched chain amylopectin and releases linear starch chains, thus reducing reorientation capacity during retrogradation.

1.7.3 Impact Ultrasonication on Structural Characteristics

In general, ultrasounds are insufficient to cause disruption or distortion to the starch granule crystallites. However, it will alter the amorphous region and the breakdown of the double helical structure in both amorphous and crystalline lamella (Bonto et al. 2021). Studies reported that ultrasonicated starches maintained their intact crystalline pattern, but it changed the RC. The ultrasonication of cassava and corn starches at 40 kHz for a different time decreased the peak intensity and RC (Rahaman et al. 2021). Similarly, ultrasonication (20 kHz) of potato and maize starches reduced their RC (Wang et al. 2022). The unstable layered arrangement of starch could reduce the RC of starch by ultrasound. Further, the breakdown of the double helix can also facilitate a reduction in the RC of ultrasonicated starches.

1.7.4 Impact of Ultrasonication on Starch Digestibility

Ultrasonication of starch significantly impacts its in vitro digestibility, and it varies with the starch origin and treatment conditions. The sonication of partially gelatinized high-amylose starch at different temperatures potentially increased its digestibility. A long duration at a high temperature (50 °C) decreased the digestibility (Chan et al. 2021). However, low-temperature (40 °C) sonication decreased the RS content but did not alter the SDS content. Similarly, the sonication of (140 W, 40 °C) cassava starch decreased its digestibility (Khurshida et al. 2021). A higher-temperature sonication enhances the long-range ordered molecular structure and degree of order and double helix in short-range molecular structure. It gives stability to the granules against hydrolysis. Likewise, the ultrasonication of retrograded starch increased the RS content at lower power (200 W) and then decreased at higher power (300–600 W) (Ding et al. 2020). The sonication of talipot starch (24 kHz, 400 W) increased the RDS and decreased the RS content. Besides, it did not cause any significant changes in SDS content (Aaliyaa et al. 2021).

1.7.5 Impact of Ultrasonication on Granule Morphology

The ultrasonication of starch causes obvious changes in its morphology, and it varies with the type of starch, the concentration of starch slurry, the temperature of the system, and treatment conditions, i.e., frequency, energy input, and duration of treatment (Zhu 2015). An exposure of rice starch in aqueous ethanolic suspension at 20 kHz caused the formation of fissures and depressions (Bonto et al. 2021).

Likewise, deep pitting and cracks were formed on the maize and potato starch granules (Wang et al. 2022). The ultrasound treatment of cassava (40 kHz) and corn starch (40 kHz) resulted in roughness on the granular surface, and its intensity increased with an increase in treatment time (Krishnakumar and Sajeev 2018; Rahaman et al. 2021). Sujka (2017) studied the influence of the different aqueous mediums on the morphology of potato starches. The researcher observed that the sonication in water resulted in surface erosion, cracks, and depression on the surface, whereas sonication in ethanol did not show any symptoms of surface erosion. Chan et al. (2021) studied the impact of different treatment temperatures on granule morphology and reported that the low-and high-temperature treatment sonication increased the number of pores and cracks on the granule surface. In addition, extended treatment at high temperatures caused wrinkles and fragmentation of granules.

1.7.6 Impact of Ultrasonication on Starch Functional Properties

Sonication of starch causes physical damage to the granules, disrupting the ordered arrangement and crystalline structure. It facilitates the easy bonding between H_2O and the OH group of starch chains, and hence it allows a higher amount of water to enter into starch granules. This gives an increased SP to ultrasonicated starch granules (Wang et al. 2019). The SP of different starches, such as cassava (Krishnakumar and Sajeev 2018), plantain and taro (Carmona-García et al. 2016), sweet potato (Wang et al. 2019), etc. was increased after sonication. The extent of increased SP of ultrasonicated starch may vary with its type, composition, and sonication conditions.

The solubility of starch increases with sonication and increases with an increase in the intensity of ultrasonication. Starch chains' increased mobility during sonication enhances the hydration of amylose molecules, and it results in higher solubility. In addition, the fragmentation that occurs during ultrasonication increases the amount of amylose, and surface damages like pores and cracks also promote higher solubility of ultrasonicated starches (Wang et al. 2019).

The enhanced hydration and solubility and reduced amounts of remnants of the granules in starch give a higher LT and paste clarity to the ultrasonicated starch gels (Zhu 2015). The ultrasound probe and ultrasound bath treatment of oat starch significantly increased its LT and paste clarity (Reza et al. 2019). On the contrary, the LT of cassava starch was decreased with an increase in time, temperature, and sonication intensity (Krishnakumar and Sajeev 2018). The decreased paste clarity was explained by the higher disintegration of starch granules by sonication. It leads to the development of a more viscous paste and thus decreases the LT.

The texture profile analysis of ultrasonicated starch shows that sonication significantly decreases the gel hardness, and it depends on the treatment duration and intensity. The gel hardness of potato starch (Nie et al. 2019) and oat starch (Reza et al. 2019) decreased with an increase in sonication power.

The ultrasonication of cassava starch at 30 kHz with different treatment times and temperatures significantly decreased the syneresis value (Krishnakumar and

Sajeev 2018). However, ultrasonication of oat starch increased its syneresis value at the initial storage days, and later it remained constant over time (Reza et al. 2019). Sonication-induced breakage and reordering of the starch chain enhance the number of hydrogen bonds with water molecules, and it could hold more water, reducing the syneresis.

1.8 Plasma Treatment

As the fourth state of matter, plasma is an ionized gas made up of electrons, atoms, and molecules with a net neutral charge. In general, gas or gas flowing between two charged electrodes is given electrical energy in order to create plasma. Due to the collision between a free electron and gas molecules caused by the potential difference between two electrodes, gas gets ionized. High-temperature plasma, also known as fusion plasma, and low-temperature plasma, also known as gas discharge, are the two basic categories of plasmas. The former is not suited for food processing because it has high temperatures. The latter has a low temperature and won't harm the food's thermo-labile components (Laroque et al. 2022). Thus, it has great utility in foods, such as shelf-life extension, seed germination, cooking optimization, and starch modification. The plasma treatment of starch may lead to physical or chemical impacts, and it varies with the sources and type of plasma and starch used (Kalaivendan et al. 2022). Both surface etching and morphological alterations are both possible outcomes of the physical effects of plasma. Plasma's chemical reactions could cause the molecular chain to depolymerize and cross-link (Thirumdas et al. 2017a). The physicochemical and functional characteristics of starch are greatly impacted by changes in its molecular chains brought on by plasma. The pictorial representation of plasma treatment on starch granules is given in Fig. 1.8. The effect of plasma treatment on the physicochemical and functional properties of starch is discussed below.

1.8.1 Impact of Plasma Treatment on Gelatinization Characteristics

The gelatinization characteristics of starch were altered by plasma treatment, and it may either increase, have little effect, or reduce the gelatinization attributes. The cross-linking, depolymerization, and oxidation induced by plasma treatment significantly impact its gelatinization properties. The degree of cross-linking can also affect the gelatinization temperature, and highly cross-linked starches exhibited a high gelatinization temperature. Similarly, excessive depolymerization and degradation decrease the gelatinization temperature (Zhu 2017). The gelatinization characteristics of plasma-treated starches are given in Table 1.7. The plasma treatment of corn and tapioca starch using low-pressure radio frequency air plasma (LPRFP) decreased the gelatinization temperature and ΔH value of corn starch. In contrast, it was increased in tapioca starch (Banura et al. 2017). A low-pressure argon plasma

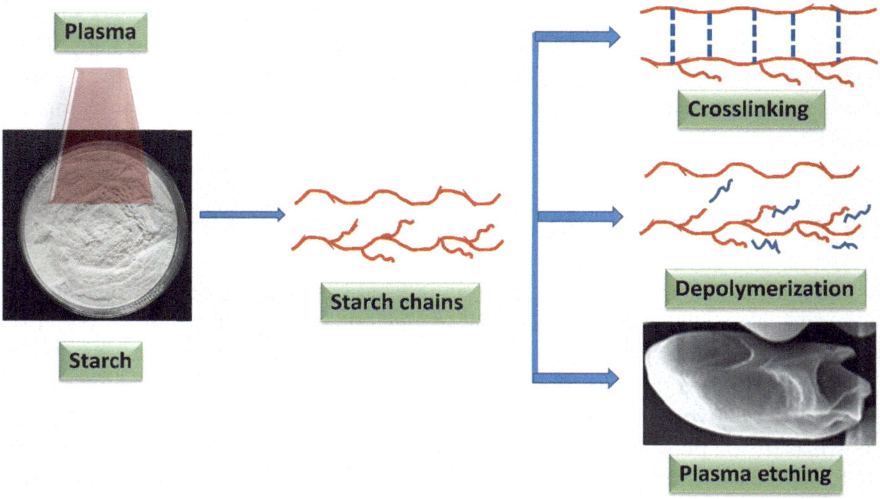

Fig. 1.8 The pictorial representation of plasma treatment on starch granule

(LPAP) treatment reduced the gelatinization properties of talipot starch (Navaf et al. 2022b), whereas the exposure of tapioca starch to argon plasma did not significantly affect its gelatinization temperatures; however, the ΔH value was slightly increased (Wongsagonsup et al. 2014). The authors also stated that the ΔH value increases with the degree of cross-linking while the gelatinization temperatures are not affected. Sudheesh et al. (2019b) noticed a contrary observation when kithul starch was treated with energetic neutral nitrogen atoms created by glow discharge air plasma (EN air plasma). The authors reported that the gelatinization parameters were decreased at a lower plasma power, whereas they increased at a higher plasma power. A gelatinization temperature and enthalpy increase was reported in dielectric discharge plasma-treated buckwheat, sorghum, wheat, and quinoa starches (Gao et al. 2019). The atmospheric pressure jet plasma (APJP) treatment decreased the gelatinization temperature of both waxy and normal maize starches (Zhou et al. 2019). The decreased gelatinization characteristics of plasma-treated starch are due to the disruption and depolymerization of starch chains, and the increased gelatinization parameters could be due to the existence of more stable crystallites after depolymerization.

1.8.2 Impact of Plasma Treatment on Pasting and Rheological Properties

Plasma treatment of starch significantly affects the pasting properties. A reduced pasting profile was observed when potato starch was treated with glow discharge helium and nitrogen plasma (Zhang et al. 2015), wheat and sorghum starches were treated with dielectric glow discharge plasma (Gao et al. 2019), and corn starch was

Table 1.7 The gelatinization characteristics of plasma-treated starches

Starch	Types of plasma and power	T_O (°C)	T_P (°C)	T_C (°C)	ΔT (°C)	ΔH (J/g)	References
Cassava	Oxygen glow plasma, 30 min	66.49	70.57	76.27	9.78	13.19	Bie et al. (2016)
	Oxygen glow plasma, 60 min	66.96	71.49	76.46	9.5	12.61	
	Helium glow plasma, 30 min	65.77	69.32	76.11	10.34	15.00	
	Helium glow plasma, 60 min	66.47	70.57	76.49	10.02	14.46	
Sorghum	Dielectric barrier discharge plasma	67	71.7	84.6	17.6	11.5	Gao et al. (2019)
Wheat		62	67.6	80.2	18.2	10.2	
Quinoa		51.4	64.1	79.2	27.8	13.2	
Tapioca	Jet atmospheric argon plasma, 50 W	64.4	70.4	79.6	15.2	14.9	Wongsagonsup et al. (2014)
	Jet atmospheric argon plasma, 100 W	64.1	70.4	79.0	14.9	13.0	
Corn	Low-pressure radio frequency air plasma, 40 W, 10 min	66.61	71.98	75.68	9.07	12.12	Banura et al. (2017)
	Low-pressure radio frequency air plasma, 60 W, 10 min	66.13	71.70	75.79	9.66	12.42	
Rice	Cold plasma, 40 W, 10 min	64.7	73.2	78.1	13.4	14.6	Thirumdas et al. (2017a)
	Cold plasma, 60 W, 10 min	59.9	73.1	77.9	18	13.4	
Waxy maize	Radio frequency cold plasma, 120 W, 30 min	64.5	70.8	78.7	14.2	12	Okyere et al. (2019)
Waxy rice		60.5	67.4	75.2	14.7	9.2	
Waxy potato		61.7	66.8	72.6	10.9	4.4	

treated with APJP (Wu et al. 2018). On the contrary, an increase in paste viscosities was reported in corn and tapioca starches when treated with LPRFP (Banura et al. 2017).

The effect of plasma treatment on pasting properties varies with the types of starch and plasma and treatment conditions. The reactive plasma species may weaken or break down the glycosidic linkage of starch molecules. Granular corrosion and molecular depolymerization caused by the destabilization of starch granules could lower viscosity (Wu et al. 2018). Plasma treatment would also reduce the viscosity of starches by disrupting granular structure, reducing amylose content, and increasing RC (Navaf et al. 2022b). Banura et al. (Banura et al. 2017) stated that simple sugars formed after depolymerization held a higher amount of water, resulting in increased starch viscosity. Moreover, plasma-treated starch may leach out a greater amount of amylose through the weakening or breaking of bonds, and it may also increase the PV (Sudheesh et al. 2019b).

The modification of starch with plasma causes molecular changes in the granules. The rheological characteristics of plasma-treated starches vary depending on the type of changes, i.e., cross-linking or depolymerization. In general, treating starch with low power of plasma leads to cross-linking, resulting in a strong gel formation and increased G' and G" values. Contrarily, the treatment of starch with higher plasma power levels results in depolymerization and disruption of starch molecules, forming a weak gel. The exposure of kithul starch with an EN air plasma at a power of 5 W for 30 increased the G' and G" values (Sudheesh et al. 2019b). However, a further increase in treatment time and plasma power decreased the G' and G" values. The treatment of tapioca starch with low-pressure air, CO_2, and helium plasma increased the G' value; however, it was decreased by the treatment of LPAP (Zhang et al. 2022b). The exposure of tapioca starch to atmospheric jet argon plasma with a power of 50 W increased the G' and G" values, whereas it decreased with an increase in power level (100 W) (Wongsagonsup et al. 2014). Similarly, a weak gel formation was observed in LPAP-treated talipot starch (Navaf et al. 2022b). The authors added that, as the plasma power and treatment time increased, the reduction in G' and G" was increased.

1.8.3 Impact of Plasma Treatment on Structural Characteristics

The amorphous region of starch granules is highly sensitive to active species; thus, the exposure of starch to plasma results in a greater reduction in the amorphous region than the crystalline region. In general, like other physical treatments, starch had less or no effect on its polymorphic structure or crystalline pattern. However, it alters the crystalline peak intensity. The reduction in RC of plasma-treated starch highly depends on the mc of the starch since the water molecules are highly prone to molecular degradation by secondary electrons formed by the plasma. The formation of reactive hydroxyl species also reduced the RC. Sudheesh et al. (Sudheesh et al. 2019b) reported that the exposure of EN air plasma did not alter the crystalline pattern of kithul starch; however, the RC was significantly decreased with an increase in plasma power and duration of treatment. A similar observation was made in argon plasma-treated talipot starch (Navaf et al. 2022b). The treatment of potato starch with glow discharge plasma also maintained its intact B-type crystalline pattern, decreasing the RC of starch granules (Zhang et al. 2015). Likewise, the exposure of cassava starch to oxygen and helium plasma did not affect its intact A-type crystalline pattern, whereas the RC was decreased (Bie et al. 2016).

In other contexts, a few studies reported an increase in the RC of plasma-treated starches, which depends on the nature of the starch treated. The treatment of wheat, sorghum, and quinoa starches with dielectric barrier discharge plasma increased their RC (Gao et al. 2019). It could be due to the reorganization of smaller fragmented molecules formed from amylose molecules into a more ordered crystalline structure.

1.8.4 Impact of Plasma Treatment on Starch Digestibility

The impact of plasma treatment on the digestibility of starch varies with the structural changes made by plasma treatment in the starch granules. The cross-linking formed by plasma treatment strictly hinders the enzyme susceptibility of starch granules, thus decreasing the starch digestibility. In contrast, the molecular breakdown and granular corrosion of starches may increase starch digestibility (Zhu 2017). The exposure of wheat and quinoa starches to dielectric barrier discharge plasma (DBDP) treatment did not impact their digestibility, whereas it slightly increased the RDS content of sorghum starch (Gao et al. 2019). On the contrary, Shen et al. (2022) observed an increase in RDS and RS content of wheat starch with DBDP treatment. Similarly, an increase in RDS content was observed in dielectric barrier atmospheric cold plasma-treated aria starch (Carvalho et al. 2021). A short period (1 min) of cold plasma exposure increased the digestibility of adzuki bean starch. However, an extended treatment time (5 and 10 min) decreased the digestibility (Ge et al. 2021).

1.8.5 Impact of Plasma Treatment on Granule Morphology

Plasma treatment alters the granule morphology to various extents and varies with the type of starch, plasma, and treatment condition. However, if the treatment conditions are mild, plasma treatment may not cause any changes in morphology. A LPAP treatment caused the development of holes and cracks in talipot starch (Navaf et al. 2022b). The cold plasma treatment (60 W-10 min) of rice starch develops fissures or cavities in the granule surface. However, it did not affect the morphology of waxy rice, corn, and potato starches (Okyere et al. 2019). Nevertheless, DBDP treatment caused the formation of larger and deeper fissures or holes in sorghum and quinoa starches (Gao et al. 2019). However, in wheat starch, it caused the formation of indentations, corrosion, and microdeposition on the granule surface (Shen et al. 2022). Similar observations were made in neutral nitrogen plasma-treated kithul starch (Sudheesh et al. 2019b).

1.8.6 Impact of Plasma Treatment on Starch Functional Properties

The plasma treatment of starches significantly alters their functional properties. The effect of plasma on the functional properties of starch varies with the type of starch, plasma, and treatment condition. The exposure of APJP to corn starch did not alter its SP significantly, but it increased solubility (Wu et al. 2018). DBDP-treated buckwheat, sorghum, wheat, and quinoa starches showed increased solubility and decreased SP (Gao et al. 2019).

Similarly, the cold plasma treatment of red adzuki starch also shows increased solubility and decreased SP (Ge et al. 2021). The SP and solubility of kitul starch were increased by exposure to an EN air plasma (Sudheesh et al. 2019b). Similarly, the low-pressure radio frequency air plasma treatment of corn and tapioca starches showed an increase in SP with an increase in plasma power and treatment time (Banura et al. 2017).

The molecular depolymerization, degradation, and oxidation of starch molecules by plasma treatment lead to smaller fragments with higher solubility (Wu et al. 2018). The degradation of the starch chain during plasma treatment negatively affects the water absorption capacity of swollen granules, thus resulting in a decreased swelling index (Gao et al. 2019). Banura et al. (2017) stated that the higher SP of plasma-treated starch could be due to a reduction in amylose content by plasma treatment since the high amylose starch exhibited a lower swelling index than low amylose starches. In addition, Sudheesh et al. (Sudheesh et al. 2019b) stated that surface damage and molecular depolymerization of starch by plasma treatment facilitate high hydration, resulting in a higher swelling index.

The LT of plasma-treated starch varies with the treatment conditions. The plasma-induced molecular depolymerization and cross-linking may increase LT and paste clarity (Zhu 2017). For example, the LPAP-treated talipot starch (Navaf et al. 2022b), APJP-treated corn starch (Wu et al. 2018), etc. exhibited an increased LT and paste clarity. The plasma-induced depolymerization of starch granules is the primary cause of an increase in paste clarity. Depolymerization results in the formation of starch fragments with short chain lengths. These have higher dispersibility and solubility during gelatinization resulting in higher LT. Contrarily, a decrease in paste clarity and LT was reported in corona electrical discharge plasma-treated corn starch (Nemtanu and Minea 2006). The atmospheric argon jet plasma treatment of tapioca starch decreased LT at a low plasma power (50 W). In contrast, it further increased the LT at a higher plasma power (100 W) (Wongsagonsup et al. 2014). The authors stated that the reduced LT of plasma-treated starches could be due to the predominant effect of cross-linking.

The plasma-induced molecular depolymerization and breakdown of glycosidic linkage may increase the syneresis value of plasma-treated starch. The DBDP treatment of buckwheat, wheat, sorghum, and quinoa starches increases the syneresis value compared to the native counterpart (Gao et al. 2019). Similarly, a slight increase in syneresis value was reported in cold plasma-treated rice starch (Thirumdas et al. 2017b).

1.9 Conclusion

The physical modifications of starches can be done by various methods, i.e., thermal and nonthermal methods. Thermal methods include heat moisture treatment, annealing, etc. Nonthermal methods include cold plasma treatment, ultrasound treatment, etc. The starch modification by these recent techniques is gaining popularity since these are highly efficient in producing starches with desired functionalities. Physical modifications alter the physicochemical and functional properties of starch without affecting its chemical structure. In addition, the impact of different physical modifications on starch functionalities varies with the starch source, type, and treatment conditions. Depending on the method and intensity of treatment, the physical modifications lead to a change in morphology, thermal properties, pasting properties, digestibility, etc. By modifying these properties, physical modifications can tailor starch for specific applications in food, pharmaceutical, and other industries.

References

Aaliya B, Sunooj KV, Navaf M, Akhila PP, Sudheesh C, Sabu S et al (2022) Influence of plasma-activated water on the morphological, functional, and digestibility characteristics of hydrothermally modified non-conventional talipot starch. Food Hydrocoll 130:130

Aaliyaa B, Sunooj KV, Navaf M, ParambilAkhila P, Sudheesh C, Sabu S et al (2021) Effect of low dose γ-irradiation on the structural and functional properties, and in vitro digestibility of ultrasonicated stem starch from Corypha umbraculifera L. Appl food Res 1:1000013

Adewale P, Yancheshmeh MS, Lam E (2022) Starch modification for non-food, industrial applications: market intelligence and critical review. Carbohydr Polym 291:119590

Ahmed J, Thomas L, Taher A, Joseph A (2016) Impact of high pressure treatment on functional, rheological, pasting, and structural properties of lentil starch dispersions. Carbohydr Polym 152:639–647

Ahmed J, Thomas L, Arfat YA, Joseph A (2018) Rheological, structural and functional properties of high-pressure treated quinoa starch in dispersions. Carbohydr Polym 197:648–657

Almeida RLJ, Santos NC, Feitoza JVF, de Alcântara Ribeiro VH, de Alcântara Silva VM, de Figueiredo MJ et al (2022) The impact of the pulsed electric field on the structural, morphological, functional, textural, and rheological properties of red rice starch (Oryza sativa). J Food Process Eng 45(11):1–11

Ariyantoro AR, Katsuno N, Nishizu T (2018a) Effect of annealing process on physicochemical, morphological and gelatinization properties of cereal starches. Rev Agric Sci 6:81–92

Ariyantoro AR, Katsuno N, Nishizu T (2018b) Effects of dual modification with succinylation and annealing on physicochemical, thermal and morphological properties of corn starch. Food Secur 7(9):133

Arns B, Bartz J, Radunz M, do Evangelho JA, Pinto VZ, da Zavareze ER et al (2015) Impact of heat-moisture treatment on rice starch, applied directly in grain paddy rice or in isolated starch. LWT Food Sci Technol 60(2):708–713

Atrous H, Benbettaieb N, Hosni F, Danthine S, Blecker C, Attia H et al (2015) Effect of γ-radiation on free radicals formation, structural changes and functional properties of wheat starch. Int J Biol Macromol 80:64–76

Bajaj R, Singh N, Ghumman A, Kaur A, Mishra HN (2022) Effect of high pressure treatment on structural, functional, and in-vitro digestibility of starches from tubers, cereals, and beans. Starch/Staerke 74(1–2):74

Banura S, Thirumdas R, Kaur A, Deshmukh RR, Annapure US (2017) Modification of starch using low pressure radio frequency air plasma. LWT Food Sci Technol 89:719–724

Bao J, Ao Z, Jane JL (2005) Characterization of physical properties of flour and starch obtained from gamma-irradiated white rice. Starch/Staerke 57(10):480–487

Barroso AG, del Mastro NL (2019) Physicochemical characterization of irradiated arrowroot starch. Radiat Phys Chem 158(February):194–198

Bashir K, Aggarwal M (2017) Physicochemical, thermal and functional properties of gamma irradiated chickpea starch. Int J Biol Macromol 97:426–433

Bashir K, Swer TL, Prakash KS, Aggarwal M (2017) Physico-chemical and functional properties of gamma irradiated whole wheat fl our and starch. LWT Food Sci Technol 76:131–139. https://doi.org/10.1016/j.lwt.2016.10.050

Bauer BA, Knorr D (2005) The impact of pressure, temperature and treatment time on starches: pressure-induced starch gelatinisation as pressure time temperature indicator for high hydrostatic pressure processing. J Food Eng 68:392–334
Bharti I, Singh S, Saxena DC (2018) Exploring the influence of heat moisture treatment on physicochemical, pasting, structural and morphological properties of mango kernel starches from Indian cultivars. LWT Food Sci Technol 2019(110):197–206
Bie P, Li X, Xie F, Chen L, Zhang B, Li L (2016) Supramolecular structure and thermal behavior of cassava starch treated by oxygen and helium glow-plasmas. Innov Food Sci Emerg Technol 34:336–343
Bonto AP, Tiozon RN, Sreenivasulu N, Camacho DH (2021) Impact of ultrasonic treatment on rice starch and grain functional properties: a review. Ultrason Sonochem 71:105383. https://doi.org/10.1016/j.ultsonch.2020.105383
Brahma B, Sit N (2020) Physicochemical properties and digestibility of heat moisture–treated potato starches for different treatment conditions. Potato Res 63(3):367–383
Cahyana Y, Wijaya E, Halimah TS, Marta H, Suryadi E, Kurniati D (2017) The effect of different thermal modifications on slowly digestible starch and physicochemical properties of green banana flour (Musa acuminata colla). Food Chem 2019(274):274–280
Carmona-García R, Bello-Pérez LA, Aguirre-Cruz A, Aparicio-Saguilán A, Hernández-Torres J, Alvarez-Ramirez J (2016) Effect of ultrasonic treatment on the morphological, physicochemical, functional, and rheological properties of starches with different granule size. Starch/Staerke 68(9–10):972–979
Carvalho APMG, Barros DR, da Silva LS, Sanches EA, da Pinto CC, de Souza SM et al (2021) Dielectric barrier atmospheric cold plasma applied to the modification of Ariá (Goeppertia allouia) starch: effect of plasma generation voltage. Int J Biol Macromol 182:1618–1627
Castanha N, Miano AC, Sabadoti VD, Augusto PED (2019) Irradiation of mung beans (Vigna radiata): a prospective study correlating the properties of starch and grains. Int J Biol Macromol 129:460–470
Castro LMG, Alexandre EMC, Saraiva JA, Pintado M (2020) Impact of high pressure on starch properties: a review. Food Hydrocoll 106:105877
Chakraborty I, Govindaraju I, Rongpipi S, Mahato KK, Mazumder N (2021) Effects of hydrothermal treatments on physicochemical properties and in vitro digestion of starch. Food Biophys 16(4):544–554
Chan CH, Wu RG, Shao YY (2021) The effects of ultrasonic treatment on physicochemical properties and in vitro digestibility of semigelatinized high amylose maize starch. Food Hydrocoll 119:106831
Chen L, McClements DJ, Yang T, Ma Y, Ren F, Tian Y et al (2021) Effect of annealing and heat-moisture pretreatments on the oil absorption of normal maize starch during frying. Food Chem 353:129468
Chi C, Li X, Lu P, Miao S, Zhang Y, Chen L (2019) Dry heating and annealing treatment synergistically modulate starch structure and digestibility. Int J Biol Macromol 137:554–561
Chung HJ, Liu Q (2010) Molecular structure and physicochemical properties of potato and bean starches as affected by gamma-irradiation. Int J Biol Macromol 47(2):214–222
Chung HJ, Liu Q, Hoover R (2009a) Impact of annealing and heat-moisture treatment on rapidly digestible, slowly digestible and resistant starch levels in native and gelatinized corn, pea and lentil starches. Carbohydr Polym 75(3):436–447
Chung HJ, Hoover R, Liu Q (2009b) The impact of single and dual hydrothermal modifications on the molecular structure and physicochemical properties of normal corn starch. Int J Biol Macromol 44(2):203–210
Chung HJ, Liu Q, Hoover R (2010) Effect of single and dual hydrothermal treatments on the crystalline structure, thermal properties, and nutritional fractions of pea, lentil, and navy bean starches. In: Food research international, vol 43. Elsevier, Amsterdam, p 501. https://doi.org/10.1016/j.foodres.2009.07.030

Colussi R, Kringel D, Kaur L, da Rosa ZE, Dias ARG, Singh J (2020) Dual modification of potato starch: effects of heat-moisture and high pressure treatments on starch structure and functionalities. Food Chem 318:126475
Cui L, Pan Z, Yue T, Atungulu GG, Berrios J (2010) Effect of ultrasonic treatment of Brown Rice at different temperatures on cooking properties and quality. Cereal Chem 87(5):403–408
Dar MZ, Deepika K, Jan K, Swer TL, Kumar P, Verma R et al (2018) Modification of structure and physicochemical properties of buckwheat and oat starch by γ-irradiation. Int J Biol Macromol 108:1348–1356
de Almeida MCBM, dos Costa S, Cavalcanti MT, Almeida EL (2020) Characterization of Prata Banana (Musa AAB-Prata) starch: native and modified by annealing. Starch 72:1900137
Deng Y, Jin Y, Luo Y, Zhong Y, Yue J, Song X et al (2014) Impact of continuous or cycle high hydrostatic pressure on the ultrastructure and digestibility of rice starch granules. J Cereal Sci 60(2):302–310
Dhull SB, Punia S, Kumar M, Singh S, Singh P (2021) Effect of different modifications (physical and chemical) on morphological, pasting, and rheological properties of black Rice (Oryza sativa L. Indica) starch: a comparative study. Starch/Staerke 73(1–2):1–8
Dias ARG, da Rosa ZE, Spier F, de Castro LAS, Gutkoski LC (2010) Effects of annealing on the physicochemical properties and enzymatic susceptibility of rice starches with different amylose contents. Food Chem 123(3):711–719
Dikkala PK, Shirisha SDSN (2018) Irradiation of starch and its effect on pasting properties: a review. Int J Curr Microbiol Appl Sci 7(02):1618–1625
Ding Y, Liang Y, Luo F, Ouyang Q, Lin Q (2020) Understanding the mechanism of ultrasonication regulated the digestibility properties of retrograded starch following vacuum freeze drying. Carbohydr Polym 228:115350
Ezekiel R, Rana G, Singh N, Singh S (2007) Physicochemical, thermal and pasting properties of starch separated from γ-irradiated and stored potatoes. Food Chem 105(4):1420–1429
Gani A, Bashir M, Wani SM, Masoodi FA (2012) Modification of bean starch by γ-irradiation: effect on functional and morphological properties. LWT Food Sci Technol 49(1):162–169
Gao S, Liu H, Sun L, Liu N, Wang J, Huang Y et al (2019) The effects of dielectric barrier discharge plasma on physicochemical and digestion properties of starch. Int J Biol Macromol 138:819–830
Ge X, Shen H, Su C, Zhang B, Zhang Q, Jiang H et al (2021) the improving effects of cold plasma on multi-scale structure, physicochemical and digestive properties of dry heated red adzuki bean starch. Food Chem 349:129159
Genkina NK, Wasserman LA, Noda T, Tester RF, Yuryev VP (2004) Effects of annealing on the polymorphic structure of starches from sweet potatoes (Ayamurasaki and Sunnyred cultivars) grown at various soil temperatures. Carbohydr Res 339:1093–1098
Gomes AMM, Mendes Da Silva CE, Ricardo NMPS, Sasaki JM, Germani R (2004) Impact of annealing on the physicochemical properties of unfermented cassava starch ("polvilho doce"). Starch 56:419–423
Gul K, Singh AK, Sonkawade RG (2016) Physicochemical, thermal and pasting characteristics of gamma irradiated rice starches. Int J Biol Macromol 85:460–466
Guo Z, Zeng S, Lu X, Zhou M, Zheng M, Zheng B (2015) Structural and physicochemical properties of lotus seed starch treated with ultra-high pressure. Food Chem 186:223–230
Han Z, Zeng XA, Zhang BS, Yu SJ (2009a) Effects of pulsed electric fields (PEF) treatment on the properties of corn starch. J Food Eng 93(3):318–323
Han Z, Zeng XA, Yu SJ, Zhang BS, Chen XD (2009b) Effects of pulsed electric fields (PEF) treatment on physicochemical properties of potato starch. Innov Food Sci Emerg Technol 10(4):481–485
Han Z, Zeng XA, Fu N, Yu SJ, Chen XD, Kennedy JF (2012) Effects of pulsed electric field treatments on some properties of tapioca starch. Carbohydr Polym 89(4):1012–1017
Hoover R (2010) The impact of heat-moisture treatment on molecular structures and properties of starches isolated from different botanical sources. Crit Rev Food Sci Nutr 50(9):835–847

Hoover R, Manuel H (1997) Effect of heat-moisture treatment on the structure and physicochemical properties of legume starches. Food Res Int 29(8):731–750
Jayakody L, Hoover R (2008) Effect of annealing on the molecular structure and physicochemical properties of starches from different botanical origins—a review. Carbohydr Polym 74:691–703
Ji X, Rong L, Yu Q, Chen Y, Li J, Xie J (2022) Gallic acid and heat moisture treatment improve pasting, rheological, and microstructure properties of Chinese yam starch-chitosan gels: a comparative study. Int J Biol Macromol 222:114–120
Jiang B, Li W, Hu X, Wu J, Shen Q (2015) Rheology of mung bean starch treated by high hydrostatic pressure. Int J Food Prop 18(1):81–92
Jyothi AN, Sajeev MS, Sreekumar J (2010) Hydrothermal modifications of tropical tuber starches. 1. Effect of heat-moisture treatment on the physicochemical, rheological and gelatinization characteristics. Starch-Starke 62:28–40
Kalaivendan RGT, Mishra A, Eazhumalai G, Annapure US (2022) Effect of atmospheric pressure non-thermal pin to plate plasma on the functional, rheological, thermal, and morphological properties of mango seed kernel starch. Int J Biol Macromol 196:63–71
Kaur M, Singh S (2019) Influence of heat-moisture treatment (HMT) on physicochemical and functional properties of starches from different Indian oat (Avena sativa L.) cultivars. Int J Biol Macromol 122:312–319
Khurshida S, Das MJ, Deka SC, Sit N (2021) Effect of dual modification sequence on physicochemical, pasting, rheological and digestibility properties of cassava starch modified by acetic acid and ultrasound. Int J Biol Macromol 188:649–656
Kim HS, Kim BY, Baik MY (2012) Application of ultra high pressure (UHP) in starch chemistry. Crit Rev Food Sci Nutr 52(2):123–141
Kong X, Kasapis S, Bao J, Corke H (2009) Effect of gamma irradiation on the thermal and rheological properties of grain amaranth starch. Radiat Phys Chem 78:954–960
Kong X, Zhou X, Sui Z, Bao J (2016) Effects of gamma irradiation on physicochemical properties of native and acetylated wheat starches. Int J Biol Macromol 91:1141–1150
Krishnakumar T, Sajeev MS (2018) Effect of ultrasound treatment on physicochemical and functional properties of cassava starch. Int J Curr Microbiol Appl Sci. 7(10):3122–3135
Kumar P, Prakash KS, Jan K, Swer TL, Jan S, Verma R et al (2017) Effects of gamma irradiation on starch granule structure and physicochemical properties of brown rice starch. J Cereal Sci 77:194–200
Laroque DA, Seó ST, Valencia GA, Laurindo JB, Carciofi BAM (2022) Cold plasma in food processing: design, mechanisms, and application. J Food Eng 312:110748
Leite TS, de Jesus ALT, Schmiele M, Tribst AAL, Cristianini M (2017) High pressure processing (HPP) of pea starch: effect on the gelatinization properties. LWT Food Sci Technol. 76:361–369
Li G, Zhu F (2018) Effect of high pressure on rheological and thermal properties of quinoa and maize starches. Food Chem 241:380–386
Li W, Zhang F, Liu P, Bai Y, Gao L, Shen Q (2011) Effect of high hydrostatic pressure on physicochemical, thermal and morphological properties of mung bean (Vigna radiata L.) starch. J Food Eng 103(4):388–393
Li W, Bai Y, Mousaa SAS, Zhang Q, Shen Q (2012) Effect of high hydrostatic pressure on physicochemical and structural properties of Rice starch. Food Bioprocess Technol 5(6):2233–2241
Li W, Tian X, Liu L, Wang P, Wu G, Zheng J et al (2015) High pressure induced gelatinization of red adzuki bean starch and its effects on starch physicochemical and structural properties. Food Hydrocoll 45:132–139
Li Q, Wu QY, Jiang W, Qian JY, Zhang L, Wu M et al (2019) Effect of pulsed electric field on structural properties and digestibility of starches with different crystalline type in solid state. Carbohydr Polym 207:362–370
Liu T, Ma Y, Xue S, Shi J (2012) Modifications of structure and physicochemical properties of maize starch by γ-irradiation treatments. LWT Food Sci Technol 46(1):156–163
Liu H, Lv M, Wang L, Li Y, Fan H, Wang M (2016a) Comparative study: how annealing and heat-moisture treatment affect the digestibility, textural, and physicochemical properties of maize starch. Starch/Staerke 68(11–12):1158–1168

Liu H, Guo X, Li Y, Li H, Fan H, Wang M (2016b) In vitro digestibility and changes in physicochemical and textural properties of tartary buckwheat starch under high hydrostatic pressure. J Food Eng 189:64–71
Liu H, Wang L, Cao R, Fan H, Wang M (2016c) In vitro digestibility and changes in physicochemical and structural properties of common buckwheat starch affected by high hydrostatic pressure. Carbohydr Polym 144:1–8
Liu M, Wu NN, Yu GP, Zhai XT, Chen X, Zhang M et al (2018) Physicochemical properties, structural properties, and in vitro digestibility of pea starch treated with high hydrostatic pressure. Starch/Staerke 70(1–2):1–9
Marboh V, Mahanta CL (2021) Physicochemical and rheological properties and in vitro digestibility of heat moisture treated and annealed starch of sohphlang (Flemingia vestita) tuber. Int J Biol Macromol 168:486–495
Martins A, Beninca C, Bet CD, Bisinella RZB, de Oliveira CS, Hornung PS et al (2020) Ultrasonic modification of purple taro starch (Colocasia esculenta B. Tini): structural, psychochemical and thermal properties. J Therm Anal Calorim 142(2):819–828
Mason WR (2009) Starch use in foods. In: BeMiller J, Whistler R (eds) , vol 4. Academic, New York, pp 775–795
Monroy Y, Rivero S, García MA (2017) Microstructural and techno-functional properties of cassava starch modified by ultrasound. Ultrason Sonochem 2018(42):795–804
Navaf M, Sunooj KV, Krishna NU, Aaliya B, Sudheesh C, Akhila PP et al (2021) Effect of different hydrothermal treatments on pasting, textural, and rheological properties of single and dual modified Corypha umbraculifera L. Starch 74:2100236
Navaf M, Sunooj KV, Aaliya B, Sudheesh C, Akhila PP, Sabu S et al (2022a) Impact of gamma irradiation on structural, thermal, and rheological properties of talipot palm (Corypha umbraculifera L.) starch: a stem starch. Radiat Phys Chem 201:110459
Navaf M, Sunooj KV, Aaliya B, Akhila PP, Sudheesh C, Sinha SK et al (2022b) Impact of low-pressure argon plasma on structural, thermal, and rheological properties of Corypha umbraculifera L. starch: a non-conventional source of stem pith starch. Starch 75:2200165
Nemtanu MR, Minea R (2006) Functional properties of corn starch treated with corona electrical discharges. Macromol Symp 245–246:525–528
Nie H, Li C, Liu P, Lei C, Li J (2019) Retrogradation, gel texture properties, intrinsic viscosity and degradation mechanism of potato starch paste under ultrasonic irradiation. Food Hydrocoll 95:590–600. https://doi.org/10.1016/j.foodhyd.2017.08.035
Nwaogazie FO, Akinwande BA, Oyeyinka SA (2022) Physicochemical properties of Bambara groundnut (Vigna subterranea) starch annealed at different temperatures. J Food Process Preserv 46:e17183
Ogundiran OO, Adeniyi Afolabi T (2018) Effects of annealing and heat moisture treatments on the physicochemical properties of white sorghum (Sorghum bicolor). Starch J Chem Soc 43(3):463–470. http://www.journals.chemsociety.org.ng/index.php/jcsn/article/view/177
Okyere AY, Bertoft E, Annor GA (2019) Modification of cereal and tuber waxy starches with radio frequency cold plasma and its effects on waxy starch properties. Carbohydr Polym 223(July):115075
Othman Z, Hassan O, Hashim K (2015) Physicochemical and thermal properties of gamma-irradiated sago (Metroxylon sagu) starch. Radiat Phys Chem 109:48–53
Oyeyinka SA, Adegoke R, Oyeyinka AT, Salami KO, Olagunju OF, Kolawole FL et al (2018) Effect of annealing on the functionality of Bambara groundnut (Vigna subterranea) starch–palmitic acid complex. Int J Food Sci Technol 53(2):549–555
Park EY, Ma JG, Kim J, Lee DH, Kim SY, Kwon DJ et al (2018) Effect of dual modification of HMT and crosslinking on physicochemical properties and digestibility of waxy maize starch. Food Hydrocoll 75:33–40
Piecyk M, Domian K (2021) Effects of heat–moisture treatment conditions on the physicochemical properties and digestibility of field bean starch (Vicia faba var. minor). Int J Biol Macromol 182:425–433

Piecyk M, Drużyńska B, Ołtarzewska A, Wołosiak R, Worobiej E, Ostrowska-Ligęza E (2018) Effect of hydrothermal modifications on properties and digestibility of grass pea starch. Int J Biol Macromol 118:2113–2120

Polesi LF, Sarmento SBS, De Moraes J, Franco CML, Canniatti-Brazaca SG (2016) Physicochemical and structural characteristics of rice starch modified by irradiation. Food Chem 191:59–66

Pratiwi M, Faridah DN, Lioe HN (2018) Structural changes to starch after acid hydrolysis, debranching, autoclaving-cooling cycles, and heat moisture treatment (HMT): a review. Starch 1700028:1–13

Puelles-Román J, Barroso NG, Cruz-Tirado JP, Tapia-Blácido DR, Angelats-Silva L, Barraza-Jáuregui G et al (2021) Annealing process improves the physical, functional, thermal, and rheological properties of Andean oca (Oxalis tuberosa) starch. J Food Process Eng 44(6):1–10

Rahaman A, Kumari A, Zeng XA, Adil Farooq M, Siddique R, Khalifa I et al (2021) Ultrasound based modification and structural-functional analysis of corn and cassava starch. Ultrason Sonochem 80:105795

Reza S, Maghsoudlou Y, Rostamabadi H (2019) Preparation of physically modified oat starch with different sonication treatments. Food Hydrocoll 89:311–320

Rocha TS, Felizardo SG, Jane J, lin, Franco CML. (2012) Effect of annealing on the semicrystalline structure of normal and waxy corn starches. Food Hydrocoll 29(1):93–99

Schafranski K, Ito VC, Lacerda LG (2021) Impacts and potential applications: a review of the modification of starches by heat-moisture treatment (HMT). Food Hydrocoll 117:106690

Sharma M, Yadav DN, Singh AK, Tomar SK (2015a) Effect of heat-moisture treatment on resistant starch content as well as heat and shear stability of pearl millet starch. Agric Res 4(4):411–419

Sharma M, Yadav DN, Singh AK, Tomar SK (2015b) Rheological and functional properties of heat moisture treated pearl millet starch. J Food Sci Technol 52(10):6502–6510

Shen H, Ge X, Zhang Q, Zhang X, Lu Y, Jiang H et al (2022) Dielectric barrier discharge plasma improved the fine structure, physicochemical properties and digestibility of α-amylase enzymatic wheat starch. Innov Food Sci Emerg Technol 78:78

Sindhu R, Devi A, Khatkar BS (2019) Morphology, structure and functionality of acetylated, oxidized and heat moisture treated amaranth starches. Sci Total Environ 706:135907. https://doi.org/10.1016/j.scitotenv.2019.135907

Singh H, Chang YH, Lin JH, Singh N, Singh N (2011a) Influence of heat-moisture treatment and annealing on functional properties of sorghum starch. Food Res Int 44(9):2949–2954

Singh H, Ho Y, Lin J, Singh N, Singh N (2011b) Influence of heat—moisture treatment and annealing on functional properties of sorghum starch. Food Res Int 44:2949–2954

Singh S, Singh N, Ezekiel R, Kaur A (2011c) Effects of gamma-irradiation on the morphological , structural , thermal and rheological properties of potato starches. Carbohydr Polym 83:1521–1528

Sofi BA, Wani IA, Masoodi FA, Saba I, Muzaffar S (2013) Effect of gamma irradiation on physicochemical properties of broad bean (Vicia faba L.) starch. LWT Food Sci Technol 54(1):63–72

Song HY, Lee SY, Choi SJ, Kim KM, Kim JS, Han GJ et al (2014) Digestibility and physicochemical properties of granular sweet potato starch as affected by annealing. Food Sci Biotechnol 23(1):23–31

Sudheesh C, Sunooj KV, George J, Kumar S, Vikas SVA (2019a) Impact of γ− irradiation on the physico-chemical, rheological properties and in vitro digestibility of kithul (Caryota urens) starch; a new source of nonconventional stem starch. Radiat Phys Chem 162(April):54–65

Sudheesh C, Sunooj KV, Sinha SK, George J, Kumar S, Murugesan P et al (2019b) Impact of energetic neutral nitrogen atoms created by glow discharge air plasma on the physico-chemical and rheological properties of kithul starch. Food Chem 294(April):194–202

Sudheesh C, Sunooj KV, Navaf M, Bhasha SA, George J, Mounir S et al (2020a) Hydrothermal modifications of nonconventional kithul (Caryota urens) starch: physico-chemical, rheological properties and in vitro digestibility. J Food Sci Technol 57(8):2916–2925

Sudheesh C, Sunooj KV, Alom M, Kumar S, Sajeevkumar VA, George J (2020b) Effect of dual modification with annealing, heat moisture treatment and cross-linking on the physico-

chemical, rheological and in vitro digestibility of underutilised kithul (Caryota urens) starch. J Food Meas Charact 14(3):1557–1567
Sujka M (2017) Ultrasonic modification of starch—impact on granules porosity. Ultrason Sonochem 37:424–429
Thakur Y, Thory R, Singh K, Maninder S (2021) Effect of selected physical and chemical modifications on physicochemical , pasting , and morphological properties of underutilized starch from rice bean (Vigna umbellata). J Food Sci Technol 04:4974
Thirumdas R, Kadam D, Annapure US (2017a) Cold plasma: an alternative technology for the starch modification. Food Biophys 12(1):129–139
Thirumdas R, Trimukhe A, Deshmukh RR, Annapure US (2017b) Functional and rheological properties of cold plasma treated rice starch. Carbohydr Polym 157:1723–1731
Trung PTB, Ngoc LBB, Hoa PN, Tien NNT, Van HP (2017) Impact of heat-moisture and annealing treatments on physicochemical properties and digestibility of starches from different colored sweet potato varieties. Int J Biol Macromol 105:1071–1078
Varatharajan V, Hoover R, Liu Q, Seetharaman K (2010) The impact of heat-moisture treatment on the molecular structure and physicochemical properties of normal and waxy potato starches. Carbohydr Polym 81(2):466–475. https://doi.org/10.1016/j.carbpol.2010.03.002
Verma R, Jan S, Rani S, Jan K, Swer TL, Prakash KS (2017) Physicochemical and functional properties of gamma irradiated buckwheat and potato starch. Radiat Phys Chem 2018(144):37–42
Verma K, Jan K, Bashir K (2019) γ Irradiation of Cowpea and potato starch: effect on physicochemical functional and rheological properties. J Food Process Technol 10:1000810
Waduge RN, Hoover R, Vasanthan T, Gao J, Li J (2006) Effect of annealing on the structure and physicochemical properties of barley starches of varying amylose content. Food Res Int 39:59–77
Wang S, Wang J, Yu J, Wang S (2014) A comparative study of annealing of waxy , normal and high-amylose maize starches : the role of amylose molecules. Food Chem 164:332–338
Wang S, Wang J, Wang S, Wang S (2017a) Annealing improves paste viscosity and stability of starch. Food Hydrocoll 62:203–211
Wang L, Zhang C, Chen Z, Wang X, Wang K, Li Y et al (2017b) Effect of annealing on the physico-chemical properties of rice starch and the quality of rice noodles. J Cereal Sci 2018(84):125–131
Wang H, Liu Y, Chen L, Li X, Wang J, Xie F (2018) Insights into the multi-scale structure and digestibility of heat-moisture treated rice starch. Food Chem 242:323–329
Wang H, Xu K, Liang Y, Zhang H, Chen L, Xu K et al (2019) Impact of ultrasonication on the aggregation structure and physicochemical characteristics of sweet potato starch. Ultrason Sonochem 63:104868
Wang Q, Li L, Zheng X (2021) Recent advances in heat-moisture modified cereal starch: structure, functionality and its applications in starchy food systems. Food Chem 344:128700
Wang L, Wang M, Zhou Y, Wu Y, Ouyang J (2022) Influence of ultrasound and microwave treatments on the structural and thermal properties of normal maize starch and potato starch: a comparative study. Food Chem 377:131990
Wani IA, Jabeen M, Geelani H, Masoodi FA, Saba I, Muzaffar S (2014) Effect of gamma irradiation on physicochemical properties of Indian horse chestnut (Aesculus indica Colebr.) starch. Food Hydrocoll 35:253–263. https://doi.org/10.1016/j.foodhyd.2013.06.002
Wongsagonsup R, Deeyai P, Chaiwat W, Horrungsiwat S, Leejariensuk K, Suphantharika M et al (2014) Modification of tapioca starch by non-chemical route using jet atmospheric argon plasma. Carbohydr Polym 102(1):790–798
Wu TY, Chang CR, Chang TJ, Chang YJ, Liew Y, Chau CF (2018) Changes in physicochemical properties of corn starch upon modifications by atmospheric pressure plasma jet. Food Chem 2019(283):46–51
Wu C, Wu QY, Wu M, Jiang W, Qian JY, Rao SQ et al (2019) Effect of pulsed electric field on properties and multi-scale structure of japonica rice starch. LWT Food Sci Technol. 116:108515

Wu Z, Qiao D, Zhao S, Lin Q, Zhang B, Xie F (2022) Nonthermal physical modification of starch: an overview of recent research into structure and property alterations. Int J Biol Macromol 203:153–175

Xie Y, Li MN, Chen HQ, Zhang B (2018a) Effects of the combination of repeated heat-moisture treatment and compound enzymes hydrolysis on the structural and physicochemical properties of porous wheat starch. Food Chem 2019(274):351–359

Xie H, Gao J, Xiong X, Gao Q (2018b) Effect of heat-moisture treatment on the physicochemical properties and in vitro digestibility of the starch-guar complex of maize starch with varying amylose content. Food Hydrocoll 83:213–221

Xu M, Saleh ASM, Gong B, Li B, Jing L, Gou M et al (2018) The effect of repeated versus continuous annealing on structural, physicochemical, and digestive properties of potato starch. Food Res Int 111(May):324–333

Yadav BS, Guleria P, Yadav RB (2013a) Hydrothermal modification of Indian water chestnut starch: influence of heat-moisture treatment and annealing on the physicochemical, gelatinization and pasting characteristics. LWT Food Sci Technol 53(1):211–217

Yadav BS, Guleria P, Yadav RB (2013b) Hydrothermal modification of Indian water chestnut starch: Influence of heat-moisture treatment and annealing on the physicochemical, gelatinization and pasting characteristics. LWT Food Sci Technol. 53:211–217

Yan Y, Peng B, Niu B, Ji X, He Y, Shi M (2022) Understanding the structure, thermal, pasting, and rheological properties of potato and pea starches affected by annealing using plasma-activated water. Front Nutr 9:1–13

Yang Q, Lu X, Chen Y, Luo Z, Xiao Z (2019) Fine structure, crystalline and physicochemical properties of waxy corn starch treated by ultrasound irradiation. Ultrason Sonochem 51:350–358

Yang S, Dhital S, Zhang MN, Wang J, Chen ZG (2022) Structural, gelatinization, and rheological properties of heat-moisture treated potato starch with added salt and its application in potato starch noodles. Food Hydrocoll 131:107802

Yeum JH, Choi HW, Yun KH, Oh SM, Bae JE, Ye SJ et al (2021) Effect of hydrothermal treatment on physicochemical properties of amorphous granular potato starch (AGPS). Int J Biol Macromol 168:816–822

Yoon HS, Yoo JY, Kim JH, Lee JW, Byun MW, Baik BK et al (2010) In vitro digestibility of gamma-irradiated corn starches. Carbohydr Polym 81(4):961–963

Yu S, Zhang Y, Ge Y, Zhang Y, Sun T, Jiao Y et al (2013) Effects of ultrasound processing on the thermal and retrogradation properties of nonwaxy rice starch. J Food Process Eng 36(6):793–802

Yu B, Li J, Tao H, Zhao H, Liu P, Cui B (2021) Physicochemical properties and in vitro digestibility of hydrothermal treated Chinese yam (Dioscorea opposita Thunb.) starch and flour. Int J Biol Macromol 176:177–185

Zavareze EDR, Dias ARG (2011) Impact of heat-moisture treatment and annealing in starches: a review. Carbohydr Polym 83(2):317–328. https://doi.org/10.1016/j.carbpol.2010.08.064

Zeng F, Ma F, Kong F, Gao Q, Yu S (2015) Physicochemical properties and digestibility of hydrothermally treated waxy rice starch. Food Chem 172:92–98

Zeng F, Gao QY, Han Z, Zeng XA, Yu SJ (2016) Structural properties and digestibility of pulsed electric field treated waxy rice starch. Food Chem 194:1313–1319

Zeng F, Li T, Gao Q, Liu B, Yu S (2018) Physicochemical properties and in vitro digestibility of high hydrostatic pressure treated waxy rice starch. Int J Biol Macromol 120:1030–1038

Zhang B, Chen L, Li X, Li L, Zhang H (2015) Understanding the multi-scale structure and functional properties of starch modulated by glow-plasma: a structure-functionality relationship. Food Hydrocoll 50:228–236

Zhang F, Zhang YY, Thakur K, Zhang JG, Wei ZJ (2018) Structural and physicochemical characteristics of lycoris starch treated with different physical methods. Food Chem 2019(275):8–14

Zhang B, Saleh ASM, Su C, Gong B, Zhao K, Zhang G et al (2020) The molecular structure, morphology, and physicochemical property and digestibility of potato starch after repeated and continuous heat–moisture treatment. J Food Sci 85(12):4215–4224

Zhang J, Zhang M, Bai X, Zhang Y, Wang C (2022a) The impact of high hydrostatic pressure treatment time on the structure, gelatinization and thermal properties and in vitro digestibility of oat starch. Grain Oil Sci Technol 5(1):1–12
Zhang K, Zhang Z, Zhao M, Milosavljević V, Cullen PJ, Scally L et al (2022b) Low-pressure plasma modification of the rheological properties of tapioca starch. Food Hydrocoll:125
Zhao K, Zhang B, Su C, Gong B, Zheng J, Jiang H et al (2020) Repeated heat-moisture treatment: a more effective way for structural and physicochemical modification of mung bean starch compared with continuous way. Food Bioprocess Technol 13(3):452–461
Zheng Y, Chai Z, Kong X, Chen S, Ye X, Tian J (2023) Effect of annealing treatment on the physicochemical properties and enzymatic hydrolysis of different types of starch. Food Chem 403:134153
Zhou Y, Yan Y, Shi M, Liu Y (2019) Effect of an atmospheric pressure plasma jet on the structure and physicochemical properties of waxy and normal maize starch. Polymers (Basel) 11(8):1–15
Zhu F (2015) Impact of ultrasound on structure, physicochemical properties, modifications, and applications of starch. Trends Food Sci Technol 43:1–17
Zhu F (2016) Impact of γ-irradiation on structure, physicochemical properties, and applications of starch. Food Hydrocoll 52:201–212
Zhu F (2017) Plasma modification of starch. Food Chem 232:1–7
Zhu F (2018) Modifications of starch by electric field based techniques. Trends Food Sci Technol 75:158–169
Zhu ZM, Xiao Y, Yang S, Min LH, Hong LM, Yaqoob S et al (2020) Effects of heat–moisture, autoclaving, and microwave treatments on physicochemical properties of proso millet starch. Food Sci Nutr 8(2):735–743
Zuo JY, Knoerzer K, Mawson R, Kentish S, Ashokkumar M (2009) The pasting properties of sonicated waxy rice starch suspensions. Ultrason Sonochem 16(4):462–468

Chemical Crosslinking, Acid Hydrolysis, Oxidation, Esterification, and Etherification of Starch

2

Tahira Mohsin Ali, Salman Haider, Marium Shaikh, Natasha Abbas Butt, and Nusrat Zehra

Abstract

Chemically modified starches are one of the most commonly used additives in food products. Different chemical reactions like esterification, etherification, hydrolysis, oxidation, crosslinking, and bleaching have been employed to develop chemically modified starches globally. The most commercially exploited sources for the development of modified starches are waxy corn, normal and high-amylose corn, tapioca, potato, and rice starch. These are used as bulking agent, thickening agent, coating agent, anti-crystallizing agent, and texturizing agent in wide range of products. Their fat replacement potential in different ready-to-eat products has also been a known starch application. This chapter in-depth discussed preparation, applications, and safety assessment of crosslinked, esterified, etherified, oxidized/bleached, and acid-thinned starches along with the impact of these modifications on the structural, functional, and digestibility characteristics of starches.

Keywords

Chemically modified · Starches · Crosslinking · Oxidation · Bleaching · Etherification · Acid hydrolysis

T. M. Ali (✉) · S. Haider · M. Shaikh
Department of Food Science and Technology, University of Karachi, Karachi, Pakistan
e-mail: tahira.mohsin@uok.edu.pk

N. A. Butt · N. Zehra
Department of Biomedical Engineering, Faculty of Engineering, Science, .Technology and Management, Ziauddin University, Karachi, Pakistan

N. Mazumder, M. H. Rahman (eds.), *Advanced Research in Starch*,
https://doi.org/10.1007/978-981-99-9527-1_2

2.1 Introduction

Starch extraction is carried out commercially from different sources including cereals, tubers, fruits, and roots. However, this form of starch finds limited utility in foods owing to lack of process tolerance against high shear, temperature, and low pH conditions. These problems are resolved by the food industries by using modified starches that are tailor made according to specific product requirement. The term starch modification refers to the alteration that is brought about in the structural arrangement of starch chains through physical, chemical, and enzymatic means. The chemical modifications of starch are highly complex reactions that might result in low or very high DS products or may result in promotion of side reactions if the process conditions related to temperature, catalyst addition, pH, stirring, overhead sealing, and total solids in slurry are not properly controlled. The mechanism of chemical modification is either depolymerization of long starch chains or substitution of hydroxyl groups with different functional groups arising from the chemical compounds used for modification technique or due to crosslinking (di-esterification). The resultant chemically modified starch is characterized by having additional new functional groups like acetyl, succinyl, hydroxypropyl, carboxyl, or octenyl on the starch chains. Different e-numbers are designated by Codex alimentarius to starch additives based on the single or dual modification, and based on the modification reaction, the polymer has undergone. The Code of Federal regulations USA also strictly regulates the reagent to be used and their maximum quantities that are allowed during modification reactions. These functional groups alter the physicochemical, thermal, textural, rheological, and nutritional behavior of native starch. This chapter will give deep insight regarding the types of chemical modifications, safety assessment of chemically modified starches, and effects of chemical modifications on the structural and functional starch characteristics with their specific food applications.

2.1.1 Types of Chemical Modifications

The chemical modification types currently employed on starches are described in Fig. 2.1.

2.1.2 CrossLinked Starches

The importance of crosslinked starches is growing equally in both food and non-food sectors as it provides process tolerance more than any other modified starch. The crosslinked starches produced from corn however carry the lion share in the commercial market. The crosslinking reaction is dependent on various factors like reagents used, pH of the reaction, percent solids in starch slurry, temperature of modification, starch source, amylose content, and total duration of reaction (Ashogbon and Akintayo 2014). Different reagents like STMP (sodium

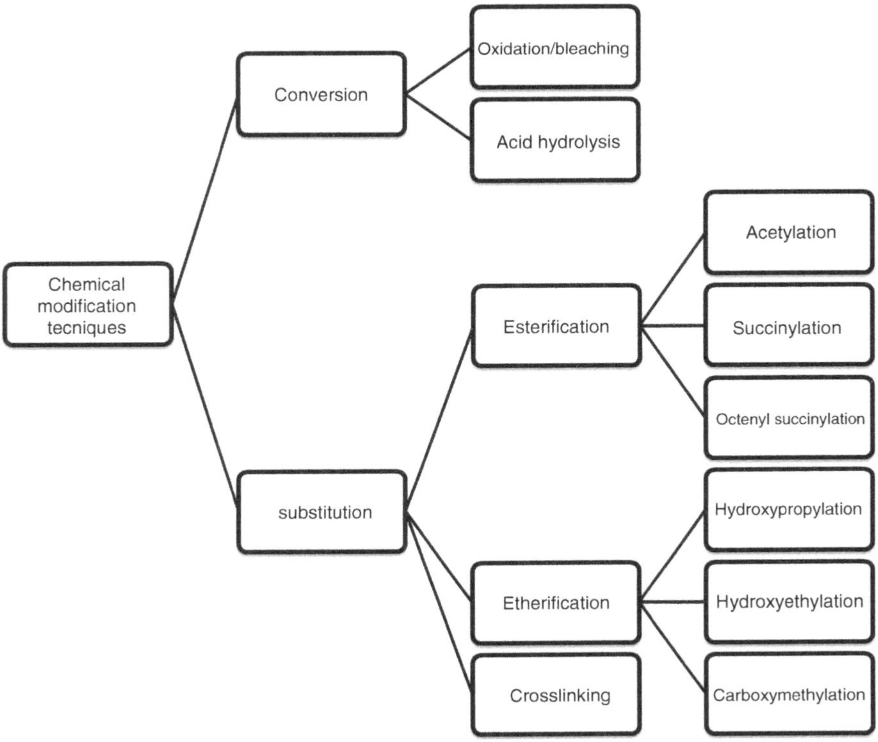

Fig. 2.1 Types of starch-based chemical modifications

trimetaphosphate), STPP (sodium tripolyphosphate), epichlorohydrin, phosphorus oxychloride ($POCl_3$), adipic acid/adipic anhydride, citric acid, etc. have been used for crosslinking of different starches like pearl millet starch (Dhull et al. 2021), corn starch (Dong and Vasanthan 2020), waxy potato starch (Luo et al. 2009), sweet potato starch (Zhao et al. 2012), cassava starch (Jyothi et al. 2006), barely starch (Mehfooz et al. 2020), sorghum starch (Mehboob et al. 2020), and many others.

2.1.2.1 Types of Crosslinking Reagents (Conventional and Non-conventional)

The various types of conventional (commercially implemented) and non-conventional reagents used for crosslinking modification hitherto are detailed in Fig. 2.2. Depending on the type of reagent used, the starches crosslinked may be called (a) distarch phosphates, (b) acetylated-distarch adipates, (c) distarch citrates, (d) distarch glycerol, and (e) phosphated distarch phosphate (Jyothi et al. 2006; Ashogbon and Akintayo 2014).

Table 2.1 shows the reagents used for the aforementioned chemically crosslinked starches along with their E-numbers available in Codex. Except distarch glycerol which is ether, all other crosslinks formed are di-esters. The reagents used for this

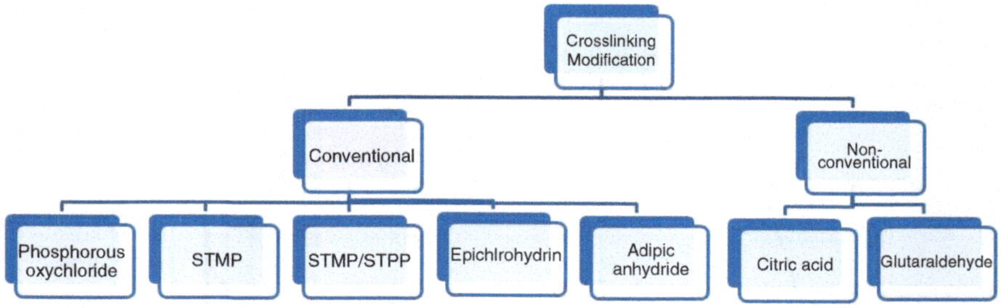

Fig. 2.2 Conventional and non-conventional reagents used for starch crosslinking

Table 2.1 Types of crosslinked starches and the reagents used for their preparation

Chemical name of crosslinked starch	Crosslinking reagents	E-numbers
Distarch phosphate	(a) Sodium trimetaphosphate (STMP) or (b) phosphorus oxychloride	E1412
Phosphated distarch phosphate	Mixture of STMP and STPP (sodium tripolyphosphate)	E1413
Distarch-citrate	Citric acid anhydrous	–
Acetylated distarch adipate	Mixture of acetic and adipic anhydrides	E1422
Distarch glycerol	Epichlorohydrin	–

modification are polyfunctional groups capable of reacting with at least two hydrophilic groups on starch leading to the formation of a "crosslink" (Luo et al. 2009). Crosslinking does not take place in amylose chains but is usually formed between two amylopectin chains or between amylose and amylopectin molecules. Therefore, crosslinking reactions are known to increase crystallinity and reduce amylose content in starch (Kou and Gao 2018).

The crosslinking reaction reinforces the whole starch structure by introducing covalent bonds which gives more strength to the starch structure and also enhances the hydrogen bonds bridging between the starch chains (Jyothi et al. 2006). The different methods used for crosslinking of starch are described below.

Distarch Phosphates

The distarch phosphates can be prepared via phosphorus oxychloride or by using sodium trimetaphosphate.

Using Phosphorus Oxychloride

Phosphorus oxychloride is a fast-acting crosslinking agent that leads to diesterification at pH ≥ 10 preferably. The phosphorus oxychloride contains three chloride ions in its molecule. On addition of $POCl_3$ in water, it forms phosphorus dichloride which is the primary crosslinking agent as two starch chains get esterified to this phosphorus dichloride molecule to form an intermolecular bridge or crosslink. The sodium sulfate anhydrous is also added (2–15%) based on starch weight dry basis which increases the rate of reaction and also prevents starch gelatinization under

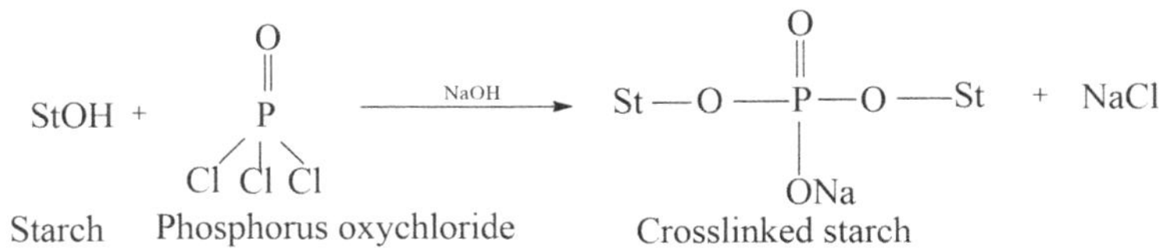

Fig. 2.3 Crosslinking of starch using phosphorus oxychloride

highly alkaline conditions used during the reaction. This reaction is carried out for 1–2 h at 25 °C (Woo and Seib 2002; Delley et al. 2012; Shah et al. 2016). As per CFR (2023), the amount of phosphorus oxychloride based on starch content must be limited to $\leq 0.1\%$. The typical formation of distarch phosphate using $POCl_3$ is shown in Fig. 2.3 (Singh et al. 2007).

Using Sodium Trimetaphosphate (STMP)

For crosslinking with STMP, a starch suspension having 30–50% solids is prepared in the presence of sodium sulfate anhydrous (5%, starch db). No cut-off limit is set by the Code of Federal Regulations regarding reagent addition during the starch modification. Compared to phosphorus oxychloride, the reaction of sodium trimetaphosphate with starch is slow and usually takes 2–5 h to complete in the presence of sodium chloride, sodium carbonate, or sodium sulfate anhydrous which act as catalysts. The reaction is carried out from 40 to 50 °C (Gao et al. 2014; Sugih et al. 2019; Xie et al. 2019). The distarch phosphate is formed using STMP between C6 and C2/C3 of different glucose monomers in starch (Desam et al. 2018).This chemical is cheaper and safer compared to phosphorus oxychloride and has therefore become a commercial choice for crosslinking modification of starch. Also, STMP is safe for human consumption (Woo and Seib 1997). The starches modified by both phosphorus oxychloride and sodium trimetaphosphate are labeled as E1412. The typical distarch phosphate formed using STMP is shown in Fig. 2.4 (Singh et al. 2007).

Using Mixture of STMP and STPP

Phosphated distarch phosphate (E1413) is produced using a mixture of STMP and STPP (sodium tripolyphosphate). Studies have reported two reaction regimes to produce this modified starch which includes (a) semi-dry method and (b) slurry method. However, the latter is more extensively reported in literature. In the slurry method, STMP/STPP is added to a 35–45% starch suspension under basic conditions. The reaction is terminated after 1–3 h by adjusting pH to 6.5. However, in semi-dry method, starch is treated with the mixture of STMP and STPP such that starch is equilibrated to less than 20% moisture content. This semi-dry starch is then heated to 120–140 °C for 2 hours in oven (Lim and Seib 1993; Sang and Seib 2006; Park et al. 2018; Dong and Vasanthan 2020). The dry roasting method results both in heat-moisture treatment and crosslinking and was found to have more thermal stability compared to slurry method (Dong and Vasanthan 2020). Deetae et al. (2008) found that STPP results in mono-esterification, i.e., substitution of

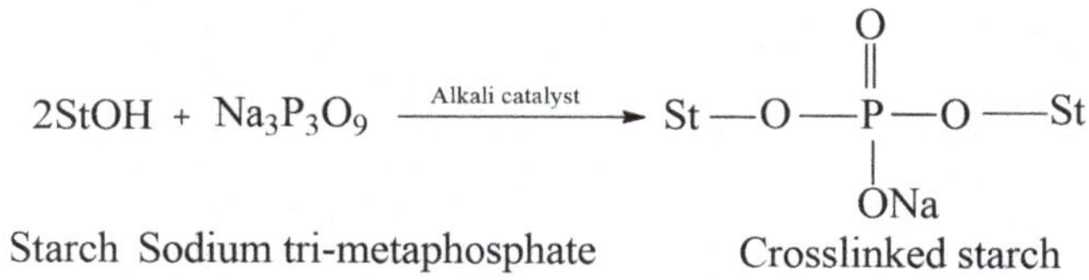

Fig. 2.4 Crosslinking of starch with sodium trimetaphosphate

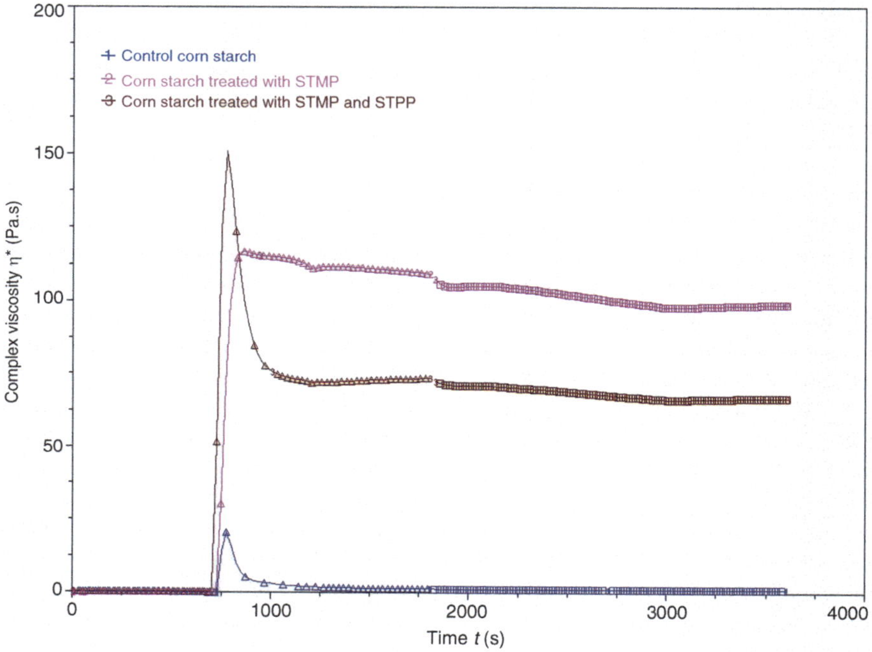

Fig. 2.5 Pasting properties of corn starch treated with STMP and STMP/STPP

phosphate groups on starch chains, whereas STMP results in the formation of intermolecular bridge through distarch esterification resulting in crosslinks. Therefore, the starches when modified using a mixture of STMP and STPP result in both crosslinking and substitution reactions, and that is why it is called phosphated distarch phosphate. Also, according to Sugih et al. (2019), the STMP and STPP together result in the addition of more phosphate groups in a particular period of time, then when these two chemicals are separately used for starch modification. Also, another advantage is that the max phosphorus group allowed when using STMP is 0.04% which increases to 0.4% when STMP and STPP are used together (CFR 2023). Also the dry method of crosslinking results in higher-resistance starch content when used at 5–12% based on starch weight (db) (Park et al. 2018). Figure 2.5 shows the pasting profile of native STMP and STPP/STMP added waxy corn starch at 2.5% level using slurry method.

Distarch Citrate

Distarch citrate though a chemically modified starch is prepared from a nutritionally harmless and green chemical as it is present in the Generally regarded as safe (GRAS) list of FDA (Food and Drug Administration, USA). The citric acid modified starch hasn't been assigned any E-number as yet. Citric acid on reaction with starch can lead to both hydrolysis and esterification depending upon the temperature of the reaction. At temperatures less than gelatinization one, hydrolysis is dominant however at temperatures >100 °C; the citric acid molecule dehydrates to form a reactive anhydride which esterifies with alcoholic hydroxyl group of starch to form a monoester. The same citric acid molecule when esterifies to another starch chain results in the formation of a crosslink or distarch citrate as two starch molecules get esterified to a single citric acid bearing three carboxyl groups (Zhang et al. 2023). Typically the starches on dry basis are treated with 5–40% citric acid based on starch weight followed by adjustment of pH to 3.0. Water and citric acid are allowed to equilibrate with starch for 6–8 hours under ambient conditions and are eventually dried at 50 °C for 24 h. The dried starch is then heated at 130 °C for 2–5 h. The washing of starch as a final step is essential to remove the unreacted acid (You et al. 2020; Zhang et al. 2023). The mechanism of formation of distarch citrate is outlined in Fig. 2.6 (Chowdary et al. 2011).

Di-Acyl Glycerol

Epichlorohydrin (EPI) acts as a crosslinking reagent to produce distarch glycerols. This reagent is not approved by FAO (Codex) as a crosslinking agent but is approved by the Code of Federal Regulations, USA. As per (CFR 2023), epichlorohydrin

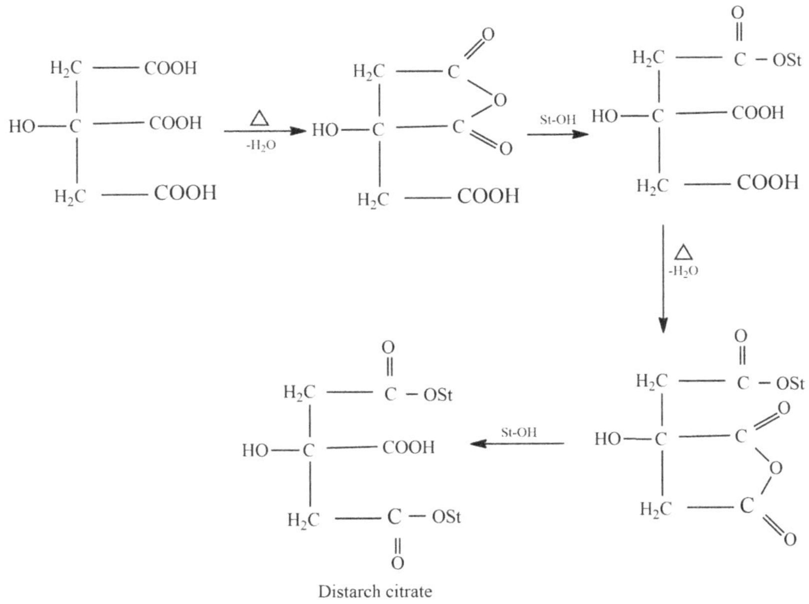

Fig. 2.6 Formation of distarch citrate

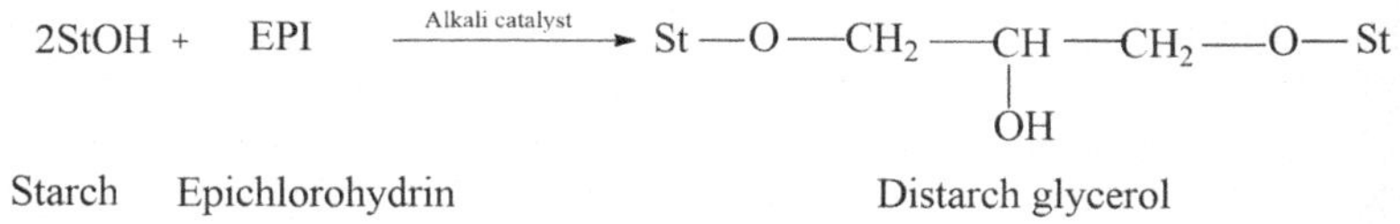

Fig. 2.7 Formation of distarch glycerol

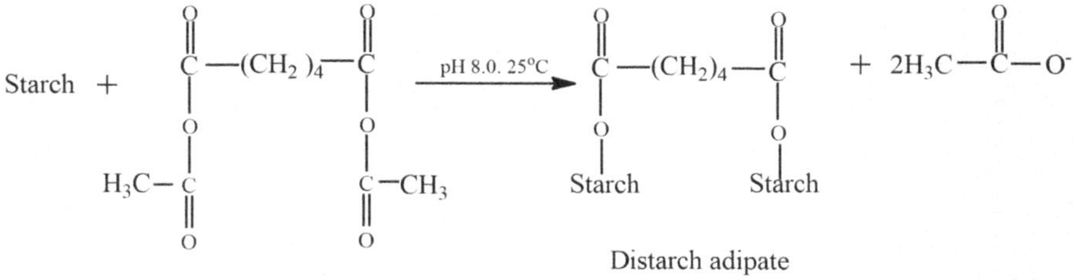

Fig. 2.8 Formation of crosslinked distarch adipate

must not be used greater than 0.3%. However, if it is used in combination with propylene oxide (max 10%) in any sequence, then the amount should not exceed 0.1%. Tetrabutylammonium bromide (TBAB) and DMF (*N,N*-dimethylformamide) are used as catalyst to accelerate the rate of chemical reaction with EPI. Also, if EPI is used in combination with succinic anhydride or acetic anhydride, then also its amount should be less than or equal to 0.3%. The reaction begins with the formation of starch anion under highly alkaline conditions. TBAB and DMF act as a phase transfer catalyst and are expected to create a sodium-saturated environment around starch which facilitates the formation of starch anion. This starch anion then reacts with epichlorohydrin to produce distarch glycerols (Jyothi et al. 2006). The reaction is typically carried out at 40 °C for 17 hours and is terminated by maintaining pH to approximately 5.0. A single crosslink in starch can consume one or two molecules of EPI (Shah et al. 2016). The reaction for the formation of distarch glycerol is shown in Fig. 2.7 (Singh et al. 2007).

Distarch Adipate

Distarch adipate is designated E1422 additive number by FAO. It is made by using the mixture of acetic anhydride and adipic anhydride but is also made by using vinyl acetate and adipic anhydride. Similar to phosphorus oxychloride, it also quickly forms a crosslink. However, instead of phosphate group, an adipate group crosslinks the two starch molecules. Addition of acetic anhydride results in substitution (esterification) reaction by insertion of acetyl groups in parallel to crosslinking reaction (Shah et al. 2016). Therefore, this starch is termed as a dual-modified starch. This reaction is also carried out under alkaline conditions for around 2 h at 35 °C. The reaction is ended by maintaining 6.5 pH (Luo et al. 2009). According to the Code of Federal Regulations, no upper limit of reagent has been set for vinyl acetate and acetic anhydride. However, no more than 0.12% (starch weight dry basis) adipic anhydride must be used in the reaction (CFR 2023). The formation of distarch adipate is shown in Fig. 2.8 (Shah et al. 2016).

Crosslinking with Glutaraldehyde

Glutaraldehyde is a five carbon di-aldehyde with no hydrophilic groups; hence on crosslinking it reduces the hydrophilicity of starch. In contrast to multifunctional reagents like STMP, phosphorus oxychloride, adipic anhydride, epichlorohydrin, and citric acid form crosslinks through the formation of diesters and diethers. But using this crosslinking agent, hemi-acetal ring is formed as nucleophilic (-OH) group is added to the carbonyl group as shown in Fig. 2.9 (Gonenc and Us 2019) leading to the development of a crosslink.

Glutaraldehyde is considered less toxic compared to phosphorus oxychloride and epichlorohydrin; however, scarce research is present on the utilization of glutaraldehyde as starch crosslinking agent. According to MSDS list, STMP, citric acid, and boric acid have a rating of 1 in health hazard risk, "2" for glutaraldehyde, and "3" for epichlorohydrin and phosphorus oxychloride (Reddy and Yang 2010; Gonenc and Us 2019). According to research conducted by Gonenc and Us (2019), glutaraldehyde was used under both alkaline and acidic medium, separately at 5–20% level based on starch weight (db). The reaction time varied from 1 to 4 h. For acidic and alkaline mediums, the reaction was terminated by the addition of ethanol and HCl (1 N), respectively. This chemical however hasn't been approved by CFR or Codex for production of food and pharmaceutical grade starches.

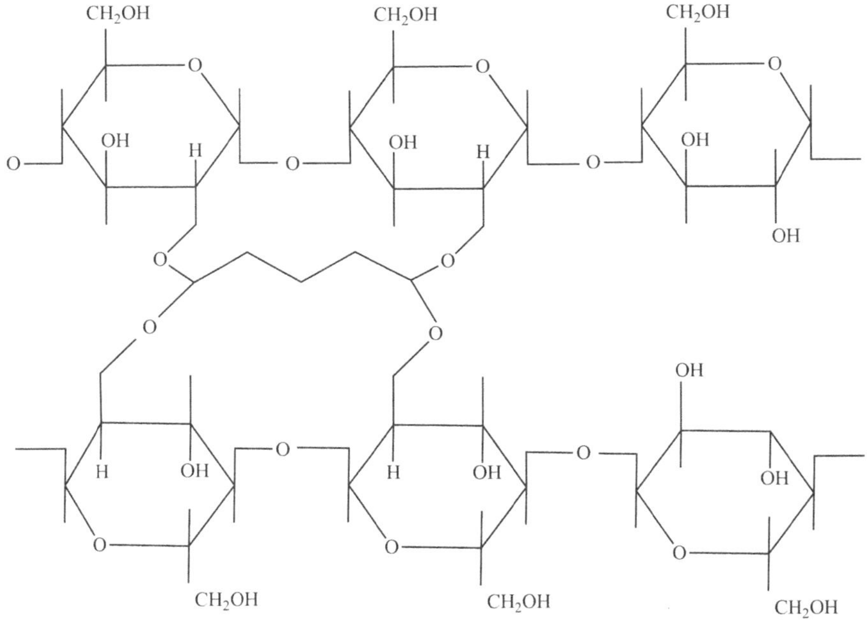

Fig. 2.9 Crosslinking of starch through glutaraldehyde crosslinking agent

2.1.2.2 Safety Assessment as per CFR and Codex (FAO) for CrossLinked Starches

The chemical modification process for starches to be used in foods is mainly regulated by Codex that works under the umbrella of Food and Agriculture Organization (FAO) and the Code of Federal Regulations, USA. The permissible limits of chemical reagents to be added and their safety assessment are elaborated in Table 2.2.

2.1.2.3 Impact of Crosslinking on Structural Properties of Starches

Crosslinking of native starch does not bring any drastic change on the surface morphology, and the granular structure remains intact as the induction of crosslinks occurs at the molecular level both for distarch phosphates and distarch adipates which are more popularly available commercially (Luo et al. 2009; Kou and Gao 2018). The results of microscopy found that too much level of disphosphate crosslinks (STMP/STTP addition 0.25–6%) resulted in reduced swelling of granules with lesser disintegration after gelatinization suggesting shear and thermal stability of starch granules so much so that starch granules crosslinked with 6% STMP/STPP showed Maltese cross after heat treatment under polarized microscopy which completely disappeared in native and slightly crosslinked starch granules (Wongsagonsup et al. 2014). Dong and Vasanthan (2020) observed a decrease in FTIR absorbance at (1047/1022) cm^{-1} which is the ratio of crystalline to amorphous areas of the starch for corn, faba bean, and field pea starch after crosslinking with phosphorus oxychloride, STMP (semi-dry), and STMP (slurry) treatments suggestive of the fact that short-range molecular order was disturbed and reduced near the surface of starch granules as the IR beam can only penetrate up to 2 μm from the surface of the granules. The decrease in organization order is also seconded by XRD studies which reported decrease in starch crystallinity after induction of distarch phosphate crosslinks (Dong and Vasanthan 2020). The pNMR shows peaks between 4.0 and 3.0 ppm for distarch phosphate-based crosslinks, and the intensity of peaks increases with the increase in crosslinks (Zhao et al. 2015; Desam et al. 2018).

2.1.2.4 Impact of Crosslinking on Functional Properties of Starches

Increasing levels of modification decreases both swelling power and solubility. The higher number of crosslinks is known to form a thick crust around the granules which not only restricts entry of water into starch but also curtails leaching of chains leading to reduced solubility. Too much crosslinking results in a very low swelling starch which may not be able to contribute to texture or viscosity of the finished product after heating (Wongsagonsup et al. 2014). The settling volume (SV) decreases with the increase in crosslinking that is less granule sediment down after gelatinization (Kou and Gao 2018). Crosslinked starches are characterized by high shear, pH, and thermal stability. The presence of any chemical crosslink increases the strength within the granule and increases the molecular density which results in reduced molecular mobility. The increased molecular density due to crosslinks results in increased opacity of starch gels also. The higher gelatinization temperature and decreased weight loss in TGA analysis support the fact that crosslinking improves thermal stability of starches (Wongsagonsup et al. 2014; Gonenc and Us

Table 2.2 Safety assessment of single and dual modified crosslinked starches

Chemical name of crosslinked starch	Modification reagents	Single or dual modification	Crosslinking agent permissible limit as per CFR (starch, dry basis)	Safety assessment as per CFR (starch, dry basis)	Safety assessment as per JECFA (starch, dry basis)
Distarch phosphate	Phosphorus oxychloride ($POCl_3$)	Single	≤0.1%	–	Phosphate ≤0.5% (wheat and potato starch) and ≤ 0.4% (for all starches except potato and wheat)
Distarch phosphate	Sodium trimetaphosphate	Single	–	Residual phosphate ≤0.04%	Phosphate ≤0.5% (wheat and potato starch) and ≤ 0.4% (for all starches except potato and wheat)
Phosphated distarch phosphate	Sodium trimetaphosphate and sodium tripolyphosphate	Double	–	Residual phosphate ≤0.4%	Phosphate ≤0.5% (wheat and potato starch) and ≤ 0.4% (for all starches except potato and wheat)
Acetylated distarch phosphate	Crosslinking with phosphorus oxychloride/ sodium trimetaphosphate and acetylation using acetic anhydride or vinyl acetate	Double	$POCl_3$ ≤ 0.1, acetic anhydride ≤8%, vinyl acetate ≤7.5%	Acetyl groups ≤2.5%	Phosphate ≤0.14% (wheat and potato starch) and ≤ 0.04% (for all starches except wheat and potato). Ester groups ≤0.5%. Acetyl groups ≤2.5%
Hydroxypropyl distarch phosphate	Crosslinking with phosphorus oxychloride/ sodium trimetaphosphate and etherification with propylene oxide	Double	$POCl_3$ ≤ 0.1%, no limit for STMP, propylene oxide ≤10%	Residual propylene chlorohydrins≤5 ppm	Hydroxypropyl groups ≤7%, propylene chlorohydrins ≤1 ppm, phosphate ≤0.14% (wheat and potato) and ≤ 0.04% (for all starches except wheat and potato)

(continued)

Table 2.2 (continued)

Chemical name of crosslinked starch	Modification reagents	Single or dual modification	Crosslinking agent permissible limit as per CFR (starch, dry basis)	Safety assessment as per CFR (starch, dry basis)	Safety assessment as per JECFA (starch, dry basis)
Distarch glycerol	Epichlorohydrin	Single	Epichlorohydrin ≤0.3%	–	Epichlorohydrin is impermissible
Acetylated distarch glycerol	Epichlorohydrin and acetic anhydride	Double	Epichlorohydrin ≤0.3%	Acetyl groups ≤2.5%	Epichlorohydrin is impermissible
Hydroxypropylated distarch glycerol	Epichlorohydrin and propylene oxide in any sequence	Double	Epichlorohydrin ≤0.1%, propylene oxide ≤10%	Residual propylene chlorohydrins≤5 ppm	Epichlorohydrins impermissible
Acetylated distarch adipate	Mixture of acetic anhydride/ vinyl acetate, adipic anhydride	Double	Adipic anhydride ≤0.12%	Acetyl groups ≤ 2.5%	Acetyl groups ≤2.5%, vinyl acetate ≤0.1 mg/kg, ester groups ≤0.5%, adipate groups ≤0.135%, residual free adipic acid ≤0.025%

References: JECFA (2021), CFR (2023)

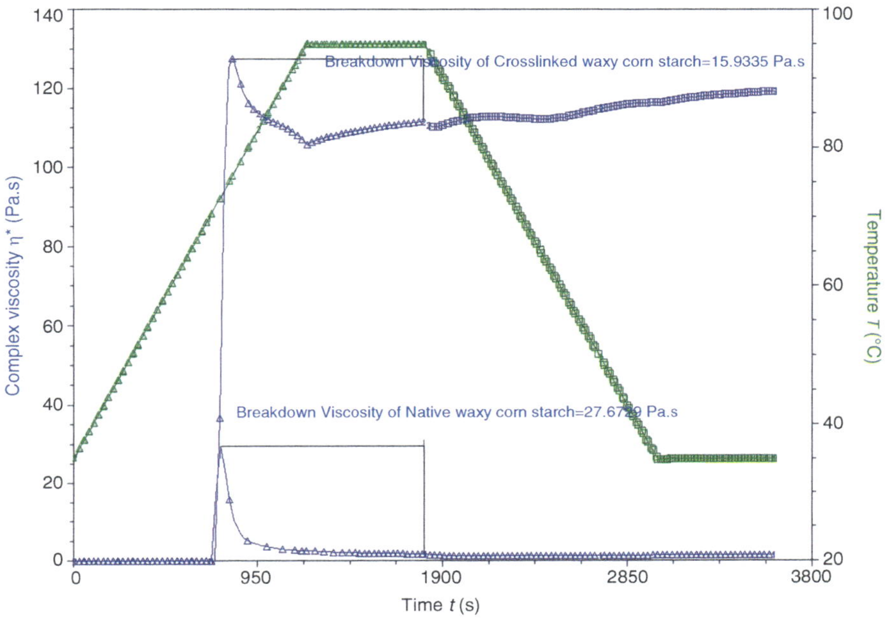

Fig. 2.10 Comparative breakdown viscosity of native and crosslinked waxy corn starches

2019; Mehfooz et al. 2019). Also breakdown viscosity observed during viscoamylography (Fig. 2.10) reduces after crosslinking, indicating shear stability of starch granules post modifications.

2.1.2.5 Impact of Crosslinking on Digestibility of Starches

The starch remaining indigested in the small intestine is called "resistant starch." The chemically modified resistant starches are known as RS4 starches. The crosslinking technique is now being explored for the development of RS4 starch as it reduces the glycemic index of the products in which it is incorporated. The crosslinking is known to increase the RS4 content mainly because of restriction of alpha-amylase to enter the starch granules. The Food and Drug Administration (FDA) intends to approve crosslinked phosphorylated starch in the list of dietary fibers (FDA 2019). Starch citrates formed by crosslinking with citric acid also showed higher RS content (Shaikh et al. 2019a). Also starch citrate when incorporated in flat bread up to 15% was found to reduce the predicted glycemic index from 57.25 to 44.97 (Shaikh et al. 2022).

2.1.2.6 Food Applications of CrossLinked Starches

Crosslinked starches single or dual modified are extensively used in food applications. It is widely used as a fat imitate in mayonnaise, soups, ice creams, and muffins (Subroto et al. 2020) to develop low-fat variants of these products. Commercial producers of modified starches like Ingredion, Roquette, Bluecraft Agro, and Tate and Lyle recommend their usage in wide-range processing conditions. Due to their high process tolerance and resistance to pH and shear, these are extensively used as texturizers and thickening agent. These are also used as anti-crystallizing agent for depression of freezing point and prevent breakdown of emulsion by raising viscosity of the continuous phase which ultimately prevents oil separation. The E1442 starch also improves freeze thaw stability of the products in which it is incorporated. Due to these reasons, crosslinked starches are widely used in muffins, puddings, cookies, sauces, noodles, marshmallows, vermicelli, etc.

2.1.3 Acid-Hydrolyzed Starches

Acid-hydrolyzed starches (E1401), commonly known as acid-thinned or thin-boiling starches, are produced by treating starches with acids such as H_2SO_4, HCl, and H_3PO_3 under controlled temperature conditions below gelatinization until the required degree of hydrolysis is achieved (Thomas and Atwell 1999; Pratiwi et al. 2018). Although the dextrin-roasted starches (E1400) are also prepared with acids such as ortho-phosphoric acid or hydrochloric acid, the difference between dextrin-roasted starches and acid-hydrolyzed starches is that the former is treated by dry heating or roasting the native starch under acidic conditions, while the latter involves use of acids (dilute) which are added to starch suspensions (JECFA 2021). The extent of hydrolysis is assessed via intrinsic viscosity, alkaline number, molecular weight, and reducing value. However, the acid-hydrolyzed starches prepared commercially use the empirical scale technique which is also known as "water fluidity." The water fluidity scale is referred to as the reciprocal of viscosity and ranges from 0 to 100 where 0 is for unmodified starches and 100 is the viscosity of water. This standardized water fluidity scale is designed based on the fact that the viscosity of starch decreases with the increase in the extent of hydrolysis (Thomas and Atwell 1999). Acid-thinning involves electrophilic attack on oxygen of α (1 $\rightarrow$ 4) glycosidic linkage by hydroxonium ion which leads to the formation of high-energy intermediate carbocation form (Lewis acid) which subsequently leads to hydrolysis of glycosidic linkage (Hoover 2000). The process of starch acid hydrolysis is dependent on time, temperature, and concentration of acid used for the treatment (Wang et al. 2003). The steps of acid hydrolysis are quite simple. Starch is first suspended in an organic or inorganic acid solution with uniform stirring, followed by neutralization and filtering. Typically, the temperature for acid modification of starches is adjusted at 50 °C or lower, with steady stirring for different intervals of time ranging from one to several hours (Wang et al. 2012; Ali and Hasnain 2014; Moin et al. 2017; BeMiller 2019). Post-neutralizing, the slurry undergoes washing and drying procedures while ensuring temperatures below gelatinization (Zhang et al. 2023).

2.1.3.1 Safety Assessment as per Code of Federal Regulations and JECFA for Acid-Thinned Starches

The HCl or H_2SO_4 may be used separately or together for acid-thinning of starches (CFR 2023), whereas, according to JECFA (2021), the dilute solution of orthophosphoric is also allowed in addition to HCl and H_2SO_4 to produce acid modified starch.

But to produce dextrin roasted starch, only HCl and phosphoric acid are allowed as per JECFA regulations.

2.1.3.2 Impact of Acid-Thinning on the Structural Properties of Starches

Starch morphology as impacted by acid-thinning varies depending on the source of starch and the degree of hydrolysis. The granular structure of acid-hydrolyzed starches is expected to remain intact after mild hydrolysis, although the physicochemical properties of starches are significantly altered after modifications with HCl, H_2SO_4, or orthophosphoric acid (Wang et al. 2003; Shujun et al. 2007; Sun et al. 2015; Zhang et al. 2023). As the hydrolysis duration progresses, the starch granules change shape from oval or spherical to cake-like or bread-like, and extensive duration might lead to complete disappearance of intact granules (Shujun et al. 2007; Wang and Copeland 2015). Essentially, starch acid hydrolysis is a two-stage process, which first takes place in amorphous region (faster reaction) and is followed by occurrence in semi-crystalline and crystalline regions (slower reaction) (Wang et al. 2012). The loose packing of starch chains in amorphous regions of starch granules is thought to make them more susceptible to acid attack than crystalline regions. Initially, hydrolysis is influenced by the amylose content, size, pores, and level of amylose-lipid complexes present. However, the second stage in which acid attacks both amorphous and crystalline regions, the double helical packing, the chain length of AP (amylopectin), and its presence between crystalline and amorphous region can impact rate of acid hydrolysis (Le Corre et al. 2010; Wang and Copeland 2015). Interestingly, the hydrolysis of starches is also thought to be a three-stage process which is based on rapid hydrolysis, slow hydrolysis, and very slow hydrolysis process occurring in amorphous, semi-crystalline, and crystalline regions, respectively (Li et al. 2007). The final characteristics of starch granules after modifications are dictated by both structure and composition of starch under consideration (Wang et al. 2003). For example, the micropores present on the granular surface of waxy starch allow easier infiltration of acids into the core of granules as these pores connect hilum of starch via hollow channels (Zhang et al. 2023), whereas high-amylose granules which have a reasonably complete and smooth surface do not allow easier penetration, and therefore they are more resistant to hydrolysis (Wang et al. 2003; Zhang et al. 2023). During the acid hydrolysis process, the amorphous regions undergo degradation faster compared to the crystalline regions because of the less ordered structure of the former (Zhang et al. 2023). Increase in crystallinity is more evident with the increase in acid thinning owing to the formation of shorter chains which undergoes further reordering to give more close packing and a distinct XRD crystalline structure (Wang et al. 2003; Shujun et al. 2007). The degree of acid hydrolysis is also observed to change the polymorph type from C type to A. Since the C-type polymorph is composed of both A- and B-type polymorphs, degradation begins first in the B-type polymorph present in the C-type polymorph during acid treatment, which seconds with the statement that hydrolysis first begins in the amorphous region of the granules due to the fact that B-type polymorphs form amorphous region and A-type are the crystalline region in the C-type starch granules (Shujun et al. 2007; Haq et al. 2019).

2.1.3.3 Impact of Acid Thinning on Functional Properties of Starches

The swelling power of starches decreases with acid hydrolysis, and the decline is more pronounced with the increase in the degree of hydrolysis owing to the more rigid structure caused by frequent entanglement and close packing of shorter AP chains. The degradation of amorphous region also improves the relative crystallinity of starch granules after acid treatment leading to restricted swelling behavior (Singh et al. 2009; Ali and Hasnain 2014; Moin et al. 2017). Solubility on the other hand increases with acid hydrolysis primarily due to the rise in low molecular weight linear fractions of the starch chains, which solubilize easily compared to high molecular weight intact chains (Sandhu et al. 2007; Singh et al. 2009; Moin et al. 2017). The rise in solubility followed by acid hydrolysis is thus due to the depolymerization and weakened structure of starch granules. Similarly, the water and oil absorption tendency of acid-thinned starches is reduced compared to their native counterparts due to the decreased amorphous region within the starch granules leading to reduced binding sites available for absorption of water and oil in starch granules (Lawal 2004). The decrease in percent transmittance of acid-hydrolyzed starches is the result of degradation of amorphous region and leaching out of amylose during the hydrolysis process which increases the retrogradation rate and decreases the transmittance of monochromatic light (Lawal 2004). The acid-hydrolyzed starches form soft gels. The lesser number of intermolecular interactions between the low molecular weight starch chains after hydrolysis may be the cause of decreased gel strength of acid-hydrolyzed starch (Ali and Hasnain 2014; Moin et al. 2017). The viscosity profile of starches also decreases during acid hydrolysis because long chains convert into shorter chains together with reduction in swelling power (Ali and Hasnain 2014; Moin et al. 2017). The peak gelatinization temperature usually increases after acid hydrolysis as attack in amorphous region at the first place leaves behind more crystalline region with higher thermal stability. Therefore, these remnants of starch crystallites melt at higher temperatures after acid-based modification (Sandhu et al. 2007; Abdorreza et al. 2012).

2.1.3.4 Impact of Acid Hydrolysis on Digestibility of Acid-Thinned Starches

Acid hydrolysis is thought to have a significant impact on starch digestibility (Miao et al. 2011; Wang and Copeland 2015). During the acid hydrolysis process, the acids initially hydrolyze the amorphous region followed by crystalline region of starch granules leading to the formation of shorter chains. These shorter chains will disorganize during the autoclaving process but produces a highly ordered double helical structure when reoriented during retrogradation which then prevents enzymatic hydrolysis (Liu et al. 2014). Furthermore, the digestibility of starches is also influenced by changes in the chain length of the starches upon acid hydrolysis, which is degraded into short A chains and long B chains in amylopectin region (Miao et al. 2011). Acid hydrolysis of starches tends to increase the RDS fraction of starches and decreases the RS fraction of starches, whereas the SDS fraction is said to be dependent upon the source of starch and degree of hydrolysis (Wang and Copeland 2015). The increase in RDS is due to increased effective surface area of small-sized granules for enzymatic attack after acid degradation. Moreover, the loss of α-(1–6) bonds and their inter-cluster chains in the amorphous region makes substrate more accessible to the enzyme and easier for α-amylase to cleave internal structures resulting in increased RDS fraction (Srichuwong et al. 2005; Zhang et al. 2006; Miao et al. 2011). However, an increase in acid concentration is known to increase the RS content, which is suggested due to the formation of short chains (Miao et al. 2011; Wang and Copeland 2015; Li et al. 2020). Furthermore, the resistant starch is a proportional factor of amylose content present in starch granules and can be developed through amylose retrogradation. The hydrolysis of starch by acid raises the short linear chains of amylose, which favors retrogradation which in turn produces resistant starch on storage of gelatinized acid-hydrolyzed starch (Dundar and Gocmen 2013; Aparicio-Saguilán et al. 2015; Pratiwi et al. 2018).

2.1.3.5 Food Applications of Acid-Hydrolyzed Starches

Acid-thinned starches exhibit reduced hot paste viscosity, enhanced gel hardness on storage, and higher water solubility. These thin boiling starches are widely used in many countries throughout the world in the making of different food and non-food products (Wang and Copeland 2015). Acid-hydrolyzed starches are widely used in confectionery operations for producing jelly and gum candies, where the hot liquid because of its low viscosity is easily poured into molds which then gel on cooling to give the desired texture and shape (Mason 2009). Also, these are frequently utilized in salad dressing and mayonnaise due to their low viscosities in both cold and hot conditions (Egharevba 2019). Acid-thinned starches, which have excellent flow and compressibility characteristics, are also frequently employed as fillers (Garcia et al. 2020). Many commercial companies, including Allwyn Chem Industries, Roquette, Grain Processing Corporation, and Nguyen starch, supply acid-thinned starches for a variety of food products, including cereal snacks, confectionery, dairy, meat, poultry, and seafood where they are used as an additive to improve textural properties, gelling ability, and for reduced hot-paste viscosity.

2.1.4 Oxidized Starches

Oxidized starches are extensively used to alter/modify structural and functional properties in food and non-food products. The starch molecules that have undergone a chemical reaction with an oxidizing agent under good manufacturing practices (GMP) are termed "Oxidized starches." These are designated as E1404 in Codex. This process modifies the properties of starch, leading to changes in its functionality and behavior. Carboxylic and carbonyl groups are formed in starches as a result of oxidation reaction. Various factors affect oxidation phenomenon including pH, reaction time, temperature, concentration, and active chlorine level. Oxidized starches can also be coupled with any acidic, basic, bleaching, and enzymatic reagent to obtain starches with desired functional properties (Yi et al. 2014; Vanier et al. 2017). Oxidation first produces carbonyl functional groups which then further oxidize to carboxyl moieties.

Therefore, the ratio of carboxyl to carbonyl moieties in oxidized starches dictates the magnitude of oxidation. Under controlled reaction conditions of pH, temperature, and pressure, specific oxidizing agents react with intermolecular linkages of starch. It thereby cleaves the polymeric chains and oxidizes aldehyde group to carboxylic one (Okekunle et al. 2020). This type of oxidation occurs at C-2, C-3, and C-6 positions of hydroxyl groups. Insertion of these functional groups in starches plays a distinctive role in film formation, binding material in textile, paper industry, and food industries. Botanical origin, molecular structure of native starches, and reaction conditions also affect the structural and physicochemical properties of oxidized starch (Vanier et al. 2017).

2.1.4.1 Process of Making Oxidized Starches

Commercially, preparation of oxidized starches utilizes sodium hypochlorite as it is the only approved chemical by FDA and Codex Alimentarius and introduces carboxyl and carbonyl moieties in starch. Carbonyl to carboxyl ratio devices functional properties of starches. The pH of reacting medium has a key role in oxidation rate. According to Sangseethong et al. (2009), near neutral pH, reaction rate becomes faster; however near acidic or alkaline conditions, it becomes slower. However, several other authors reported alkaline pH using hypochlorite solution to produce oxidized starches as it favors the production of carboxyl groups (Sánchez-Rivera et al. 2009; Sangseethong et al. 2009; Dias et al. 2011; Halal et al. 2015). After maintaining alkaline pH, sodium hypochlorite is added, and well-stirred conditions are maintained. The temperature at which oxidation is conducted varies between 30 and 40 °C (Garrido et al. 2014; Liu et al. 2018). After an appropriate amount of time, starch slurry is filtered, washed, and dried to obtain oxidized starch (Wang and Wang 2003). Insertion of carboxyl groups reduces starch retrogradation and gives low temperature stabilizing property to starch (Shi et al. 2023). The amount of amylose content and active chlorine in oxidizing agent also influences ratio of carboxyl and carbonyl groups (Zhang et al. 2022). General oxidation reaction of starch with NaOCl is presented in Fig. 2.11.

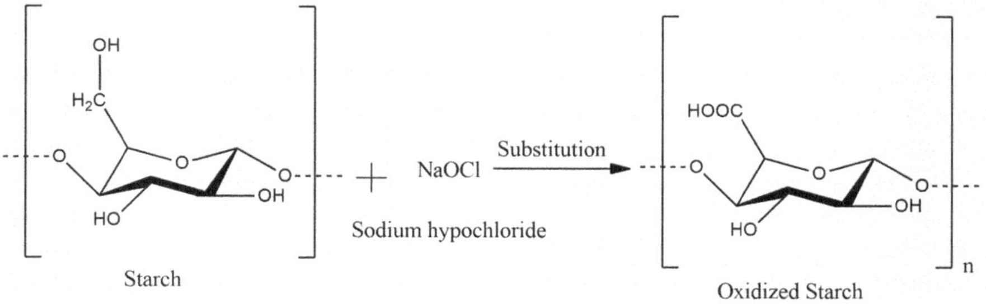

Fig. 2.11 Oxidation of starch using sodium hypochlorite

2.1.4.2 Permissible Limit and Safety Assessment of Oxidized Starches as per CFR and JECFA

According to CFR (2023), the chlorine in NaOCl must not exceed 0.055 lb. per pound of starch (dry basis). But no guidelines are available regarding safety assessment of finished oxidized starch in the Code of Federal Regulations. However, as per JECFA (2021), the hypochloride oxidized starch (E1404) should be regarded as safe if it has carboxyl groups based on starch weight dry basis ≤1.3%, and residual oxidizing substances (as H_2O_2) must be restricted to less than 180 ppm. The same regulations must be followed when oxidized starch is further acetylated to produce "acetylated oxidized starch" or (E1451). Compliance with these regulations helps ensure safety in these additives being modified chemically. It is therefore essential for food manufacturers to adhere to these guidelines and regularly monitor and verify the levels of oxidizing agents in the modified starches.

2.1.4.3 Impact of Oxidation on Structural Properties of Starches

Surface morphology remained unaffected after oxidation (Fonseca et al. 2015; Sindhu et al. 2021). Kuakpetoon and Wang (2008) and Adebowale et al. (2006), however, observed severe damage with signs of cracks and breakage which may affect physicochemical characteristics suggesting the impact of starch sources on morphological changes after oxidation. Ali and Hasnain (2014) found an increase in diameter of pores naturally present on the surface of sorghum starch when starch was dual modified via oxidation and acetylation. It may be presumed that mild oxidation with hypochlorite does not affect the surface of starch but intense oxidation singly or in combination with other modifications may cause severe erosion and holes/cavities in the center of granules as explained by Kuakpetoon and Wang (2008), Okekunle et al. (2020) and Guleria and Yadav (2022). The amorphous region of starch is first attacked by oxidation, thereby removing it from the surface and ultimately causing it to burst (Adebowale et al. 2006). In another finding Adebowale et al. (2006) revealed that the external surface of bean starch appeared rougher and irregular after 1.5% active chlorine oxidation. The oxidation reaction also causes depolymerization or hydrolysis of starch chains (James BeMiller 2009). The structural characterization has also been observed by high-performance size-exclusion chromatography (HPSEC). As the extent of oxidation by hypochlorite increases, the fraction of low molecular weight chains also increases (Wang and Wang 2003). Nuclear magnetic resonance (^{13}C NMR) showed signal at 176 ppm corresponding to carboxyl groups' presence in oxidized starch while, FTIR spectroscopy showed a characteristic peak at 1600 cm^{-1} in oxidized starches (Yi et al. 2014; Liu et al. 2018). These structural changes make oxidized starches more suitable for various food applications where enhanced stability, texture modification, and improved functionality are desired.

2.1.4.4 Impact of Oxidation on Functional Properties of Starches

Oxidation of starch causes breakage of inter- and intramolecular linkages and limited depolymerization resulting in intermediate thickening tendency or may even lead to formation of soft gels (Egharevba 2019). Oxidation by hypochlorite increases the swelling capabilities of starch granules with respect to temperature; however with respect to native counterparts, oxidized starches display lower swelling power due to depolymerization and lack of water binding sites (Rao et al. 2022). This trend is shown by various sources of starches modified by hypochlorite including common bean, potato, and lima bean (Zhou et al. 2016; Okekunle et al. 2020). The depolymerization is also responsible for increase in water solubility of oxidized starches which increases with the increase in the level of oxidation. High-amylose starches are found to be more resistant against oxidation as compared to normal and waxy corn starches (Pietrzyk et al. 2019). Thermal analysis of oxidized starches displayed lowering of ΔH as compared to native starch. Peak gelatinization temperature also exhibited reduction as the active chlorine concentration increases in NaOCl. As analyzed through RVA, peak viscosity and setback viscosity also decreased as compared to native. Pastes with low setback are potentially required in bakery industries where less retrogradation is preferred (Hornung et al. 2015; Liu et al. 2018).

2.1.4.5 Food Applications of Oxidized Starches

The carboxyl and carbonyl groups induced during oxidation reactions can significantly enhance food applications of this starch. Oxidized starches typically have improved stability, enhanced thickening and binding capabilities, and increased resistance to breakdown during processing or storage. These are commonly used in various food applications as thickeners, binders, film formers, stabilizers, and texturizing agents. Oxidized potato starch treated with varying concentrations of sodium hypochlorite displayed improved water-holding capacity and viscosity. The addition of oxidized starches improved sensory profile of yogurt and prevented separation of whey in yogurt (Zhang et al. 2022).

Due to their strong adhesive properties, oxidized starches have also been extensively used in coating applications in food products. According to Purcell et al. (2014), the oxidized starches displayed higher cold batter viscosity due to the presence of carboxyl groups. Thus, these are utilized in preparation of batters for chicken nuggets along with other ingredients (Kuakpetoon and Wang 2001). Acetylated oxidized starches also find applications as thickener, emulsifier, and bulking agent. Another important feature of this type of oxidized starch is enhancement in transparency and whiteness in food products. Oxidized starches are also used as an ingredient of aromatized alcoholic beverages, water-based flavored drinks, dietetic foods, soups and broths, batters, confectionary, cheese analogs, condensed milk, fat spreads, pasteurized cream, and fat emulsions. In sauces, gravies, and soups, oxidized starches provide enhanced viscosity and stability, ensuring desired texture and preventing phase separation. They are also employed in bakery products, where they improve moisture retention, enhance dough handling properties, and contribute to the desired crumb structure. Oxidized starches find application in dairy products like puddings, custards, and yogurts, where they help in achieving a smooth and creamy texture (Purcell et al. 2014; Ahsan et al. 2023). Moreover, hypochlorite-oxidized starches are utilized in the production of frozen foods, acting as cryoprotectants by forming a protective barrier around food particles, preventing ice crystal formation, and maintaining product quality during freezing and thawing processes. In meat products, oxidized starches can serve as binders, improving water retention, texture, and yield (Egharevba 2019). Overall, the functional properties of oxidized starches make them valuable ingredients in various food formulations, providing stability, texture enhancement, and improved overall quality.

2.1.5 Bleached Starches

Bleached starch is a type of starch that undergoes a chemical treatment process to remove impurities and also whitens the color. Bleaching is normally carried out using "sodium hypochlorite, sodium chlorite, hydrogen peroxide, potassium permanganate, peracetic acid, ammonium persulfate, sulfur dioxide," or other permitted forms of sulfites at a very low level for different starches (Iwuagwu 2000; Klang et al. 2019; CFR 2023). Enzymatic, acid, or alkali-based treatments may be followed by bleaching according to GMP (Good Manufacturing Practices). In bleaching process, the interaction of oxidizing agents with the starch molecule must be very slight so that these do not make any change in the physical properties except color (JECFA 2021). The major purpose of this treatment is mediated toward removal/bleaching of impurities (colored) linked with this polymer like phenols, carotenoids, and xanthophylls. The bleached starch has been labeled as E1403 by Codex Alimentarius.

2.1.5.1 Types of Bleaching Agents

There are two categories of food starches when treated with oxidant: (a) oxidized starches (chlorinated) normally prepared by treatment with hypochlorite and (b) bleached starches (which receive low-level treatments with some specific oxidizing chemicals). Bleaching is opted to bring changes in starch color only, whereas oxidation involves insertion of carboxyl groups which are carefully controlled during the reaction as per limit of regulatory bodies explained above. The various types of bleaching agents that have been used since long for bleaching purpose of starch are listed in Fig. 2.12.

The different methods used for bleaching of starch are described below.

2.1.5.2 Treatment with Sodium Hypochlorite/Calcium Hypochlorite or Sodium Chlorite

Hypochlorite-bleached starch is prepared under highly alkaline conditions (pH 10) by making starch slurry (40% dry solids) and treating it with sodium hypochlorite with maximum (0.82%) active Cl_2, based on dried starch mass through dropwise addition. Well-stirred conditions are maintained during the reaction. The reaction is carried out at pH 10 and at a temperature of 30 °C for 30–300 min. To finish the reaction, sodium bisulfite ($NaHSO_3$) is added followed by bringing of pH back to 6.5–7.0. This slurry is then filtered and then undergoes multiple washings. To check efficacy of washing and to confirm absence of chemical residues, it should not give any precipitates with silver nitrate ($AgNO_3$) solution. This obtained starch is then

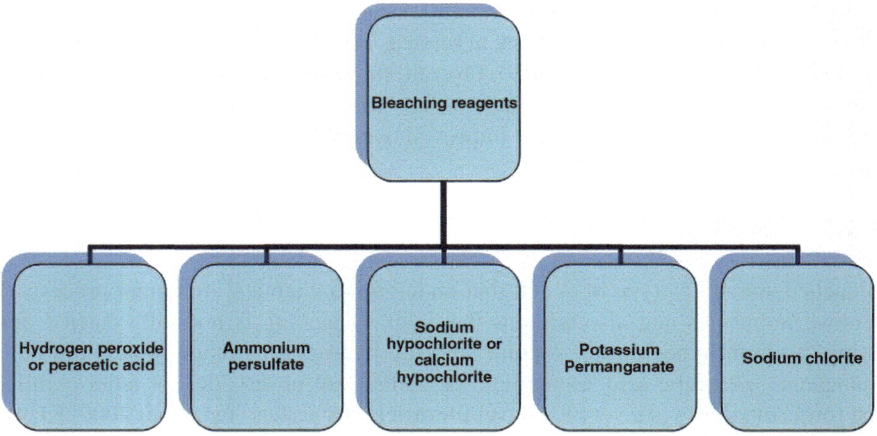

Fig. 2.12 Modification reagents used for preparation of bleached starch

dried at 50 °C for 2 h in the oven (Sangseethong et al. 2009). The pH is very crucial to this modification. The reaction rate is faster at pH 7 but decreases at pH 10 (Garrido et al. 2014). The modification conditions must be alkaline due to the fact that acidic and high temperature accelerate conversion of hypochlorite to chlorate that resist oxidation. Toxic gases with huge volumes of waste water are generated during oxidation of starch raising environmental concerns.

2.1.5.3 Treatment with Peracetic Acid and/or Hydrogen Peroxide

In contrast to sodium hypochlorite, H_2O_2 is not used much commercially in spite of the fact that it is environment friendly and generates oxygen and water as by-products which have no environmental concerns (Tethool et al. 2012). Bleaching through hydrogen peroxide is often preferred, particularly when a chlorine-free procedure is required. During bleaching process, hydrogen peroxide is typically added to a slurry of starch, where it breaks down into water and oxygen gas, which then reacts with the impurities in the starch to cause their bleaching. Hydrogen peroxide can generate lots of highly oxidizing hydroxyl radicals (OH) in the presence of a catalyst mainly copper Cu (II) or iron Fe (II) (Zhang et al. 2012; Broekman et al. 2023). Hydrogen peroxide combines with the hydroxyl groups in starch polymers in a manner similar to sodium hypochlorite, and as a result carbonyl groups are produced. These carbonyl (-CO) groups then further oxidize to carboxyl (COOH) groups which give improved stability to starch paste. In peroxide oxidation, the major functional groups produced are carbonyl (Sangseethong et al. 2010). Under alkaline conditions, the hydroperoxyl anion (HOO) is also created by the breakdown of hydrogen peroxide which causes bleaching (An et al. 2021). The hydrogen peroxide shows stable bleaching effect particularly at pH 11 (An et al. 2021). Conclusively, hydrogen peroxide and peracetic acid are known to be stable, non-toxic, odorless, chlorine free, and non-polluting bleaching agents.

2.1.5.4 Treatment with Potassium Permanganate or Ammonium Persulfate

In this treatment, the starch sample is treated with standard potassium permanganate ($KMnO_4$) for 1 hour at pH 6.0 followed by washing. The starch is then filtered for the complete removal of the oxidizing agents followed by oven drying (Lazzarotto et al. 2017). The mechanism is complicated because the subsequent oxidation using potassium permanganate or ammonium persulfate generates free radicals, i.e., the reduction of Mn^{4+} to either Mn^{3+} or Mn^{2+} or direct oxidation (Hornung et al. 2016).

2.1.5.5 Safety Assessment as per CFR and JECFA for Bleached Starches

The permissible limits of chemical reagents used for modification and safety assessment of oxidized starches in line with JECFA and CFR are listed in Table 2.3.

Under these limits the role of oxidizing agent is principally that of bleach. It must be taken into consideration that excessive use of oxidizing agents can lead to over-bleaching and synergistically affect the quality of the starch. Therefore, its usage should be carefully regulated and monitored in the food industry.

2.1.5.6 Impact of Bleaching on Structural Properties of Starch

Amorphous region is primarily attacked by hydrogen peroxide; therefore, the X-ray diffraction patterns and granular morphology remain unchanged (Vanier et al. 2017). However, starch oxidation through other chemical compounds disrupts the

Table 2.3 Safety assessment of bleached starches

Chemical reagents used for bleaching	Limit of chemical reagents used for bleaching as per CFR	Bleached starch specifications as per CFR (dry basis starch)	Bleached starch specifications as per JECFA (dry basis starch)
Peracetic acid or hydrogen peroxide	Active oxygen $\leq 0.45\%$		Carboxyl groups $\leq 0.1\%$ after applying the phosphate content correction factor for phosphorus containing starches; residual oxidizing substances <180 ppm (as H_2O_2)
Sodium hypochlorite	Chlorine $\leq 0.82\%$ based on dry starch weight		
Sodium chlorite	$\leq 0.5\%$		
Potassium permanganate	$\leq 0.2\%$	Residual manganese (Mn) ≤ 50 ppm	
Ammonium persulfate	$\leq 0.075\%$		
Calcium hypochlorite	Chlorine $\leq 0.036\%$ based on dry starch weight	Only allowed as component of batter for commercially processed foods	
Sulfur dioxide	$\leq 0.05\%$		

References: JECFA (2021), CFR (2023)

hydrogen bonding and causes hydrolysis of glycosidic linkages, thus resulting in depolymerization and weakening of starch granular structure (Chiu and Solarek 2009). Sangseethong et al. (2010) reported that the molecular size of cassava starch decreased to similar extent by both oxidants sodium hypochlorite and hydrogen peroxide.

2.1.5.7 Impact of Bleaching on Functional Properties of Starch

Relatively bleaching has mild effects on the functional properties of the starch. The main purpose is to bleach the colored impurities present within the starch. The bleaching however may affect apparent viscosity and gel strength. This modification causes decline in apparent viscosity which could be used to increase solids. Hydrogen peroxide leads to faster oxidation compared to sodium hypochlorite. Iwuagwu (2000) reported that the swelling power was not exaggerated by bleaching treatment although a slight increase was observed in hydration capacity of bleached maize starch. Klang et al. (2019) observed a decrease in swelling power and water retention capacity of bleached potato starch. Hard gels are formed by hydrogen peroxide-led bleach starches compared to those prepared with hypochlorite (Sangseethong et al. 2010). Similarly, gelatinization of peroxide-bleached starches takes place at higher temperature too. Probably, it is likely that peroxide oxidation leads to hydrolysis of amorphous region responsible for destabilizing crystalline region. Thus, in the absence of any destabilizing effects on ordered crystalline lamella, the gelatinization temperature increased (Manelius et al. 2000; Sangseethong et al. 2010). Since carboxyl anions are induced more by hypochlorite, therefore this results in lower starch melting temperatures. Thus, this weakens the starch granule by facilitating hydration and causes it to gelatinize at lower temperature. This was also seconded by DSC analysis that bleached starches show decreased gelatinization enthalpies, probably as a result of partial starch molecule breakdown in the crystalline lamella (Sangseethong et al. 2010).

2.1.5.8 Food Applications of Bleached Starches

Currently bleached starch has limited food applications. In contrast, the paper and textile industries are the major users of bleached starches. Singh et al. (2007) reported application of oxidized/bleached starch in batters and breading for coating several food stuffs, as binders and film former in confectionary, and as texturizers in dairy industries. Additionally bleached starch is widely used in preparation of various adhesive coatings in the paper industry and in the textile industry. According to CFR (2023), if the bleached starch is prepared using calcium hypochlorite, then it should only be used as a component of batter for commercially processed foods.

2.1.6 Esterified Starches

Starch esters are prepared by reacting acids or their anhydrides or chlorides with starch in the presence of sufficient amount of water for a few hours under alkaline condition. The starch ester formation with both acid and anhydride is shown in Figs. 2.13 and 2.14, respectively (Imre and Vilaplana 2020).

In aqueous medium, the former acts as acyl donor and reacts with alcoholic or (-OH) group of starch to produce starch ester. Simply esterification reaction can be defined by the following reaction:

$$\mathrm{RCOOH(acid) + ROH(alcohol) = RCOOR(ester)}$$

Based on reactants used to carry out esterification, the properties of starch esters may change and are assigned different E-numbers as presented in Table 2.4.

Nevertheless, different modification reagents actually substitute functional groups on different glucose monomers of starch chains. This esterification reaction is dependent on the structure of acids/anhydrides, time, temperature, starch concentration in suspension, and presence of catalyst (Otache et al. 2021). It has been reported that in case of long chain esterifying reagents (C8-C18), the substitution is quite slow and inefficient due to lack of solubility of reagent in aqueous medium which leads to phase separation and inefficient entry of reagents into the starch granules. The subsequent slow esterification process may lead to the hydrolysis of

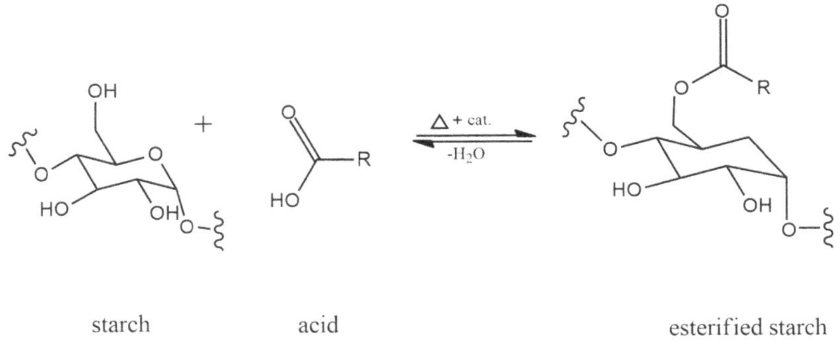

Fig. 2.13 Esterification of starch using acid

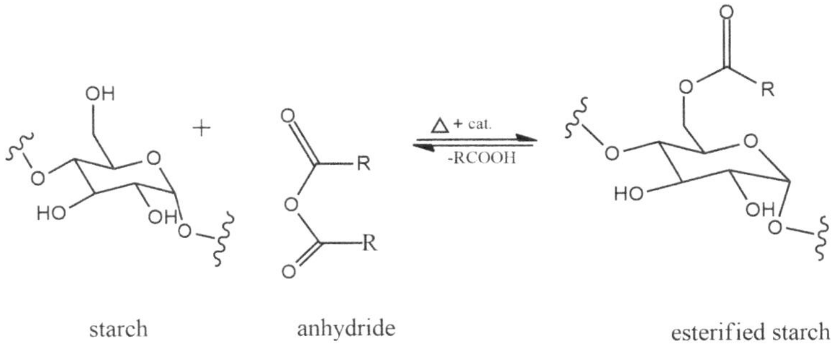

Fig. 2.14 Esterification of starch using anhydride

Table 2.4 E-numbers, modification reagents, and safety assessment of esterified starches

Esterification reagent	Chemical name of modified starch	E-number	Limit of modification chemicals permitted by CFR	Safety assessment as per JECFA (starch dry basis)	Safety assessment as per Code of Federal Regulations, USA (starch dry basis)
Sodium or potassium orthophosphate, orthophosphoric acid, sodium tripolyphosphate (STPP)	Monostarch phosphate	E1410	–	Phosphate ≤0.5% (wheat and potato starch) and ≤ 0.4% (for all starches except wheat and potato)	Residual phosphate ≤0.4 percent calculated as phosphorus
Vinyl acetate or acetic anhydride	Starch acetate	E1420	–	Acetyl groups ≤2.5% Ester groups ≤0.5%	Acetyl groups ≤2.5%
Octenyl succinic anhydride	Starch sodium octenyl succinate	E1450	(a) 1–Octenyl succinic anhydride ≤3 percent (b) When 1-octenyl succinic anhydride is used in combination with aluminum sulfate, then OSA should be ≤2%. Also aluminum sulfate should be ≤2%[a]	Octenyl succinyl groups ≤3%, residual free octenyl succinic acid ≤0.3%	–

[a]If starch is treated with beta-amylase after OSA treatment at ≤3%, then its use must be limited in beverage or beverage bases as emulsifier or stabilizer

reagent itself leading to extremely low DS products. For example, to produce a high DS octenyl succinate starch, Niu et al. (2014) carried out an esterification reaction in the presence of 50% (v/v) aqueous ethanol solution so as to increase the miscibility of the esterifying reagent in the reaction medium. The degree of substitution is explained as the "average number of hydroxyl groups replaced by other functional groups per repeating unit." The DS can range from 0 to 3 for starch where DS = 3 means that all the three (-OH) groups on the repeating anhydrous glucose units are substituted with the functional moieties. The major food-based esterified starches and their methods of preparation are described below:

2.1.6.1 Starch Acetates

Acetylated starch or E1420 is a substituted starch extensively used in food industries. The allowed reagents as per JECFA (2021) and CFR (2023) are acetic anhydride and vinyl acetate, and the maximum % acetyl groups allowed in E1420 should be ≤2.5% irrespective of source of starch and reagent (acetic anhydride/vinyl acetate) used for modification. During this modification acetyl moieties are substituted on to the starch chains as shown in Figs. 2.15 and 2.16 (Singh et al. 2007).

The addition of acetyl groups makes starch amphiphilic in nature. Different methods have been used to conduct acetylation of starches. Based on DS, the acetylated starches are divided into three and is presented in Fig. 2.17 along with their applications (Imre and Vilaplana 2020).

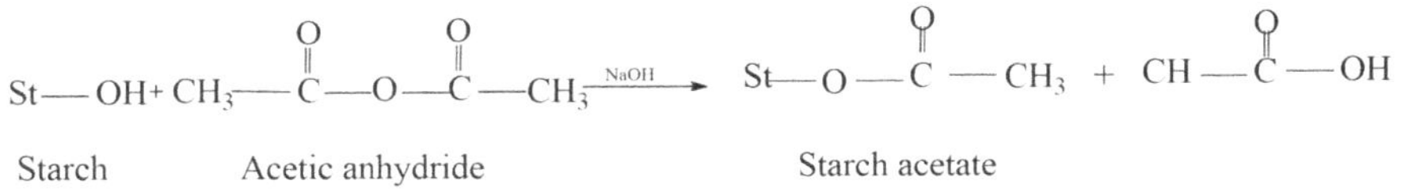

Fig. 2.15 Formation of starch acetate using acetic anhydride

Fig. 2.16 Formation of starch acetate using vinyl acetate

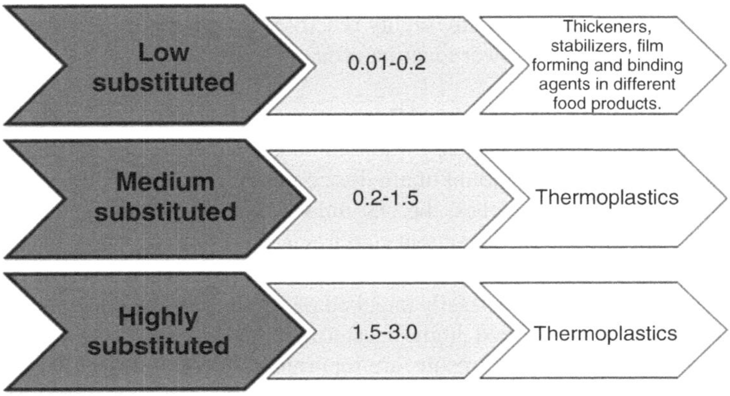

Fig. 2.17 Classification and application of acetylated starches based on their degree of substitution

The hydroxyl groups attached to C-2, C-3, and C-6 possess different magnitude of reactivity for substitution reaction. The (-OH) attached to C-6 also called primary is highly reactive and is present on the surface of the granule, while the hydroxyl groups attached to other two carbon atoms are called secondary which are buried in the interior of granules and are less reactive due to steric hindrance (Kalita et al. 2014). One of the oldest and conventional procedures is to treat starch slurry in a pH range of 8.0–9.0, while acetic anhydride is added dropwise. After the reaction, the pH is reduced to 4.5, and the reaction chemicals are washed out with repeated washings. The acetic anhydride used is in the range of 3–9% in conventional methods, while time duration varies between 10 and 60 minutes resulting in mostly low to intermediate substituted starches of DS (0.03–0.21) (Ali and Hasnain 2011; Ayucitra 2012). Thus, to prepare starches with intermediate and higher degree of substitution, many researchers have reported acetylation reaction in which starch is directly added in acetic anhydride. Other methodologies for increasing DS of acetylated starches include use of metallic catalysts, enzymes, ball milling, high-pressure treatment, pulsed electric field, and use of microwaves (J. Hong et al. 2016). The study by Imre and Vilaplana (2020) used different organocatalysts like citric acid, tartaric acid, propionic acid, etc. for conducting acetylation of starch with the aim to conduct reaction under non-alkaline conditions (which raise environmental concerns). It was found that tartaric acid acts as the best organocatalyst and was able to give the same level of DS as that of conventional method discussed above. This method involved no aqueous medium as the catalyst was directly dissolved in acetic anhydride. To dissolve the catalyst, oil bath was used to provide temperature of 120 °C. After addition of freeze-dried starch, the reaction was conducted for 1–16 h (40–50 °C). To achieve higher DS, very high levels of sodium hydroxide (20%) with increasing levels of acetic anhydride and higher temperatures are also studied (Chi et al. 2008; Luo and Shi 2012). Vinyl acetate also reacted in similar condition as that mentioned for conventional acetylation using acetic anhydride (Kalita et al. 2014). However, Huang et al. (2007) reported a higher degree of substitution for yellow pea, cow pea, and chick pea starches when reacted with vinyl acetate in contrast to acetic anhydride as the latter being highly reactive may get involved into side reactions with water leading to lower reaction efficacy.

Impact of Acetylation on Structural Properties of Starches

For acetylated starches, it has been reported that the degree of substitution was directly proportional to the amount of amylose content. In a study conducted by Luo and Shi (2012) on maize starches, the DS under the same conditions was found to be in the order of high amylose content starch > normal starch > waxy maize starch which could be due to declining amorphous regions from high amylose to waxy. The amorphous region is more easily attacked by the esterifying reagents. Imre and Vilaplana (2020) also reported highest esterification rate for high-amylose corn starch. However, contradictory results are reported by Colussi et al. (2014) for rice starches. The X-ray diffraction patterns which reflect the changes in the crystalline pattern and percent crystallinity of the starch exhibit more pronounced changes at higher acetylation levels due to disruption of intermolecular hydrogen bonds leading to decline in crystallinity. The surface erosion or lack of smoothness also occurs after acetylation as observed under scanning electron microscopy (Colussi et al. 2014). The FTIR studies also confirm the hydrophobic structure of starch as decrease in band intensity at 1650 cm^{-1} is observed which is attributed to water-starch association. Also a new peak at 1750 cm^{-1} corresponding to insertion of acetyl carbonyl group is observed after acetylation (Colussi et al. 2014; Imre and Vilaplana 2020).

Impact of Acetylation on Functional Properties of Starches

For low DS acetylated starches, it was observed that swelling is higher but decreases with progressive acetylation as the starches become more hydrophobic. Also thermal transitions take place at lower temperatures indicating weakening of starch granules (Luo and Shi 2012). Similarly pasting temperature also reduced due to acetylation, but in terms of peak and hot paste viscosity, diversified results have been reported, suggestive of the fact that viscosity profile may be dependent on source of starch, degree of substitution, location of substituent groups on starch chains, and process of acetylation (Ali and Hasnain 2011; Wani et al. 2012; Golachowski et al. 2015). Acetic anhydride-esterified starches show improved freeze-thaw stability compared to their native counterparts as the introduced functional groups create steric hindrance and prevent retrogradation of amylopectin chains and thus averts syneresis (Hong et al. 2018). The acetylated starches form soft texture gels compared to native starches (Colussi et al. 2015).

Impact of Acetylation on Digestibility of Starches

A decline in RDS (readily digestible starch), hydrolysis index, glycemic index, and increase in RS (resistant starch) is observed after acetylation suggesting increased resistance against attack of amylases and amyloglucosidases after modification (Simsek et al. 2012; Zięba et al. 2013).

2.1.6.2 Monostarch Phosphates

The monostarch phosphates (E1410) have been defined by JECFA (2021) as "the starch obtained by esterification with ortho-phosphoric acid, sodium or potassium ortho-phosphate and sodium tripolyphosphate" resulting in the formation of substituted starch or starch monoester. It is different from crosslinking which is conducted using phosphorus oxychloride or sodium trimetaphosphate that leads to the formation of diesters or distarch phosphates (E1412). In contrast E1413 with chemical name "phosphated distarch phosphate" is obtained by "crosslinking of starch with sodium trimetaphosphate or phosphorus oxychloride combined with esterification with ortho-phosphoric acid, or sodium or potassium ortho-phosphate, or sodium tripolyphosphate" (JECFA 2021). Thus, the reagents used must comply with JECFA monograph on modified starches to be termed monostarch phosphate, crosslinked (distarch phosphate), or dual modified as earlier explained in case of E1413.

Typically, monostarch phosphate is obtained by mixing starch milk with STPP in the presence of sodium sulfate anhydrous which acts as a catalyst. This mixture is then adjusted to pH (8–11), and STPP is allowed to impregnate for 1 h at 27 °C

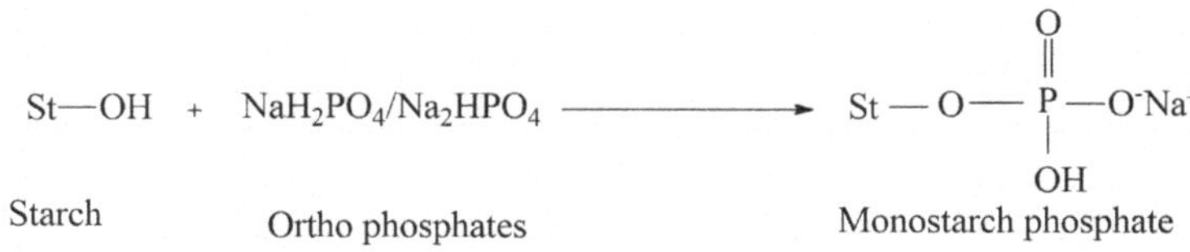

Fig. 2.18 Formation of monostarch phosphate

followed by drying starch at 40 °C to obtain a dried mixture of starch-STPP which is then heated at 120–130 °C for around 2 h so that monoesters are formed through insertion of phosphate groups, after which the starch is again slurried and pH is neutralized to 6.5 followed by excessive washing to remove the modification chemicals. The starch is then dried to 10–12% moisture content (Wongsagonsup et al. 2005; Polnaya et al. 2013). Below pH of 9.0, monoester formation takes place as STPP gets protonated and forms monometaphosphates which then esterify to hydroxyl group of glucose monomer in starch to produce singly esterified monostarch phosphates. The maximum monoester formation took place at pH = 9 for sago starch treated with STPP in the range of pH 8–11. Some studies also used a mixture of anhydrous disodium hydrogen phosphate (Na_2HPO_4) and sodium dihydrogen phosphate dehydrate ($NaH_2PO_4 \bullet 2H_2O$) (Sitohy et al. 2000; Yang et al. 2016). Monostarch phosphate formation reaction is shown in Fig. 2.18 (Singh et al. 2007).

Monostarch phosphate formation has also been reported by extruding mixture of starch/STPP equilibrated to 22% moisture content in a temperature range of 80–180 °C maintained in different extrusion zones (Sugih et al. 2019).

Impact of Phosphorylation on Structural Properties of Starches

The XRD pattern does not change much after phosphorylation (Polnaya et al. 2013; Yang et al. 2016). ^{13}C CP/MAS NMR spectra showed that the relative intensity of signals in the C4 region increased after phosphorylation (Yang et al. 2016). After phosphorylation the ^{31}P NMR spectrum shows signals in the range of δ 3.41–5.2 ppm which indicates the formation of monoesters at C-2, C-3, and C-6 hydroxyl groups of the monomeric glucose unit in starch. The signal at 3.6 ppm is associated with primary phosphate group at the C-6 position, while the signal at 4.7 ppm is due to phosphate groups at the C-2 and C-3 positions (Passauer et al. 2006; Sang et al. 2007; Polnaya et al. 2013).

Impact of Phosphorylation on Functional Properties of Starches

Water absorption index, swelling power, and percent transmittances usually increase after phosphorylation which is due to disruption of the intra- and intermolecular hydrogen bonds owing to the addition of phosphate monoester groups into the starch chains which causes weakening of the granules and allows percolation of more water into the granules which is contradictory to crosslinking where distarch esters actually strengthen the granules. This weakening of granules leads to higher solubility and peak viscosity of these starches, but enthalpy of starch gelatinization and pasting temperature decreases (Lin et al. 2009; Yang et al. 2016; Sugih et al. 2019). Wongsagonsup et al. (2005) reported increase in peak viscosity, final viscosity, and breakdown with the increasing levels of STPP used for modification, while decline in pasting temperature and time was observed suggesting that higher levels of esterification in starches cause further weakening of granules and increase hydrophilicity which leads to easier gelatinization and higher water uptake (Polnaya et al. 2013). Freeze-thaw stability also improve after phosphorylation (Passauer et al. 2006).

Impact of Phosphorylation on Digestibility of Starches

Not much work has been done on digestibility of monostarch phosphates. However, Passauer et al. (2006) observed that on reaction with STPP, high-amylose maize starches showed improvement in digestibility suggesting easier penetration of enzymes but the digestibility of waxy and normal maize starches remained largely unaffected with respect to the native starch.

2.1.6.3 Starch Sodium Octenyl Succinate

Starch sodium octenyl succinate commonly known as OSA (octenyl succinic anhydride) starch is designated as E1450 in JECFA Codex Alimentarius. It is prepared by esterification of starch using octenyl succinic anhydride. This modification turns the character of starch to hydrophobic due to insertion of octenyl succinyl groups. The commercial use of waxy corn OSA starch is the most common, whereas other sources of OSA starch production may include normal corn, potato, and tapioca. A lot of variation in process parameters has been found in studies regarding its method of preparation in terms of temperature, pH, total starch solids, and reaction time. Most commonly a 30–35% (w/w) starch slurry adjusted to pH 8–9 with sodium carbonate or sodium hydroxide solution is treated with octenyl succinic anhydride which is usually suspended in threefold absolute ethanol so as to allow better miscibility of this hydrophobic reagent into the aqueous slurry. The most commonly reported OSA concentration varies from 0.2 to 15% based on starch weight (db). However, as per CFR (2022), the amount of 1-octenyl succinic anhydride must not exceed 3% based on starch weight (db), whereas if the reaction is conducted in the presence of aluminum sulfate (not to exceed 2%), then the amount of OSA is further restricted to ≤2%. The temperature and time of reaction are usually reported as between 20–35 °C and 2–24 h. After the esterification reaction, the pH is brought to 6.5 using HCl solution, and then the starch slurry is washed with water and alcohol to remove the chemical residues of modifying reagents (Bai et al. 2011; Sharma et al. 2016; Hong et al. 2017; Zheng et al. 2017; Quintero-Castaño et al. 2020). The role of pH in the context of OSA modification carries significant importance and is an important parameter to be controlled. Usually pH between 8.0 and 8.2 is recommended. The pH above this promotes side reactions leading to decline in esterification, while below this the (OH^-) ions are not activated for nucleophilic attack by octenyl succinic anhydride (Shi and He 2012; Xu et al. 2012). Since OSA reaction efficiency is low, therefore different methods have been reported to improve the degree of substitution during the reaction like Zhang et al. (2020) used ultrasonic-assisted method to perform esterification of starch with OSA and resulted in significant improvement in DS value which were attributed to mechanochemical effects of ultrasound waves that exclude air and exert strong mechanical forces on granules' interior particularly in the amorphous region leading to better access of modification reagent. Use of ball mill to cause mechanical activation of starch surface was also found to improve the efficacy of reaction to produce this amphiphilic OSA modified starch. The reaction was conducted in a ball mill in which starch, sodium carbonate (as catalyst), and OSA reagent (8%, starch dry weight basis) were all added together (Zhang et al. 2020). This anhydrous mixture was milled from 30 to 120 min showing max DS value at 90 minutes. The higher efficacy of this reaction was due to the unveiling of the active and free hydroxyl groups of starch owing to the mechanical destruction of the crystalline structure of starch. Milling with balls also improved the area of contact between starch and long chain OSA reagent. Another non-conventional method is lipase-coupled OSA reaction conducted by Xu et al. (2012) who conducted a detailed optimization study for esterification and found that the starch when pre-heated at 65 °C for 15 min led to improvement in efficacy of lipase-coupled esterification as the gelatinization phenomenon led to less restricted entry of lipase and OSA into the starch granules. This resulted in DS = 0.0195 within 30 min of reaction. The OSA substituted starch is shown in Fig. 2.19 (Wang et al. 2010).

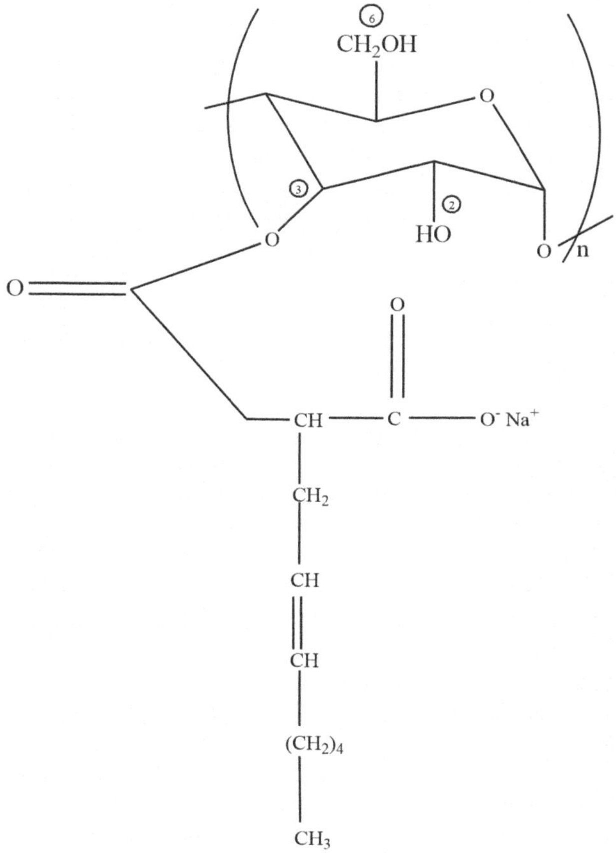

Fig. 2.19 Octenyl succinic anhydride (OSA) starch

Impact of OSA Modification on Structural Properties of Starches

Octenyl succinylation preferentially takes place in the amorphous region of the starch granules and mainly occurs on the surface of the granules as seconded by confocal laser scanning microscopy (Whitney et al. 2016). The SEM results also reported surface erosion or less smooth surface after modification, but the granular morphology largely remained intact in conventional methodology followed for OSA modification (Hu et al. 2016; Quintero-Castaño et al. 2020; J. Zhang et al. 2021). This hydrophobic modification leads to the introduction of new characteristic peak at 1713–1726) cm^{-1} that corresponds to C=O (carbonyl group) vibration due to the esterification of OSA group to hydroxyl group of repeating monomeric glucose units in starch (Xu et al. 2012; Zhang et al. 2021). The XRD patterns also largely remain unchanged during modification. However, slight decline to no change in crystallinity is observed after modification which again emphasizes the fact that modification is more pronounced in amorphous region leaving behind crystalline area largely unaffected (Won et al. 2017; Zhang et al. 2021). Some studies also reported decrease in molecular weight of starch with the increase in OSA groups' substitution (Simsek et al. 2015; Zheng et al. 2017). The ^{1}H NMR shows signals between 0.70 and 2.0 ppm which are assigned to free or bound terminal protons of the OSA groups inserted in modified starch compared to native counterpart (Simsek et al. 2015; Won et al. 2017).

Impact of OSA Modification on Functional Properties of Starches

The OSA starch is very unique as it gives the highest degree of emulsification compared to other modified starch forms. This starch has a more dominant hydrophobic character which makes it suitable for use in pickering emulsions. Since OSA ester groups' formation causes reorganization in starch granules due to steric hindrance and repulsive forces, therefore most studies reported decrease in pasting temperature and peak gelatinization temperature and decline in gelatinization enthalpy which confirms a more disorganized structure in modified starch compared to native starch which allowed easier penetration of water molecules. This disorganized structure also leads to higher swelling power, water solubility index, and peak and breakdown viscosity in OSA starch. The paste clarity was also observed to increase with the increase in degree of substitution. On the other hand, oil absorption capacity significantly increases which is expected as OSA groups being hydrophobic carry affinity for oil (Ali et al. 2015; Sharma et al. 2016; Altuna et al. 2018; Arshad et al. 2020a). The oil in water emulsions can be stabilized by using OSA starches. High DS OSA starches contribute more to emulsion stability as these tend to form smaller droplet size in emulsion which eventually results in more uniform distribution of dispersed phase (oil). Also, more OSA groups lead to higher magnitude of negative zeta potential which prevents coalescence of oil droplets in emulsions (Cheng et al. 2021; Zhang et al. 2021). Compared to native starch, OSA starch shows lower surface tension that leads to emulsion stabilizing properties. These surface active properties are measured using Wilhelmy plate technique (Zainal Abiddin et al. 2018). The emulsion stabilizing property of OSA starch is by virtue of its orientation at oil-water interface which prevents breakage of emulsion due to steric effects. Also, presence of OSA starch leads to more elastic emulsion (G′ > G″) with less dependence of storage and loss moduli on frequency (Bello-Pérez et al. 2015). Amylose content of starch also impacts emulsion stabilizing property of OSA starch. A study by Cheng et al. (2021) showed that emulsion stability of OSA waxy starch was higher compared to OSA normal corn starch which showed better stability compared to high-amylose pea starch due to higher retrogradation of high-amylose starches at oil-water interface which reduces the repulsive steric effects mainly responsible for emulsion stability.

Impact of OSA Modification on Digestibility of Starches

OSA starch can be used as a functional fiber in food products because it has a higher amount of RS content and lower RDS values. The increase in OSA substitution leads to higher resistant starch content suggesting more resistance to enzymatic degradation. The reasons associated with reduced digestibility of OSA starch compared to native starch are (a) physical or sterospecific hindrance caused by bulky OSA groups against α-amylase attack and (b) hydrophobic environment caused by OSA groups (Zhang et al. 2017; Lin et al. 2018). The pullanase treatment to debranch OSA starch further increases the RS content (Liang et al. 2021). The increase in RS content reduces the glucose yield after digestion thus could be used to design low glycemic foods for people suffering from diabetes (Zheng et al. 2017).

2.1.6.4 Permissible Limits of Esterifying Agents and Safety Assessment of Esterified Starches

The permissible modifying agents and the safety assessment of esterified starches as per the Code of Federal Regulations, USA, and JECFA are outlined in Table 2.4.

2.1.6.5 Food Applications of Esterified Starches

Esterified starches include monostarch phosphates, octenyl succinic anhydride starches and acetylated starches. Their food applications are also versatile and are summarized in Table 2.5.

Table 2.5 Applications of different starch esters

Starch esters	Applications	References
Acetylated starch	Increase thickness of cake batters, improves cake volume, binding agent in formed vegetable products, as thickening and stabilizing agents in mayonnaise and sauces, as encapsulating agent, reduces oil uptake in fried foods, and for development of biodegradable packaging material	Golachowski et al. (2015), Colussi et al. (2015), Ali and Hasnain (2016), Moin et al. (2019), Ingredion (2022)
Monostarch phosphate	Improves storage stability of refrigerated and frozen foods.	Singh et al. (2007)
Octenyl succinic anhydride starch	As wall material for encapsulating compounds and as emulsion stabilizer in Pickering emulsions	Ali et al. (2015), Arshad et al. (2018, 2020b); Wang et al. (2020)

2.1.7 Etherified Starches

Ether is a chemical compound which contains two alkyl groups on either side of an oxygen atom and is generally formed by the condensation of two alcohols. High temperature and acid conditions promote ether formation. The etherification of starch is a kind of chemical modification in which free hydroxyl groups (-OH) of glucose units are replaced by a variety of reactive functional groups donated by ether compounds to form charged (anionic and cationic), amphoteric, or nonionic starch ethers. Like other chemical modifications, the objective of starch etherification is to improve functionality of native starch. Specific etherified starches find applications in food (Shaikh et al. 2020), paper making (Hak Lae Lee et al. 2002), textile (Ragheb et al. 1997), pharmacy (Massicotte et al. 2008), and wastewater treatment (Shi et al. 2012).

2.1.7.1 Types of Etherified Starches

Based on the kind of chemical compound used for the process of etherification, there are three categories of ether modified starches, namely, charged, amphoteric, and nonionic (Fig. 2.20). The charged starch ether could be further categorized as cationic or anionic in nature. Cationic starch is characterized by having a positive ionic charge that is provided by either ammonium (NH^{4+}), sulfonium (H_3S^+), or phosphonium groups (PH^{4+}). The chemical reagents commonly used to form cationic starch include 2,3-epoxypropyltrimethylammonium chloride (ETA) or 3-chloro-2-hydroxypropyltrimethyl ammonium chloride (CTA). The applications of cationic starch are majorly targeted in paper making industry followed by their usage as flocculating agent to treat wastewater. These starches also find their functional roles in adhesives, cosmetics, textiles, and detergent-based non-food industries (Kuo and Lai 2009). Similarly, anionic starches are prepared with a chemical reagent that adds negative ions to the starch chains. The three major anion donating functional groups typically used for this type of starch production are carboxyl (-COO^-), phosphate (PO_4^{3-}), and sulfonate (SO_3^-) groups. Among anionic starch derivatives, carboxymethyl starch (CMS) is the widely studied starch ether due to its attractive functional properties and the ease of modification method with which it could be prepared. Briefly, CMS is produced by reacting starch with sodium salt of monochloroacetic acid or the acid alone in the presence of NaOH. It can be made by wet (aqueous and/or organic solvent), dry, and extrusion methods. The commercial applications of CMS are diversified including food, pharmaceutical, medical, and

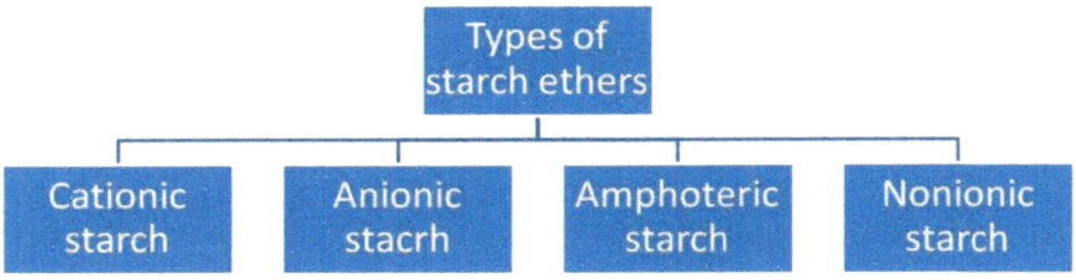

Fig. 2.20 Types of etherified starch

cosmetic industries. Amphoteric starch refers to a starch derivative that simultaneously contains both positive and negative ions derived from various functional groups as described above. The modification process could be carried out in any manner like cationic modification followed by anionic modification of native starch or vice versa. Also, the attachment of both types of ions could be on the same location or at different locations on the starch chains (Zhang et al. 2007). The utilization of amphoteric starch is limited in food products, and it finds greater acceptability in non-food applications. Nonionic starch contains no charge and is usually formed by reaction of native starch with different alkyl ethers. The two common types of alkyl ether modified starches are hydroxypropylated starch (HPS) and hydroxyethylated starch (HES). The modifying chemical reagents for HPS and HES are propylene oxide and ethylene oxide, respectively.

Among all etherified starches, only hydroxypropylated starch is permissible by the regularity authorities worldwide to be incorporated as an additive in foods.

2.1.7.2 Safety Assessment of Hydroxypropylated Starches

As mentioned above, only hydroxypropylated starch is permissible as food additive by the leading regulatory authority Joint FAO/WHO Expert Committee on Food Additives (JECFA). In the context of JECFA, two kinds of HPS are regulated: (1) hydroxpropylated starch, native starch modified with propylene oxide, and (2) hydroxypropyl distarch phosphate, native starch modified with propylene oxide and crosslinked with phosphorus oxychloride or sodium trimetaphosphate. In both cases, the hydroxyl groups of unmodified starch are replaced with hydroxypropyl groups which should not exceed 7.0%, based on dry starch weight to be safely used in food products (JECFA 2021). According to the Code of Federal Regulations (CFR), the maximum allowable amount of propylene oxide for hydroxypropylated starch is set to 25% based on dried starch weight (CFR 2023).

2.1.7.3 Impact of Hydroxypropylation on Structural Properties of Starches

Surface morphology as observed through scanning electron microscope gives an idea about the size, shape, and surface pattern of starch granules. The severity of structural change of starch granules in terms of shape and appearance change may depend on the degree of hydroxypropylation with lower MS values resulting in less obvious structural alterations. For example, Chuenkamol et al. (2007) observed that low hydroxypropylated canna starches having MS in the range of 0.01–0.11 remained intact and did not show any significant structural difference from the native starch. On the other hand, hydroxpropylated potato starches having MS range of 0.098–0.122 exhibited a great difference in morphological appearance when compared with native starch (Kaur et al. 2004). In this case, the substituted starch granules assumed somewhat doughnut-like shape because of inward movement of outer surfaces after hydroxypropylation. The difference in structural change upon hydroxypropylation at the same level may be linked with the starch source. Generally, hydroxypropylation is known to have little to no effect on the crystalline form of native starch as it did not transform the polymorphic form from one type to another (Chuenkamol et al. 2007; Kim et al. 2011; Lawal 2011; Oladebeye et al.

2013; Yang et al. 2016). However, the peak intensities were variably altered after the modification process which may arise due to difference in starch source and modification levels used. For instance, Oladebeye et al. (2013) reported no pronounced difference between the peak pattern of native and hydroxypropylated lima bean and jack bean starches, but there was a slight drop in peak intensities. Similarly, native and hydroxypropylated canna starches also showed similar peak types but with slight increment in peak intensity at 5.6° (Chuenkamol et al. 2007). When starch granules are gone through the process of hydroxypropylation, the addition of hydroxypropyl groups mainly initiates in the amorphous phase due to easy reagent accessibility followed by crystalline region, and therefore if the level of modification is low, then this phase remains unaffected (Lawal 2011). The degree of crystallinity is related with other starch characteristics like digestibility, heat stability, and retrogradation behavior. Naknaen (2014) found that the degree of crystallinity reduced after hydroxypropylation which was also dependent on the concentration of propylene oxide used. This outcome indicates that higher level of propylene oxide could interact with either amorphous or the crystalline domains of starch granule leading to reduction in crystallinity degree. Similar results of reduced degree of crystallinity are reported for hydroxypropylated jackfruit starch (Kittipongpatana and Kittipongpatana 2011) and hydroxypropylated wheat and oat starches (Majzoobi et al. 2014).

2.1.7.4 Impact of Hydroxypropylation on Functional Properties of Starches

The process of hydroxypropylation attaches hydroxypropyl functional groups that are highly hydrophilic and thereby impacts the performance of native starch. In general, hydroxypropylated starches are recognized for their increased granular swelling, better paste clarity, lower retrogradation, improved freeze-thaw stability, and greater cold water solubility (Fouladi and Nafchi 2014). According to the literature, numerous starch sources like corn (Liu et al. 1999), potato (Kaur et al. 2004), sweet potato (Lee and Yoo 2011), rice (Choi and Kim 2022), sorghum (Mehboob et al. 2021), pearl millet (Shaikh et al. 2017), and sago (Fouladi and Nafchi 2014) have been exploited to observe the impact of hydroxypropylation under different reaction conditions like concentration of propylene oxide, difference in molar substitution (MS), pH, reaction time, and temperature. The effect of hydroxypropylation on some selected functional properties is outlined in Table 2.6. Hydroxypropylated starch appears to be clearer than unmodified starch due to steric effects of ether groups. Hydroxypropylation is almost universally known to greatly enhance the granular swelling and water solubility of starch granules. The explanation to this increasing behavior may be twofold. Firstly, attachment of hydroxypropyl groups weakens the starch granules due to disruption of inter- and intrahydrogen bonding that ultimately gives easy access to water molecules. Secondly, replacement of hydroxyl groups that were less hydrophilic with highly hydrophilic hydroxypropyl groups makes the granules more water-loving, thereby causing additionally high swelling power followed by greater solubility. In terms of thermal analysis, usually HPS is known to possess lower ΔH_{gel} that means less amount of energy is required to dissociate the granular structure of such starch. Additionally, hydroxypropylated

Table 2.6 Impact of hydroxypropylation on some functional properties of starches

Starch source	Level of propylene oxide (%)	Molar substitution (MS)	Paste clarity (%)	Swelling power (g/g)	Solubility (%)	Gelatinization enthalpy (J/g)	References
Corn	0	ND[a]	ND	8.7	17.0	6.61	
	20	ND	ND	13.6	55.0	ND	
Potato	0	–	ND	30.5	3.6	11.7	Kaur et al. (2004)
	10	0.098	ND	33.8	5.1	10.7	
Rice	0	–	ND	26.6	7.78	ND	Chun and Yoo (2007)
	3	0.030	ND	35.5	8.19	ND	
	6	0.086	ND	40.6	9.01	ND	
	9	0.118	ND	48.1	9.80	ND	
	12	0.142	ND	52.8	10.7	ND	
Sweet potato	0	–	39.7	45.4	15.2	16.5	Lee and Yoo (2011)
	3	0.042	47.8	52.3	19.6	15.6	
	6	0.082	50.3	55.0	21.6	15.0	
	9	0.132	56.2	57.3	23.7	13.9	
	12	0.153	59.5	61.1	25.2	12.3	
Sorghum	0	ND	1.6	10.1	7.2	14.1	Mehboob et al. (2021)
	10	ND	28.3	11.4	33.0	11.5	
	20	ND	46.9	12.1	55.0	7.4	
	30	ND	40.4	14.2	46.0	6.8	
Tapioca	0	–	ND	ND	52.9	ND	Xiong et al. (2019)
	10	0.042	ND	ND	84.7	ND	
	20	0.082	ND	ND	81.4	ND	
	30	0.132	ND	ND	88.1	ND	

[a]Not determined

starch is universally known to exhibit less or no tendency to retrograde upon cold storage due to existence of bulky hydroxypropyl functional groups that offer steric hindrance and prevent recrystallization of amylopectin chains responsible for retrogradation.

2.1.7.5 Impact of Hydroxypropylation on Digestibility of Starches

The digestibility pattern of hydroxypropylated starches is reported to be quite variable perhaps due to difference in starch origin, level of MS, granular or gelatinized forms, and different methods of measurement. Hydroxypropylation increased RS content of waxy rice (Hwang et al. 2009), corn (Chung et al. 2008), wheat (Yousefi et al. 2015), and canna (Juansang et al. 2012) even after cooking. The existence of bulky hydroxypropyl groups on starch chains poses resistance against enzymes to access glycosidic linkage (Chung et al. 2008). The slowly digestible starches are gaining much interest these days due to their health impacts like controlling hyperglycemia. Iftikhar et al. (2022) compared the influence of hydroxypropylation on the SDS fraction of rice starches bearing different amylose composition. The authors reported that hydroxypropylated waxy and low-amylose rice starches had higher SDS content than their native counterparts. It means that hydroxypropylation is effective in increasing the SDS fraction of waxy cultivars that otherwise would have greater digestibility due to absence of amylose. Therefore, it is concluded that hydroxypropylation could be an effective modification technique to reduce RDS content with concomitant increment of SDS and RS fractions in starches.

2.1.7.6 Food Applications of Hydroxypropylated Starches

The highlighting features of hydroxypropylated starch that make it an attractive food additive are enhanced viscosity, superior clarity, reduced tendency to retrograde (no syneresis), low and high pH resistance, heat stability, and low digestibility. Two different hydroxypropylated starches are commercially available and authorized to be used as food additives, namely, hydroxypropylated starch (E1440) and hydroxypropylated distarch phosphate (E1442). The E1440 is a single modified starch prepared from propylene oxide only, whereas E1442 is a double modified hydroxypropylated distarch phosphate. Hydroxypropylated distarch phosphate in the first place is widely used as a thickener in sauces, puddings, gravies, soups, spreads, mayonnaise, and salad dressings. Apart from food applications, HPS has been used to fabricate biodegradable food packaging materials. Shaikh et al. (2019b) fabricated biodegradable films from hydroxypropylated pearl millet starches that had excellent clarity and flexibility in comparison to control film.

2.2 Conclusion

The global food resources are currently under strain due to rising world population and impacts of global warming which is negatively impacting the agriculture. In this crucial time, the starch industry is also facing problems due to lack of crops particularly corn required for starch extraction and modification. Therefore, the current situation has made the need for exploration of new crops with similar starch properties as that of corn starch crucial for the survival of starch industry.

References

Abdorreza MN, Robal M, Cheng LH, Tajul AY, Karim AA (2012) Physicochemical, thermal, and rheological properties of acid-hydrolyzed sago (Metroxylon sagu) starch. LWT Food Sci Technol 46:135–141

Adebowale KO, Afolabi TA, Olu-Owolabi BI (2006) Functional, physicochemical and retrogradation properties of sword bean (Canavalia gladiata) acetylated and oxidized starches. Carbohydr Polym 65:93–101

Ahsan M, Ali TM, Hasnain A (2023) Use of oxidized potato starch as simultaneous fat and casein replacer in analogue mozzarella cheese- : impact on functional and sensory properties of cheese. Food Hydrocoll 142:108810

Ali TM, Hasnain A (2011) Functional and morphological characterization of low-substituted acetylated white sorghum (Sorghum bicolor) starch. Int J Polym Anal Charact 16:187–198

Ali TM, Hasnain A (2014) Morphological, physicochemical, and pasting properties of modified white sorghum (Sorghum bicolor) starch. Int J Food Prop 17:523–535

Ali TM, Hasnain A (2016) Physicochemical, morphological, thermal, pasting, and textural properties of starch acetates. Food Rev Intl 32:161–180

Ali TM, Waqar S, Ali S, Mehboob S, Hasnain A (2015) Comparison of textural and sensory characteristics of low-fat mayonnaise prepared from octenyl succinic anhydride modified corn and white sorghum starches. Starch/Staerke 67:183–190

Altuna L, Herrera ML, Foresti ML (2018) Synthesis and characterization of octenyl succinic anhydride modified starches for food applications. A review of recent literature. Food Hydrocoll 80:97–110

An D, Xiao Q, Zhang C, Cai M, Zhang Y, Weng H, Chen F, Xiao A (2021) Preparation, characterization, and application of high-whiteness agar bleached with hydrogen peroxide. Food Hydrocoll 113:106520

Aparicio-Saguilán A, Valera-Zaragoza M, Perucini-Avendaño M, Páramo-Calderón DE, Aguirre-Cruz A, Ramirez-Hernández A, Bello-Pérez LA (2015) Lintnerization of banana starch isolated from underutilized variety: morphological, thermal, functional properties, and digestibility. CyTA J Food 13:3–9

Arshad H, Ali TM, Abbas T, Hasnain A (2018) Effect of microencapsulation on antimicrobial and antioxidant activity of nutmeg oleoresin using mixtures of gum Arabic, OSA, and native sorghum starch. Starch/Staerke 70:1–12

Arshad H, Ali TM, Hasnain A (2020a) Bioactive properties and oxidative stability of nutmeg oleoresin microencapsulated by freeze drying using native and OSA sorghum starches as wall materials. J Food Meas Charact 14:2559–2569

Arshad H, Ali TM, Hasnain A (2020b) Comparative study on efficiency of nutmeg microencapsulation (freeze-drying method) using native and OSA sorghum starch as wall materials in combination with gum arabic. Cereal Chem 97:589–600

Ashogbon AO, Akintayo ET (2014) Recent trend in the physical and chemical modification of starches from different botanical sources: A review. Starch-Stärke 66:41–57

Ayucitra A (2012) Preparation and characterisation of acetylated corn starches. Int J Chem Eng Appl 3:156

Bai Y, Shi YC, Herrera A, Prakash O (2011) Study of octenyl succinic anhydride-modified waxy maize starch by nuclear magnetic resonance spectroscopy. Carbohydr Polym 83:407–413

Bello-Pérez LA, Bello-Flores CA, del Carmen Nuñez-Santiago M, Coronel-Aguilera CP, Alvarez-Ramirez J (2015) Effect of the degree of substitution of octenyl succinic anhydride-banana starch on emulsion stability. Carbohydr Polym 132:17–24

BeMiller JN (2019) Corn starch modification. In: Corn: chemistry and technology (third). Elsevier, United Kingdom, pp 537–549
Broekman JOP, Genuino HC, Heeres HJ, Brinksma J, Wielema T, Deuss PJ (2023) Benign catalytic oxidation of potato starch using a homogeneous binuclear manganese catalyst and hydrogen peroxide. Catal Sci Technol 13:1233–1243
CFR (2022) CFR—Code of Federal Regulations Title 21. CFR - Code of Federal Regulations Title 21, Volume (2)
CFR (2023) CFR—Code of Federal Regulations Title 21
Cheng F, Ai Y, Ghosh S (2021) Utilization of octenyl succinic anhydride-modified pea and corn starches for stabilizing oil-in-water emulsions. Food Hydrocoll 118:106773
Chi H, Xu K, Wu X, Chen Q, Xue D, Song C, Zhang W, Wang P (2008) Effect of acetylation on the properties of corn starch. Food Chem 106:923–928
Chiu C, Solarek D (2009) Chapter 17—modification of starches. Starch, pp 629–655
Choi HW, Kim H-S (2022) Hydroxypropylation and acetylation of rice starch: effects of starch protein content. Food Sci Biotechnol 31:1169–1177
Chowdary KPR, Enturi V, Sandhya Rani A (2011) Preparation and evaluation of starch phosphate-A new modified starch as a disintegrant in tablet formulations. Int J Chem Sci 9:889–899
Chuenkamol B, Puttanlek C, Rungsardthong V, Uttapap D (2007) Characterization of low-substituted hydroxypropylated canna starch. Food Hydrocoll 21:1123–1132
Chun S, Yoo B (2007) Effect of molar substitution on rheological properties of hydroxypropylated rice starch pastes. Starch-Stärke 59:334–341
Chung HJ, Shin DH, Lim ST (2008) In vitro starch digestibility and estimated glycemic index of chemically modified corn starches. Food Res Int 41:579–585
Colussi R, El Halal SLM, Pinto VZ, Bartz J, Gutkoski LC, da Rosa Zavareze E, Dias ARG (2015) Acetylation of rice starch in an aqueous medium for use in food. LWT Food Sci Technol 62:1076–1082
Colussi R, Pinto VZ, El Halal SLM, Vanier NL, Villanova FA, e Silva RM, da Rosa Zavareze E, Dias ARG (2014) Structural, morphological, and physicochemical properties of acetylated high-, medium-, and low-amylose rice starches. Carbohydr Polym 103:405–413
Deetae P, Shobsngob S, Varanyanond W, Chinachoti P, Naivikul O, Varavinit S (2008) Preparation, pasting properties and freeze—thaw stability of dual modified crosslink-phosphorylated rice starch. Carbohydr Polym 73:351–358
Delley RJ, O'Donoghue AC, Hodgson DRW (2012) Hydrolysis studies of phosphodichloridate and thiophosphodichloridate ions. J Org Chem 77:5829–5831
Desam GP, Li J, Chen G, Campanella O, Narsimhan G (2018) Prediction of swelling behavior of crosslinked maize starch suspensions. Carbohydr Polym 199:331–340
Dhull SB, Bangar SP, Deswal R, Dhandhi P, Kumar M, Trif M, Rusu A (2021) Development and characterization of active native and cross-linked pearl millet starch-based film loaded with fenugreek oil. Food Secur 10:3097
Dias ARG, Zavareze EDR, Elias MC, Helbig E, Da Silva DO, Ciacco CF (2011) Pasting, expansion and textural properties of fermented cassava starch oxidised with sodium hypochlorite. Carbohydr Polym 84:268–275
Dong H, Vasanthan T (2020) Effect of phosphorylation techniques on structural, thermal, and pasting properties of pulse starches in comparison with corn starch. Food Hydrocoll 109:106078
Dundar AN, Gocmen D (2013) Effects of autoclaving temperature and storing time on resistant starch formation and its functional and physicochemical properties. Carbohydr Polym 97:764–771
Egharevba HO (2019) Chemical properties of starch and its application in the food industry. In: Chemical properties of starch, In Chemical properties of starch, vol 9. IntechOpen, London, UK
FDA (2019) questions-and-answers-dietary-fiber @ www.fda.gov.
Fonseca LM, Gonçalves JR, El Halal SLM, Pinto VZ, Dias ARG, Jacques AC, Zavareze EDR (2015) Oxidation of potato starch with different sodium hypochlorite concentrations and its effect on biodegradable films. Lwt 60:714–720

Fouladi E, Nafchi AM (2014) Effects of acid-hydrolysis and hydroxypropylation on functional properties of sago starch. Int J Biol Macromol 68:251–257

Gao F, Li D, Bi C, Mao Z, Adhikari B (2014) Preparation and characterization of starch crosslinked with sodium trimetaphosphate and hydrolyzed by enzymes. Carbohydr Polym 103:310–318

Garcia MAVT, Garcia CF, Faraco AAG (2020) Pharmaceutical and biomedical applications of native and modified starch: A review. Starch-Stärke 72:1900270

Garrido LH, Schnitzler E, Zortéa MEB, de Souza Rocha T, Demiate IM (2014) Physicochemical properties of cassava starch oxidized by sodium hypochlorite. J Food Sci Technol 51:2640–2647

Golachowski A, Zięba T, Kapelko-Zeberska M, Drozdz W, Gryszkin A, Grzechac M (2015) Current research addressing starch acetylation. Food Chem 176:350–356

Gonenc I, Us F (2019) Effect of glutaraldehyde crosslinking on degree of substitution, thermal, structural, and physicochemical properties of corn starch. Starch-Stärke 71:1800046

Guleria P, Yadav BS (2022) Effect of chemical treatments on the functional, morphological and rheological properties of starch isolated from pigeon pea (Cajanus cajan). Curr Res Food Sci 5:1750–1759

Halal SLME, Colussi R, Pinto VZ, Bartz J, Radunz M, Carreño NLV, Dias ARG, Zavareze EDR (2015) Structure, morphology and functionality of acetylated and oxidised barley starches. Food Chem 168:247–256

Haq F, Yu H, Wang LI, Teng L, Haroon M, Khan RU, Mehmood S, Ullah RS, Khan A, Nazir A et al (2019) Advances in chemical modifications of starches and their applications. Carbohydr Res 476:12–35

Hong J, Zeng X-A, Brennan CS, Brennan M, Han Z (2016) Recent advances in techniques for starch esters and the applications: A review. Food Secur 5:50

Hong J, Zeng XA, Buckow R, Han Z (2018) Structural, thermodynamic and digestible properties of maize starches esterified by conventional and dual methods: differentiation of amylose contents. Food Hydrocoll 83:419–429

Hong Y, Li Z, Gu Z, Wang Y, Pang Y (2017) Structure and emulsification properties of octenyl succinic anhydride starch using acid-hydrolyzed method. Starch/Staerke 69:1–9

Hoover R (2000) Acid-treated starches acid-treated starches. Food Rev Intl 370

Hornung PS, De Oliveira CS, Lazzarotto M, Da Silveira Lazzarotto SR, Schnitzler E (2016) Investigation of the photo-oxidation of cassava starch granules: thermal, rheological and structural behaviour. J Therm Anal Calorim 123:2129–2137

Hornung PS, Granza AG, de Oliveira CS, Lazzarotto M, Schnitzler E (2015) Study of the effects of ultraviolet light and sodium hypochlorite solutions on properties of cassava starch granules. Food Biophy 10:368–374

Hu H, Liu W, Shi J, Huang Z, Zhang Y, Huang A, Yang M, Qin X, Shen F (2016) Structure and functional properties of octenyl succinic anhydride modified starch prepared by a non-conventional technology. Starch/Staerke 68:151–159

Huang J, Schols HA, Jin Z, Sulmann E, Voragen AGJ (2007) Characterization of differently sized granule fractions of yellow pea, cowpea and chickpea starches after modification with acetic anhydride and vinyl acetate. Carbohydr Polym 67:11–20

Hwang D, Kim S, Kim J, Ryu J, Yoo S, Park C, Kim B, Baik M (2009) In vitro digestibility of hydroxypropylated and cross-linked waxy and non-waxy rice starches. Starch-Stärke 61:20–27

Iftikhar SA, Chakraborty S, Dutta H (2022) Effect of acetylation, hydroxypropylation and dual acetylation-hydroxypropylation on physicochemical and digestive properties of rice starches with different amylose content. Biointerface Res Appl Chem 12(2022):6788–6803

Imre B, Vilaplana F (2020) Organocatalytic esterification of corn starches towards enhanced thermal stability and moisture resistance. Green Chem 22:5017–5031

Ingredion (2022) Modified starches. https://www.ingredion.com/emea/en-uk/ingredients/ingredient-types/starches/modified-starch.html

Iwuagwu M (2000) The effect of bleaching on the disintegrant properties of maize starch. STP Pharma Sci 10(2):43–147

James BeMiller RW (ed) (2009) Starch chemistry and technology, 3rd edn. Academic Press, UK, London

JECFA (2021) Specifications Monograph prepared by the meeting of the Joint FAO/WHO Expert Committee on Food Additives (JECFA), 92nd Meeting 2021

Juansang J, Puttanlek C, Rungsardthong V, Puncha-arnon S, Uttapap D (2012) Effect of gelatinisation on slowly digestible starch and resistant starch of heat-moisture treated and chemically modified canna starches. Food Chem 131:500–507

Jyothi AN, Moorthy SN, Rajasekharan KN (2006) Effect of cross-linking with epichlorohydrin on the properties of cassava (Manihot esculenta Crantz) starch. Starch-Stärke 58:292–299

Kalita D, Kaushik N, Mahanta CL (2014) Physicochemical, morphological, thermal and IR spectral changes in the properties of waxy rice starch modified with vinyl acetate. J Food Sci Technol 51:2790–2796

Kaur L, Singh N, Singh J (2004) Factors influencing the properties of hydroxypropylated potato starches. Carbohydr Polym 55:211–223

Kim H-S, Choi H-S, Kim B-Y, Baik M-Y (2011) Ultra high pressure (UHP)-assisted hydroxypropylation of corn starch. Carbohydr Polym 83:755–761

Kittipongpatana OS, Kittipongpatana N (2011) Preparation and physicochemical properties of modified jackfruit starches. LWT Food Sci Technol 44:1766–1773

Klang JM, Tene ST, Nguemguo Kalamo LG, Boungo GT, Ndomou Houketchang SC, Kohole Foffe HA, Womeni HM (2019) Effect of bleaching and variety on the physico-chemical, functional and rheological properties of three new Irish potatoes (Cipira, Pamela and Dosa) flours grown in the locality of Dschang (West region of Cameroon). Heliyon 5:e02982

Kou T, Gao Q (2018) New insight in crosslinking degree determination for crosslinked starch. Carbohydr Res 458–459:13–18

Kuakpetoon D, Wang YJ (2001) Characterization of different starches oxidized by hypochlorite. Starch/Staerke 53:211–218

Kuakpetoon D, Wang YJ (2008) Locations of hypochlorite oxidation in corn starches varying in amylose content. Carbohydr Res 343:90–100

Kuo W-Y, Lai H-M (2009) Effects of reaction conditions on the physicochemical properties of cationic starch studied by RSM. Carbohydr Polym 75:627–635

Lawal OS (2004) Composition, physicochemical properties and retrogradation characteristics of native, oxidised, acetylated and acid-thinned new cocoyam (Xanthosoma sagittifolium) starch. Food Chem 87:205–218

Lawal OS (2011) Hydroxypropylation of pigeon pea (Cajanus cajan) starch: preparation, functional characterizations and enzymatic digestibility. LWT Food Sci Technol 44:771–778

Lazzarotto SRDS, Bet CD, Hornung PS, Lazzarotto M, Schnitzler E (2017) Induced effects by oxidation with potassium permanganate on the thermal, morphological, colorimetric and pasting properties of corn starch. Ukrainian Food J 6:197–210

Le Corre D, Bras J, Dufresne A (2010) Starch nanoparticles: a review. Biomacromolecules 11:1139–1153

Lee HL, Yoo B (2011) Effect of hydroxypropylation on physical and rheological properties of sweet potato starch. LWT Food Sci Technol 44:765–770

Lee HL, Shin JY, Koh C-H, Ryu H, Lee DJ, Sohn C (2002) Surface sizing with cationic starch: its effect on paper quality and papermaking process. TAPPI J 1:34–40

Li H, Wang R, Liu J, Zhang Q, Li G, Shan Y, Ding S (2020) Effects of heat-moisture and acid treatments on the structural, physicochemical, and in vitro digestibility properties of lily starch. Int J Biol Macromol 148:956–968

Li W, Corke H, Beta T (2007) Kinetics of hydrolysis and changes in amylose content during preparation of microcrystalline starch from high-amylose maize starches. Carbohydr Polym 69:398–405

Liang S, Hong Y, Gu Z, Cheng L, Li C, Li Z (2021) Effect of debranching on the structure and digestibility of octenyl succinic anhydride starch nanoparticles. Lwt 141:111076

Lim S, Seib PA (1993) Preparation and pasting properties of wheat and corn starch phosphates. Cereal Chem 70:137

Lin Q, Xiao H, Zhao J, Li L, Yu F (2009) Characterization of the pasting, flow and rheological properties of native and phosphorylated rice starches. Starch/Staerke 61:709–715

Lin Q, Liang R, Zhong F, Ye A, Singh H (2018) Effect of degree of octenyl succinic anhydride (OSA) substitution on the digestion of emulsions and the bioaccessibility of $β$-carotene in OSA-modified-starch-stabilized-emulsions. Food Hydrocoll 84:303–312

Liu H, Ramsden L, Corke H (1999) Physical properties and enzymatic digestibility of hydroxypropylated ae, wx, and normal maize starch. Carbohydr Polym 40:175–182

Liu H, Liang R, Antoniou J, Liu F, Shoemaker CF, Li Y, Zhong F (2014) The effect of high moisture heat-acid treatment on the structure and digestion property of normal maize starch. Food Chem 159:222–229

Liu Q, Li M, Xiong L, Qiu L, Bian X, Sun C, Sun Q (2018) Oxidation modification of debranched starch for the preparation of starch nanoparticles with calcium ions. Food Hydrocoll 85:86–92

Luo F, Huang Q, Fu X, Zhang L, Yu S (2009) Preparation and characterisation of crosslinked waxy potato starch. Food Chem 115:563–568

Luo Z-G, Shi Y-C (2012) Preparation of acetylated waxy, normal, and high-amylose maize starches with intermediate degrees of substitution in aqueous solution and their properties. J Agric Food Chem 60:9468–9475

Majzoobi M, Saberi B, Farahnaky A, Mesbahi G (2014) Comparison of physicochemical and gel characteristics of hydroxypropylated oat and wheat starches. Int J Food Eng 10:657–667

Manelius R, Buléon A, Nurmi K, Bertoft E (2000) The substitution pattern in cationised and oxidised potato starch granules. Carbohydr Res 329:621–633

Mason WR (2009) Starch use in foods. In: BeMiller JN (ed) Starch: chemistry and technology, 3rd edn. Academic Press, San Diego, California

Massicotte LP, Baille WE, Mateescu MA (2008) Carboxylated high amylose starch as pharmaceutical excipients: structural insights and formulation of pancreatic enzymes. Int J Pharm 356:212–223

Mehboob S, Ali TM, Sheikh M, Hasnain A (2020) Effects of cross linking and/or acetylation on sorghum starch and film characteristics. Int J Biol Macromol 155:786–794

Mehboob S, Mohsin Ali T, Shaikh M, Hasnain A (2021) Effects of varying levels of succinylation and hydroxypropylation on functional, thermal and textural characteristics of white sorghum starch. Cereal Chem 98:624–633

Mehfooz T, Ali TM, Hasnain A (2019) Effect of cross-linking on characteristics of succinylated and oxidized barley starch. J Food Meas Charact 13:1058–1069

Mehfooz T, Ali TM, Shaikh M, Hasnain A (2020) Characterization of hydroxypropylated-distarch phosphate barley starch and its impact on rheological and sensory properties of soup. Int J Biol Macromol 144:410–418

Miao M, Jiang B, Zhang T, Jin Z, Mu W (2011) Impact of mild acid hydrolysis on structure and digestion properties of waxy maize starch. Food Chem 126:506–513

Moin A, Ali TM, Hasnain A (2017) Influence of different molar concentrations of acid on morphological, physicochemical and pasting properties of Pakistani Basmati and Irri rice starches. Int J Biol Macromol 101:214–221

Moin A, Ali TM, Hasnain A (2019) Effect of basmati and irri acetylated rice starches on textural and sensorial characteristics of dumpling wrappers. J Food Meas Charact 13:2594–2602

Naknaen P (2014) Physicochemical, thermal, pasting and microstructure properties of hydroxypropylated jackfruit seed starch prepared by etherification with propylene oxide. Food Biophys 9:249–259

Niu CM, Li SY, Gongc YX (2014) Synthesis and properties of the cassava starch succinate. Adv Mater Res 1033–1034:346–351

Okekunle MO, Adebowale KO, Olu-Owolabi BI, Lamprecht A (2020) Physicochemical, morphological and thermal properties of oxidized starches from Lima bean (Phaseolus lunatus). Sci Afr 8:e00432

Oladebeye AO, Oshodi AA, Amoo IA, Karim AA (2013) Hydroxypropyl derivatives of legume starches: functional, rheological and thermal properties. Starch-Stärke 65:762–772

Otache MA, Duru RU, Achugasim O, Abayeh OJ (2021) Advances in the modification of starch via esterification for enhanced properties. J Polym Environ 29:1365–1379

Park EY, Ma J-G, Kim J, Lee DH, Kim SY, Kwon D-J, Kim J-Y (2018) Effect of dual modification of HMT and crosslinking on physicochemical properties and digestibility of waxy maize starch. Food Hydrocoll 75:33–40
Passauer L, Liebner F, Fischer K (2006) Synthesis and properties of novel hydrogels from cross-linked starch phosphates. In: Macromolecular symposia, vol 244, pp 180–193
Pietrzyk S, Fortuna T, Łabanowska M, Kurdziel M, Królikowska K, Juszczak L (2019) Effect of oxidation level of high-amylose corn starch on its acetylation, molecular structure, and functional properties. Starch-Stärke 71:1800097
Polnaya FJ, Haryadi, Marseno DW, Cahyanto MN (2013) Effects of phosphorylation and cross-linking on the pasting properties and molecular structure of sago starch. Int Food Res J 20:1609–1615
Pratiwi M, Faridah DN, Lioe HN (2018) Structural changes to starch after acid hydrolysis, debranching, autoclaving-cooling cycles, and heat moisture treatment (HMT): a review. Starch/Staerke 70
Purcell S, Wang YJ, Seo HS (2014) Application of oxidized starch in bake-only chicken nuggets. J Food Sci 79:C810–C815
Quintero-Castaño VD, Castellanos-Galeano FJ, Álvarez-Barreto CI, Lucas-Aguirre JC, Bello-Pérez LA, Rodríguez-Garcia ME (2020) Starch from two unripe plantains and esterified with octenyl succinic anhydride (OSA): partial characterization. Food Chem 315:126241
Ragheb AA, El-Sayiad HS, Hebeish A (1997) Preparation and characterization of carboxymethyl starch (CMS) products and their utilization in textile printing. Starch-Stärke 49:238–245
Rao H, Sindhu R, Panwar S (2022) Morphology and functionality of dry heat-treated and oxidized quinoa starches. J Food Process Preserv 46:1–10
Reddy N, Yang Y (2010) Citric acid cross-linking of starch films. Food Chem 118:702–711
Sánchez-Rivera MM, Méndez-Montealvo G, Núñez-Santiago C, De La Rosa-Millan J, Wang YJ, Bello-Péreza LA (2009) Physicochemical properties of banana starch oxidized under different conditions. Starch/Staerke 61:206–213
Sandhu KS, Singh N, Lim S-T (2007) A comparison of native and acid thinned normal and waxy corn starches: physicochemical, thermal, morphological and pasting properties. LWT Food Sci Technol 40:1527–1536
Sang Y, Prakash O, Seib PA (2007) Characterization of phosphorylated cross-linked resistant starch by 31P nuclear magnetic resonance (31P NMR) spectroscopy. Carbohydr Polym 67:201–212
Sang Y, Seib PA (2006) Resistant starches from amylose mutants of corn by simultaneous heat-moisture treatment and phosphorylation. Carbohydr Polym 63:167–175
Sangseethong K, Lertphanich S, Sriroth K (2009) Physicochemical properties of oxidized cassava starch prepared under various alkalinity levels. Starch - Stärke 61:92–100
Sangseethong K, Termvejsayanon N, Sriroth K (2010) Characterization of physicochemical properties of hypochlorite- and peroxide-oxidized cassava starches. Carbohydr Polym 82:446–453
Shah N, Mewada RK, Mehta T (2016) Crosslinking of starch and its effect on viscosity behaviour. Rev Chem Eng 32:265–270
Shaikh F, Ali TM, Arif S, Raza L, Hasnain A (2022) Rheological, textural and digestibility characteristics of chapatti as affected by RS4 starch prepared from corn and sorghum. J Food Meas Charact:1–8
Shaikh F, Ali TM, Mustafa G, Hasnain A (2019a) Comparative study on effects of citric and lactic acid treatment on morphological, functional, resistant starch fraction and glycemic index of corn and sorghum starches. Int J Biol Macromol 135:314–327
Shaikh M, Haider S, Ali TM, Hasnain A (2019b) Physical, thermal, mechanical and barrier properties of pearl millet starch films as affected by levels of acetylation and hydroxypropylation. Int J Biol Macromol 124:209
Shaikh M, Ali TM, Hasnain A (2017) Utilization of chemically modified pearl millet starches in preparation of custards with improved cold storage stability. Int J Biol Macromol 104:360–366
Shaikh M, Ali TM, Hasnain A (2020) Comparative study on the application of chemically modified corn and pearl millet starches in white sauce. J Food Process Preserv 44:e14393

Sharma M, Singh AK, Yadav DN, Arora S, Vishwakarma RK (2016) Impact of octenyl succinylation on rheological, pasting, thermal and physicochemical properties of pearl millet (Pennisetum typhoides) starch. LWT 73:52–59

Shi R, Tang N, Zhang X, Cheng Y (2023) Characterization of physicochemical properties of hypochlorite oxidized starches and their suitability for making starch noodles. Starch - Stärke 75:2200212

Shi SS, He GQ (2012) Process optimization for cassava starch modified by octenyl succinic anhydride. Procedia Eng 37:255–259

Shi Y, Ju B, Zhang S (2012) Flocculation behavior of a new recyclable flocculant based on pH responsive tertiary amine starch ether. Carbohydr Polym 88:132–138

Shujun W, Jinglin Y, Jiugao Y, Haixia C, Jiping P (2007) The effect of acid hydrolysis on morphological and crystalline properties of Rhizoma Dioscoreae starch. Food Hydrocoll 21:1217–1222

Simsek S, Ovando-Martinez M, Marefati A, Malin S, Rayner M (2015) Chemical composition, digestibility and emulsification properties of octenyl succinic esters of various starches. Food Res Int 75:41–49

Simsek S, Ovando-Martinez M, Whitney K, Bello-Pérez LA (2012) Effect of acetylation, oxidation and annealing on physicochemical properties of bean starch. Food Chem 134:1796–1803

Sindhu R, Devi A, Khatkar BS (2021) Morphology, structure and functionality of acetylated, oxidized and heat moisture treated amaranth starches. Food Hydrocoll 118:106800

Singh H, Sodhi NS, Singh N (2009) Structure and functional properties of acid thinned sorghum starch. Int J Food Prop 12:713–725

Singh J, Kaur L, McCarthy OJ (2007) Factors influencing the physico-chemical, morphological, thermal and rheological properties of some chemically modified starches for food applications-A review. Food Hydrocoll 21:1–22

Sitohy MZ, Labib SM, El-Saadany SS, Ramadan MF (2000) Optimizing the conditions for starch dry phosphorylation with sodium mono- and dihydrogen orthophosphate under heat and vacuum. Starch/Staerke 52:095–100

Srichuwong S, Isono N, Mishima T, Hisamatsu M (2005) Structure of lintnerized starch is related to X-ray diffraction pattern and susceptibility to acid and enzyme hydrolysis of starch granules. Int J Biol Macromol 37:115–121

Subroto E, Indiarto R, Djali M, Rosyida HD (2020) Production and application of crosslinking-modified starch as fat replacer: A review. Int J Eng Trends Technol 68:26–30

Sugih AK, Christabella L, Kristianto H, Prasetyo S (2019) Effect of different types of phosphorylating reagent on the synthesis of modified tapioca starch. IOP Conf Ser Mater Sci Eng 673:12001

Sun Q, Zhu X, Si F, Xiong L (2015) Effect of acid hydrolysis combined with heat moisture treatment on structure and physicochemical properties of corn starch. J Food Sci Technol 52:375–382

Tethool EF, Jading A, Santoso B (2012) Characterization of physicochemical and baking expansion properties of oxidized sago starch using hydrogen peroxide and sodium hypochlorite catalyzed by UV irradiation. Food Sci Qual Manag 10:1–11

Thomas DJ, Atwell WA (1999) Handbook series—starches. American Association of Cereal Chemists

Utomo P, Nizardo NM, Saepudin E (2020) Crosslink modification of tapioca starch with citric acid as a functional food. AIP Conference Proceedings. Vol. 2242. No. 1. AIP Publishing

Vanier NL, El Halal SLM, Dias ARG, da Rosa Zavareze E (2017) Molecular structure, functionality and applications of oxidized starches: a review. Food Chem 221:1546

Wang J, Su L, Wang S (2010) Physicochemical properties of octenyl succinic anhydride-modified potato starch with different degrees of substitution. J Sci Food Agric 90:424–429

Wang PP, Luo ZG, Chun-Chen X-F, Tamer TM (2020) Effects of octenyl succinic anhydride groups distribution on the storage and shear stability of Pickering emulsions formulated by modified rice starch. Carbohydr Polym 228:115389

Wang S, Blazek J, Gilbert E, Copeland L (2012) New insights on the mechanism of acid degradation of pea starch. Carbohydr Polym 87:1941–1949

Wang S, Copeland L (2015) Effect of acid hydrolysis on starch structure and functionality: a review https://doi.org/10.1080/10408398.2012.684551, 55, 1081–1097
Wang YJ, Truong VD, Wang L (2003) Structures and rheological properties of corn starch as affected by acid hydrolysis. Carbohydr Polym 52:327–333
Wang YJ, Wang L (2003) Physicochemical properties of common and waxy corn starches oxidized by different levels of sodium hypochlorite. Carbohydr Polym 52:207–217
Wani IA, Sogi DS, Gill BS (2012) Physicochemical properties of acetylated starches from some Indian kidney bean (Phaseolus vulgaris L.) cultivars. Int J Food Sci Technol 47:1993–1999
Whitney K, Reuhs BL, Martinez MO, Simsek S (2016) Analysis of octenylsuccinate rice and tapioca starches: distribution of octenylsuccinic anhydride groups in starch granules. Food Chem 211:608–615
Won C, Jin Y, Kim M, Lee Y, Chang YH (2017) Structural and rheological properties of potato starch affected by degree of substitution by octenyl succinic anhydride. Int J Food Prop 20:3076–3089
Wongsagonsup R, Pujchakarn T, Jitrakbumrung S, Chaiwat W, Fuongfuchat A, Varavinit S, Dangtip S, Suphantharika M (2014) Effect of cross-linking on physicochemical properties of tapioca starch and its application in soup product. Carbohydr Polym 101:656–665
Wongsagonsup R, Shobsngob S, Oonkhanond B, Varavinit S (2005) Zeta potential (ζ) and pasting properties of phosphorylated or crosslinked rice starches. Starch/Staerke 57:32–37
Woo K, Seib PA (1997) Cross-linking of wheat starch and hydroxypropylated wheat starch in alkaline slurry with sodium trimetaphosphate. Carbohydr Polym 33:263–271
Woo KS, Seib PA (2002) Cross-linked resistant starch: preparation and properties. Cereal Chem 79:819–825
Xie Y, Zhang B, Li M-N, Chen H-Q (2019) Effects of cross-linking with sodium trimetaphosphate on structural and adsorptive properties of porous wheat starches. Food Chem 289:187–194
Xiong Y, Zhang P, Warner RD, Fang Z (2019) Sorghum grain: from genotype, nutrition, and phenolic profile to its health benefits and food applications. Compr Rev Food Sci Food Saf 18:2025–2046
Xu J, Zhou CW, Wang RZ, Yang L, Du SS, Wang FP, Ruan H, He GQ (2012) Lipase-coupling esterification of starch with octenyl succinic anhydride. Carbohydr Polym 87:2137–2144
Yang L, Zhou Y, Wu Y, Meng X, Jiang Y, Zhang H, Wang H (2016) Preparation and physicochemical properties of three types of modified glutinous rice starches. Carbohydr Polym 137:305–313
Yi X, Zhang S, Ju B (2014) Preparation of water-soluble oxidized starch with high carbonyl content by sodium hypochlorite. Starch - Stärke 66:115–123
Yousefi AR, Razavi SMA, Norouzy A (2015) In vitro gastrointestinal digestibility of native, hydroxypropylated and cross-linked wheat starches. Food Funct 6:3126–3134
Zainal Abiddin NF, Yusoff A, Ahmad N (2018) Effect of octenylsuccinylation on physicochemical, thermal, morphological and stability of octenyl succinic anhydride (OSA) modified sago starch. Food Hydrocoll 75:138–146
Zhang B, Mei JQ, Chen B, Chen HQ (2017) Digestibility, physicochemical and structural properties of octenyl succinic anhydride-modified cassava starches with different degree of substitution. Food Chem 229:136–141
Zhang G, Venkatachalam M, Hamaker BR (2006) Structural basis for the slow digestion property of native cereal starches. Biomacromolecules 7:3259–3266
Zhang J, Ran C, Jiang X, Dou J (2021) Impact of octenyl succinic anhydride (OSA) esterification on microstructure and physicochemical properties of sorghum starch. Lwt 152:112320
Zhang JY, Xiong JW, Chen ZP, Cui N (2022) Physicochemical properties of potato starches oxidised by different concentrations of sodium hypochlorite, and their application in yogurt. Int Food Res J 29:582–592
Zhang M, Zhang S, Ju B, Yang J (2007) A review of amphoteric starches in preparation and application. In The Proceedings of The 3rdInternational Conference on Functional Molecules (pp. 31–35). Citeseer
Zhang M, Jia H, Wang B, Ma C, He F, Fan Q, Liu W (2023) A prospective review on the research Progress of citric acid modified starch. Food Secur 12:458

Zhang Y, Dai Y, Hou H, Li X, Dong H, Wang W, Zhang H (2020) Ultrasound-assisted preparation of octenyl succinic anhydride modified starch and its influence mechanism on the quality. Food Chem: X 5:100077
Zhang YR, Wang XL, Zhao GM, Wang YZ (2012) Preparation and properties of oxidized starch with high degree of oxidation. Carbohydr Polym 87:2554–2562
Zhao J, Chen Z, Jin Z, de Waard P, Buwalda P, Gruppen H, Schols HA (2015) Level and position of substituents in cross-linked and hydroxypropylated sweet potato starches using nuclear magnetic resonance spectroscopy. Carbohydr Polym 131:424–431
Zhao J, Schols HA, Chen Z, Jin Z, Buwalda P, Gruppen H (2012) Substituent distribution within cross-linked and hydroxypropylated sweet potato starch and potato starch. Food Chem 133:1333–1340
Zheng Y, Hu L, Ding N, Liu P, Yao C, Zhang H (2017) Physicochemical and structural characteristics of the octenyl succinic ester of ginkgo starch. Int J Biol Macromol 94:566–570
Zhou F, Liu Q, Zhang H, Chen Q, Kong B (2016) Potato starch oxidation induced by sodium hypochlorite and its effect on functional properties and digestibility. Int J Biol Macromol 84:410–417
Zięba T, Kapelko M, Szumny A (2013) Effect of preparation method on the properties of potato starch acetates with an equal degree of substitution. Carbohydr Polym 94:193–198

Enzymatic Modification of Starch: Amylases and Pullulanase

3

Abu Saeid, Farzana Akter, Md. Aslam Ali, and Md. Hafizur Rahman

Abstract

Starch-modifying enzymes are crucial biocatalysts with unique properties, versatile modes of action, and wide-ranging applications. These enzymes, often sourced from microorganisms, plants, or animals, operate under mild conditions, demonstrating high catalytic efficiency, substrate specificity, and eco-friendliness. In terms of mode of action, amylases (α-amylase and β-amylase) hydrolyze starch by cleaving α-1,4-glycosidic bonds or removing maltose units. Glucoamylase converts maltose and isomaltose into glucose. Starch phosphorylases introduce phosphate groups, yielding phosphorylated products, while debranching enzymes like pullulanase facilitate access to branched starch regions. Unlike non-food applications span numerous industries, modified starch is now employed in several food application such as bakery products, snacks, soups, sauces and gravies, batters and coatings, and meat products due to their numeral functional properties. However, starch-modifying enzymes, characterized by their properties, mode of action, and applications, are indispensable biocatalysts that continue to drive innovation across diverse sectors and sustainable industrial practices. This chapter focused on the major properties, mechanisms of action, and industrial applications of starch-modified enzymes.

A. Saeid · F. Akter
Department of Nutrition and Food Engineering, Daffodil International University, DSC, Birulia, Savar, Dhaka, Bangladesh
e-mail: abusaeid.nfe@diu.edu.bd; farzana.nfe@diu.edu.bd

M. A. Ali
Department of Agro-Processing, Faculty of Agriculture, Bangabandhu Sheikh Mujibur Rahman Agricultural University, Gazipur, Bangladesh

M. H. Rahman (✉)
Department of Quality Control and Safety Management, Faculty of Food Sciences and Safety, Khulna Agricultural University, Khulna, Bangladesh

N. Mazumder, M. H. Rahman (eds.), *Advanced Research in Starch*,
https://doi.org/10.1007/978-981-99-9527-1_3

Keywords

Starch · Starch-modifying enzyme · Functional properties · Amylase · Glucoamylase

3.1 Introduction

Starch, after cellulose, is the most abundant organic compound in nature as a whole and is an α-linked polysaccharide labeled as glucan which is produced and preserved as a reserve of carbohydrate by higher plants (Otache et al. 2021). Plants store it in granules or cells of varying sizes and shapes in their tubers and roots (such as corn, maize, wheat, rice, sorghum, barley, etc.) and seeds (such as corn, maize, wheat, rice, sorghum, barley, etc.) (Wang et al. 2018). Both amylose and amylopectin are main elements of starch, with amylose consisting primarily of α-(1—4) D-glucopyranosyl units and amylopectin of many shorter chains accompanied at their reducing ends by α-(1—6) linkage (Kaur et al. 2012) which demonstrates a major distinction between their characteristics and the functions they perform. There are three types of granules as large, medium, and small are coated with these polymers, and their concentric layers of translucent and amorphous materials are structured in disk-like and spherical shapes (Cornejo-Ramírez et al. 2018).

Some of the ways in which starch modifies products' utility are through its effects on their texture, viscosity, gel-forming capability, adhesiveness, binding, moisture absorption, film forming, and overall uniformity. Unlike non-food applications, starch is now employed in several food applications, such as soups, sauces, and gravies; bakery products; dairy confectionery; snacks; batters and coatings; and meat products, due to its various functional properties (Kaur et al. 2012).

Several industrial processes have shown interest in using starch polymers, but starch has been reported to have drawbacks including hygroscopicity, brittleness, acid and enzyme sensitivity, acid and organic solvent insolubility, etc. (Otache et al. 2021). As a result of the necessity of addressing these shortcomings, advancements in other techniques of modification emerged.

Starch is modified to improve upon its already stellar qualities and do away with any drawbacks present in its natural form. In the process of modification, the semi-transparent starch granules can be broken up, making the amylopectin polymers' reactive hydroxyl groups available to electrophilic reactants (Ojogbo et al. 2020). Cooking can cause native starches to retrograde, while pastes have a tendency to gel and can also undergo syneresis with relative ease. Since starch transformation lessens retrogradation, paste gelling tendencies, and gel syneresis while simultaneously enhancing paste glossiness and clarity, texture of paste and gel, film formation, and adhesion, it is clear that starch modification has many positive effects on pastes (Kaur et al. 2012). Improvements in gelatinization, thermal stability, solubility, swelling, pasting, rheology, structure, retrogradation, morphology, etc. are all the outcome of starch modification (Fig. 3.1) (Bangar et al. 2022).

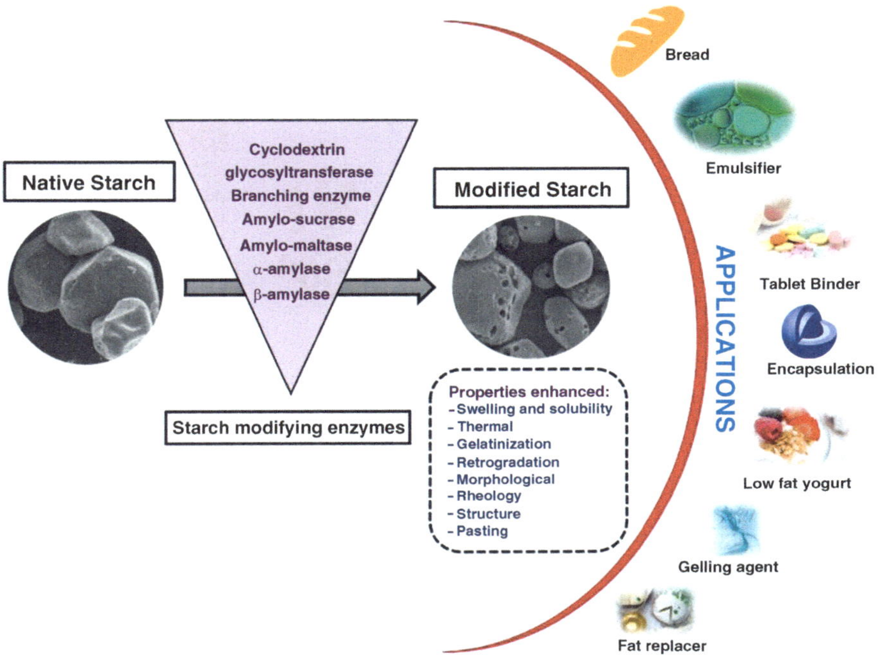

Fig. 3.1 Starch modifying enzymes and their application (Bangar et al. 2022)

Starch has been modified in several ways over the past few decades to make it more useful in a variety of industrial contexts. Utilizing physical (thermal and non-thermal), chemical, and/or enzymatic methods and modification is the process of altering the properties of native starch. Enzymatic modification of starch, as opposed to physical or chemical modification, has recently gained interest from business and academia for its ability to impart the appropriate functional properties while having minimal impact on the environment (Punia et al. 2022).

However, this chapter highlighted the enzymatic modifications starch can employ to produce novel molecules that have substantial applications in the food industries.

3.2 Starch Modifying Enzymes

Starch, which accumulates as a complex granular type structure made of α-1,4-linked and α-1,6-branched glucans, is one of the most abundant natural substances. A wide variety of technically important enzymes are broadly categorized into glycosyl hydrolases and transglycosylases (Fig. 3.2) which are employed for starch functionalization including α- and β-amylase, amylosucrase, cyclodextrin glycosyltransferase, branching enzyme (i.e., glucan transferase), de-branching enzymes

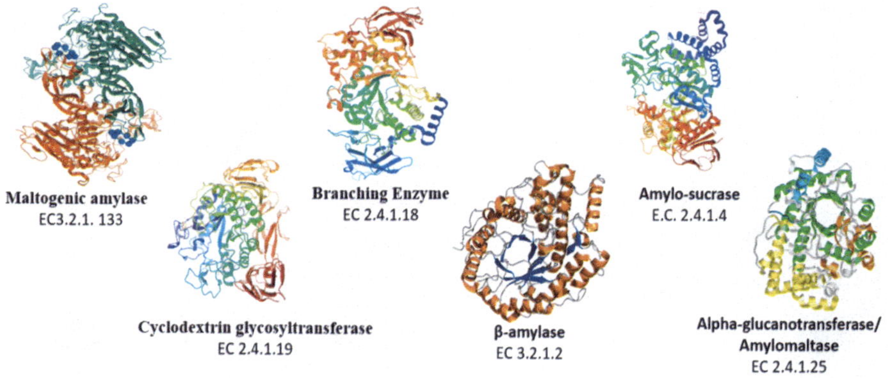

Fig. 3.2 Various enzymes responsible for starch modification

(i.e., pullulanase, isoamylase), and 4-α-glucanotransferase amylomaltase which modify or convert starch into various products (Bangar et al. 2022).

Enzymes alter starches through debranching activity, phosphate substitution, and disproportionation when applied to them. This is accomplished by modifying the starch chain length and the development of branch points (Park et al., 2008). In the following section of this chapter, the most common enzymes associated with starch transformation are covered in details.

3.3 Starch Modified Enzymes and their Properties

3.3.1 α-Amylase

The α-amylase (EC 3.2.1.1) is the member of endo-amylase family, which triggers the first starch digestion by disrupting the α-D-(1–4) glycosidic linkages, resulting in shorter oligosaccharides. It's an extremely common endoamylase, found in everything from archaea and bacteria to plants (Kandra 2003). The branched oligosaccharides that result from α-amylase action have α-configuration and α-limit dextrin as one of their branches (Van der Maarel et al. 2002).

3.3.1.1 Mode of Action

The α-(1–4) glycosidic bonds inside starch polymers are hydrolyzed by endo-acting amylases, commonly referred to as α-amylases. There are two suggested mechanisms for amylase's action: randomly and with multiple attacks. The term "random action" can refer to either a single attack or a series of attacks. In the former, the enzyme-substrate complex first thoroughly hydrolyzes the polymer molecule, and only then does it dissociate. In contrast, merely a single bond is hydrolyzed per effective encounter in the latter case. After the initial (random) hydrolytic attack, the enzyme cleaves many glycosidic bonds consecutively before dissociating from the substrate, a process known as multiple attack (Bijttebier et al., 2008). The amylase's three-dimensional structure allows it to bind to substrates and promote the breaking of the glycosidic bonds via the aid of highly specialized reactive sites (Iulek et al., 2000).

3.3.1.2 Properties

α-Amylases are thermostable and capable of catalyzing the digestion of 1,4-glucosidic linkage in the molecule of starch to liberate maltose sugar (Miao et al. 2018). Below 35 °C, this enzyme exhibits its exo-activity and multiple attack action to greatest effect (Auh et al. 2006). Temperature increment (to around 70 °C, as an example) causes MGA to primarily perform endo-hydrolysis (Grewal et al. 2015). It hydrolyzes amylose more effectively than amylopectin. The active site geometry of alpha-amylase provides insight into size discrimination by setting constraints on the substrate molecule. The use of both amyloglucosidase and α-amylase resulted in the formation of porous, amylose-rich starch, which has new properties. Short amylopectin chains were particularly advantageous by the α-amylase's solubilizing effects (Punia et al. 2022).

3.3.2 β-Amylases

The β-amylases (E.C.3.2.1.2) are another very crucial member of amylolytic enzymes that invert the anomeric shape of starch by removing β-maltose units in succession from starch molecules with the non-reducing end (Zhang et al. 2010). These enzymes are classified into the glycosides' hydrolase class of GH14, which are predominantly available in plants and microbes.

3.3.2.1 Mode of Action

The β-amylase is saccharifying exo-amylase enzyme which digests α-1,4 glucoside linkages in starch molecules and related substrates in an exo-fashion to generate β-maltose and β-limit dextrins. As a result of β-amylase's breakdown of starch, reducing sugar is rapidly released, while iodine staining capacity is gradually diminished. While the decreasing value was plotted against the iodine staining capability, an exo-pattern emerged in every case (Ray 2004). Some research explains that enzyme adsorption on starch granules is thought to occur via a C-terminal interaction site, but the exact pathway is still not well understood (Sarikaya et al. 2000).

3.3.2.2 Properties

Both higher plants and certain bacteria, such certain strains of *Bacillus* and *Clostridium thermosulfurogenes*, generate β-amylases; the latter is also discovered to create amylase with thermostable properties. β-amylase is a sulphydryl enzyme with a high turnover rate (approximately 250,000 bonds hydrolyzed every minute) and no need for a cofactor. The stable pH range for β-amylase activity is 4.0–9.0 at 20 °C for at least 24 h, with an optimum pH range of 5.0–6.0. The heat resistance of b-amylase varies by source. For instance, malted barley b-amylase is inactivated at 70 °C for 10 min, while soybean β-amylase requires 30 minutes at the same temperature (Vitolo 2020).

3.3.3 Glucoamylase

The exo-acting enzyme, glucoamylase (EC 3.2.1.3), is triggered by hydrolyzing the α-1,4 and α-1,6 glucoside bonds from starch and same as poly- and oligosaccharides with the non-reducing sites into simple glucose units, namely, saccharides. Glucoamylases from molds, as opposed to bacteria, archaea, or yeasts, are of enormous commercial interest since they are employed to generate glucose using starch and dextrins as substrate (Riaz et al. 2012).

3.3.3.1 Mode of Action

Glucoamylase's key mechanism inside the starch molecules is the digestion of α-1,4-glycosidic bonds; this leads to molecules of glucose being formed. Glucoamylase (GA) hydrolyzes the starch from the non-reducing sites in a sequential fashion, releasing glucose in the process. It can hydrolyze α-1,6 linkages to a lesser extent, yielding glucose as a by-product (Pavezzi et al. 2008). The slower catalytic action of glucoamylase can be attributed to its single-displacement mechanism, which requires the release of the shortened oligosaccharide chain following chain cleavage and then glucose discharge before the next bond disruption.

3.3.3.2 Properties

Extensive use of *Aspergillus* spp. glucoamylase in commercial starch processing is hampered by the enzyme's poor thermal stability from 55 to 60 °C; pH levels between 4 and 4.5 which are suitable for most activities; very slow activity; and the yield of by-products that diminish the final glucose production and result in millions of dollars' worth of wasted product (Kumar and Satyanarayana 2009).

3.3.4 Cyclodextrin Glycosyltransferase

The endoenzyme cyclodextrin glycosyltransferase (CGTase) is a member of the family α-amylase. It has the capability of producing cyclodextrins by transforming starch. Cyclodextrins, also known as cycloamylase dextrins (CD), are oligomers that are both cyclic and non-reducible obtained from the cyclization and disruption of α-D-glucopyranosyl α-(1, 4) glycosidic linkage (Punia et al. 2022). The cyclodextrin glycosyltransferase obtained from *Bacillus* spp. is mostly used for producing non-reducing cyclic maltooligosaccharides, called cyclodextrins (Menocci et al. 2008).

3.3.4.1 Mode of Action

CGTase has poor hydrolyzing action and primarily digests the three transglycosylation reactions: cyclization-hydrolysis, part of a linear oligosaccharide is transferred to a water molecule; coupling, the opening of a CD molecule followed by transfer to a linear oligosaccharide; and disproportionation, transfer of a portion of one linear oligosaccharide to another linear oligosaccharide (Rashid et al. 2002).

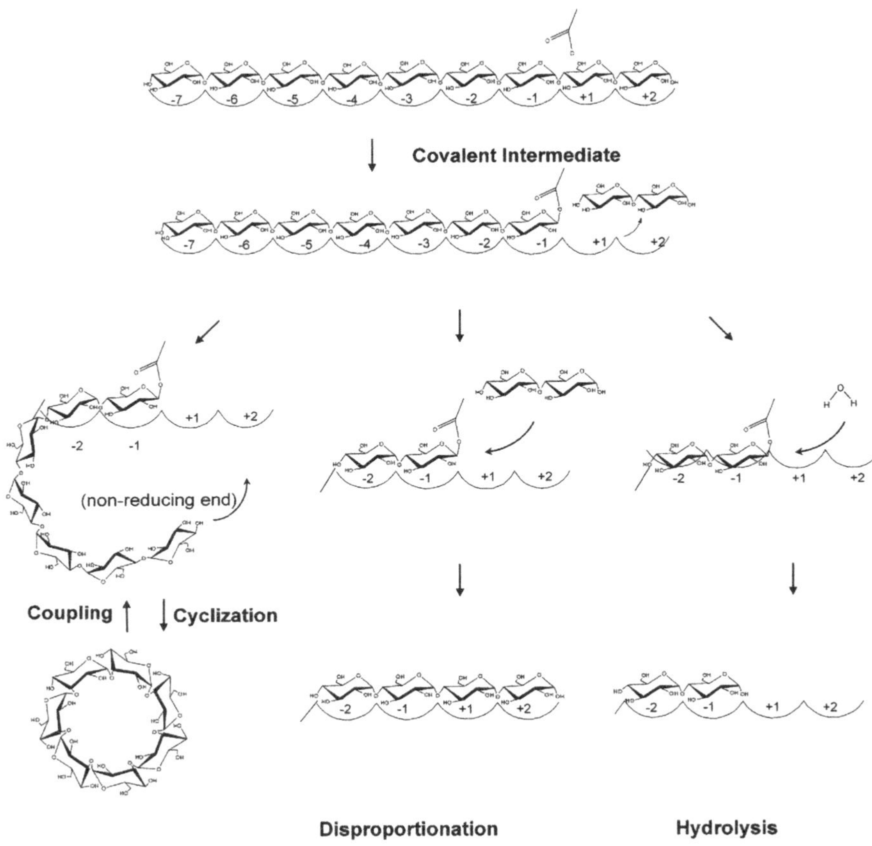

Fig. 3.3 Graphical representation of catalytic reaction of cyclodextrin glycosyltransferase (Kelly et al. 2009)

The coupling reaction catalyzed by CGTases involves the binding and ring breakage of cyclodextrins at the active site, followed by their transfer to an acceptor molecule, where they are hydrolyzed to linear products (Fig. 3.3). The three most common cyclodextrins produced by these enzymes are α-, β -, and γ-cyclodextrin, and they are distinguished by their varying degrees of polymerization (Kelly et al. 2009).

3.3.4.2 Properties

Starch is broken down primarily by the enzyme cyclodextrin glycosyltransferase (CGTase), which transforms starch into cyclodextrins (Rashid et al. 2002). The thermostability features of CGTases are heavily influenced by the host organism's environment, which can range from meso- to hyperthermophilic. Apparently evolved from a common ancestor, it acquired distinct mutations at substrate binding sites to provide a wide range of product specificities (Kelly et al. 2009).

3.3.5 Debranching Enzyme

Most of debranching enzymes are glycoside hydrolases belonging to families 13 and 57, which are found in plants and microorganisms. They fall into two main categories distinguished by the sequences of the peptides they contain and the substrates to which they bind: isoamylase (EC 3.2.1.68) and pullulanase (EC 3.2.1.41) (Wangpaiboon et al. 2023). For industrial application, most of the debranching enzymes (pullulanase) are isolated from microbial sources, for instance, *Bacillus acidopullulyticus*, *Fervidobacterium pennavorans*, *Pseudomonas amyloderamosa*, etc. (Xia et al. 2021).

3.3.5.1 Mode of Action

Debranching enzymes primarily work by disrupting the α-1, 6 glucosidic bonds. DBEs of the pullulanase type, also known as R-enzymes or limit-dextrinases, breakdown the α-1, 6 pullulan and amylopectin connections with relative ease but only have modest action toward the α-1, 6 bonds in the glycogen (Fig. 3.4). In contrast, isoamylase-type DBEs hydrolyze α-1, 6 linkages in pullulan but not in amylopectin or glycogen (Luo et al. 2019).

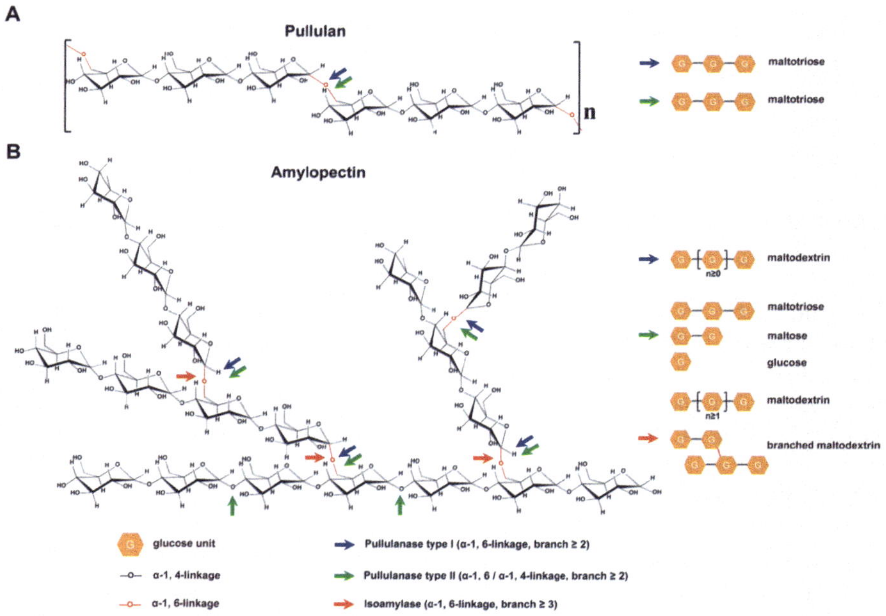

Fig. 3.4 Starch debranching enzyme (SDBE) debranching patterns on (**a**) pullulan and (**b**) amylopectin shown schematically (Xia et al. 2021)

3.3.5.2 Properties

Debranching enzyme is a monomer consisting of 1532 amino acids and has a molecular mass of about 165 kDa and has two unique triggering actions: oligo-1,4-1,4-glucantransferase and amylo-1,6-glucosidase, which are localized in a separate protein site (Lucchiari et al. 2007). The glycoside debranching enzyme pullulanase functions best in cassava starch at pH and temperature levels of 6.0 and 50 °C, respectively (Wangpaiboon et al. 2023).

3.3.6 Other Amylase

3.3.6.1 Amylomaltase

Among the α-amylase enzymes is amylomaltase (EC 2.4.1.25), which belongs to the 4-glucanotransferase (4-GTase) subfamily (Ayudhaya et al. 2016). The three facets of existence—Archaea, Eukarya, and bacteria—contain these enzymes. Amylomaltases are enzymes often employed to modify starches (Fig. 3.5), but they are also being investigated as a plant-based, chemical-free substitute for gelatin in the food sector (Kaur et al. 2012).

Mode of Action

A thermoreversible gel is formed by the action of α-1, 4—α-1, 4 glucosyl transferase enzyme called amylomaltases from starch. This is accomplished by disrupting a α-1, 4 linkage between two glucose molecules to subsequently form a novel α-1, 4 link (Kaur et al. 2012). This causes starch to undergo transglucosylation both within and between molecules, resulting in cyclodextrins containing large-ring and unbranched maltooligosaccharides (Ayudhaya et al. 2016).

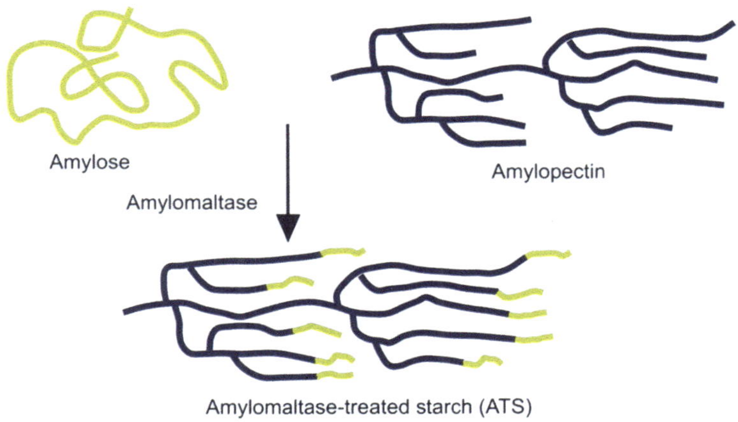

Fig. 3.5 Amylomaltase-treated starch from potato amylose and amylopectin (Alting et al. 2009)

Properties

Amylomaltase is an enzyme that acts on starch, changing it so that amylose is fully depolymerized and branch amylopectin chains are lengthened. The longer amylopectin chains produced by this approach are employed in the commercial manufacturing of thermoreversible starch gel due to their superior gel-forming capabilities (Korompokis et al. 2021). Additionally, this amylomaltase exhibits an inadequate amount of cyclization action, leading to cycloamyloses containing 8 to 32 units of glucose. At temperatures above which starch becomes gelatinous, *Thermus thermophilus* amylomaltase begins to work (the optimal temperature is 72 °C) (Van Der Maarel et al. 2005).

3.3.6.2 Amylosucrase

The enzyme amylosucrase (E.C. 2.4.1.4) is a member of the glycosidic hydrolase family (GH13) and has multiple roles in the digestion of sucrose. This enzyme is exceptional in that it can produce an amylose-like polysaccharide derived from sucrose, without having to incorporate a α-D-glucosyl-nucleotide-diphosphate, for example, ADP-glucose or UDP-glucose (Seo et al., 2020). The amylosucrase was first obtained from microbial sources especially from *Neisseria perflava* (Potocki de Montalk et al. 2000).

Mode of Action

Amylosucrase hydrolyzed the α-(1,4)- and α-(1,6)-glucosidic links of starch as its primary mechanism of action. In this case of only sucrose as a substrate, the catalytic pathway of pure amylosucrase is extremely intricate, as a result of non-Michaelian kinetic profile. Further, through the polymerization process (which formed exclusively α-1, 4-bonded glucan), the following reactions took place: hydrolysis of sucrose and the synthesis of oligosaccharide by means of acceptor reactions on the liberated glucose and fructose units (Potocki de Montalk et al. 2000).

Properties

The capacity of the enzyme amylosucrase from *Neisseria polysaccharea* to produce starches with high concentrations of fractions with an inferior glycemic index, slowly digested starch, and resistant starch has garnered a lot of attention. When treated with sucrose at 37 °C, its transglycosylation activity extends the non-reducing site of the amylose and amylopectin external chains by as much as 13 to 19 glucose units (Lee and Park 2020).

3.4 Applications of Starch Modified Enzymes

α-Amylase, β-amylase, cyclodextrin glycosyltransferase, debranching enzymes, and branching enzyme are the technically significant enzymes employed for starch modification (Wu et al. 2020). When native starch is treated by these specific enzymes to achieve desired modification, it is called single enzyme-modified starch. A notable development in the realm of food science and technology is single

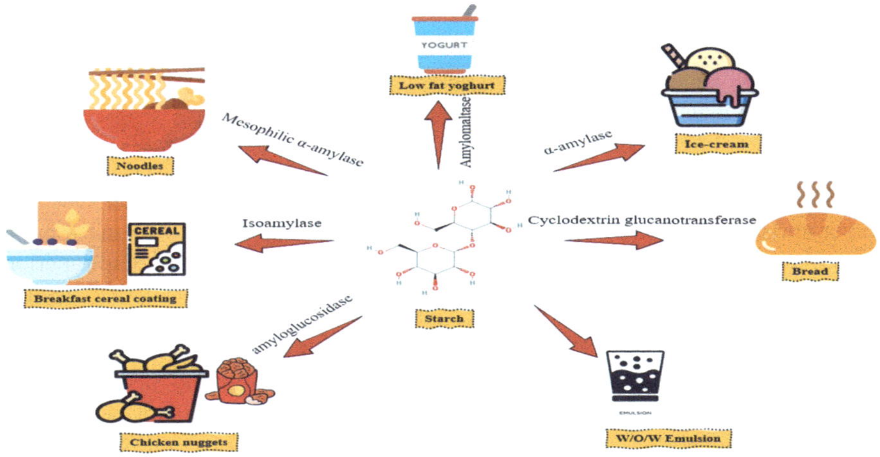

Fig. 3.6 Starch-modifying enzymes in utilization in food (Bangar et al. 2022)

enzyme-modified starch. It is feasible to acquire desired functionality by choosing the right enzyme and altering circumstances. This adaptability gives food producers new options for developing goods with superior quality and customer appeal (Fig. 3.6). Due to the fact that it streamlines the modification process, uses less chemicals, and enables focused alterations to achieve particular functionalities, this technology has significant advantages over conventional methods. The following sections will provide compiling, understanding, and synthesizing information on the application of significant single enzymes in starch modification and their applications in different food industries.

3.4.1 Application of α-Amylase in Starch Modification

α-Amylase transforms 1,4-glucosidic bonds in the starch molecules by releasing maltose as the primary product from the non-reducing sites of the chain (Miao et al. 2018). It attacks repeatedly with a great degree of activity. At 35 °C, this enzyme exhibits exo-activity. The peculiar loop that this enzyme's dimeric active site forms, which is tapering and cavernous, is thought to be the reason for the enzyme's affinity for carbohydrates (Auh et al. 2006). At rising temperature (for instance, to 70 °C), α-amylase mostly completes endo-hydrolysis (Grewal et al. 2015). The active site geometry of α-amylase, which restricts the substrate's molecular size and shape, can be used to explain how it discriminates between different sizes. Different functions are added by combining amyloglucosidase and α-amylase to create porous, highly amylose-rich starch. α-Amylase enhances the solubility of starch by shortening the amylopectin chain length. It increases the gelatinization temperature while lowering pasting characteristics like setback, breakdown, and peak viscosity. It is discovered that peas, faba beans, and lentil starches transformed by α-amylase exhibit lower pasting viscosity due to the shortening of the amylose and amylopectin chain lengths (Li et al. 2021a, b). Additionally, rice starch is treated with these enzymes, resulting in rice starch with superior physicochemical characteristics to gelatinization of starch molecules (Keeratiburana et al. 2020).

3.4.1.1 Application of α-Amylase Treated Modified Starch in Food

Retrogradation occurs more quickly in starch with a higher amylose content and longer amylopectin chains than in starch with a lower amylose amount and shorter amylopectin chains (Liu et al. 2019). Starch treated with α-amylase has superior cold-storage stability, diminished viscosities, and late retrogradation that's making it a popular anti-staling agent in baking to extend shelf life (Miao et al. 2014b). These enzymes are utilized to convert the starch in flour into fermentable dextrins by the yeast in bread dough. The final products' quality in terms of volume and texture are enhanced after adding α-amylase into the dough, which decreases the dough viscosity and hastens the process of fermentation. In addition, it raises the dough's sugar content which improves the bread's flavor, crust color, and toasting abilities. A study found that α-amylase to modify the brown rice flour made with germinated starch enhances the slow-digesting starch and the release of bioactive substances, which improves the durability of the glycemic index in type 2 diabetics. The potato starch granule's structure is altered by the addition of α-amylase, leading to the development of maltose build-up along with short outer chains and visible pores. Thus, 0.5 mg/kg at low dosage, it exhibits effective anti-staling activities and improves retrogradation qualities (Chen et al. 2019).

Thermally stable α-amylase and mesophilic α-amylase enhance extruded noodles' quality by modifying starch derived from wheat. The extruded noodles' excellent porous shape and starch gelatinization help them to rehydrate and taste better than noodles without enzymes (Rashed et al., 2019) (Table 3.1). It is also crucial in brewing industries. In the brewing business, alpha-amylase is essential for turning the starches in malted grains (like barley) into fermentable sugars. Alpha-amylase transforms complicated starch molecules into maltose and other fermentable sugars during the mashing procedure. During the fermentation process, the yeast uses these sugars as food, which produces alcohol and develops the flavor of the beer.

3.4.2 Application of β-Amylase in Starch Modification

Hydrolysis of the maltose molecule within the non-reducing portion of linear or amylopectin is the primary role of exohydrolase β-amylase. Specifically, β-amylase can cleave the glucan chain at the non-reducing site, where the α-1,4 link of the D-anhydroglucose unit is located, but it stops when there is a α-1,6 linkage or substituent group. As a means of shortening the length of the starch branches and trimming the outside branches, the starch is partially hydrolyzed to produce to β-maltose and β-limit dextrin. Research found that amylopectin prevents amylose and amylopectin from recrystallizing, suggesting it has anti-aging properties (Li et al. 2019b; Miao et al. 2015; Shi et al. 2014; Guerra and Castro 2012). Within 48 hours, the β-amylase found in hemp leaves significantly slows down the retrogradation of meals

Table 3.1 The incorporation of enzyme-modified starches into a wide variety of food products

Starch modifying enzymes	Starch	Food items	Characteristics	References
α-Amylase	Corn starch	High fructose corn syrup	Make shorter oligosaccharide chains from the long starch polymers	Bangar et al. (2022)
	Taro starch	Ice cream	Modified starch performs better as an emulsion stabilizer in ice cream than guar gum in terms of foam stability, viscosity, and overrun	More et al. (2017)
	Corn starch	Bread dough	α-Amylase treated modified starch has superior cold-storage durability, diminished viscosities, and delayed retrogradation that's making it a popular anti-staling agent in baking to extend shelf life. When α-amylase is added to the dough, the fermentation process is accelerated, and the volume and texture of the final item are improved	Miao et al. (2014b)
	Maize starch	Ice cream	Accelerate the ice cream's expansion rate, sensory score, and emulsification potential	Zou and Liu (2011)
Thermostable and mesophilic α-amylases (i.e., TS-αA and MS-αA)	Wheat starch	Extruded noodles	Utilizing modified starches TS-αA and MS-αA significantly reduces the cooking time. If anyone wants to make extrusion-based noodles, MS-αA modified starch with a concentration of 1.6% is a better option than TS-αA	Rashed et al. (2019)
A β-amylase and transglucosidase mixed form	Pea starch	Baking and brewing industry	Shorter amylopectin chains and improved branching are seen in modified starch. Improve the digestibility of modified starch	Shi et al. (2014)
β-Amylase, transglucoamylase		Bread	Retrogradation decreases, reducing amylose concentration from 14% to 9%, which contributes to the creation of fluffy bread with a highly specific volume	Le Loan et al. (2021)
		Baking industry	Ramie leaf β-amylase exhibits intermediate temperature stability, suitable for the baking industry at 65 °C. The enzyme treatment inhibited staleness without causing gumminess in baked products	He et al. 2017

(continued)

Table 3.1 (continued)

Starch modifying enzymes	Starch	Food items	Characteristics	References
Debranching enzymes	Corn starch	Frozen dough and bread	5% modified starch-containing frozen dough lost 19% less water after three freeze-thaw cycles than control dough. The retrogradation peak and storage hardness of the bread prepared with modified starch were both 14% and 37% lower than those of a control loaf	Bangar et al. (2022); Woo et al. (2021)
	Corn starch, waxy corn starch	Breakfast cereal	In comparison to uncoated and glucose-coated breakfast cereal, coating it with debranched corn starches reduced milk absorption and preserved crunchiness after 3 minutes of milk soaking. DBS-based coating materials can also increase the nutritional fiber content	Luckett and Wang (2012), Chiu and Mason (1998)
	Sweet potato starch	Ice cream	Enzyme-modified sweet potato starch can replace 20% of fat in low-calorie ice cream, and it also reduces ice cream fat from 9.85% to 7.9%, proving its efficacy as a fat replacer	Kale et al. (2020)
Cyclodextrin glucanotransferase	Corn starch	Bread, frozen dough	After three freeze-thaw cycles, 5% modified starch-containing frozen dough lost 19% less water than the control dough. Modified starch bread exhibited 14% lower retrogradation peak and 37% lower storage hardness than a control loaf	Woo et al. (2021)

containing starches, and shorter branch chains (DP < 15) of the amylopectin molecule rise, while longer chains (DP > 16) decrease (Nguyen et al. 2015). The structural alterations of starch caused by β-amylase and the formation of maltose and dextrin are effective at preventing retrogradation. When compared to wheat starch, it is clear that the crystallinity, hardness, and retrogradation enthalpy of enzymatically treated samples decrease over time. β-amylase treatment period is strongly correlated with the anti-retrogradation effect. Higher enzymatic activity in a treatment results in a stronger anti-retrogradation impact when processing time is the same.

According to research (Ichikawa 2017; Yao et al. 2003), adding β-amylase to products enhances both the product's regeneration and its qualitative attributes, which are mostly related to the hydrolysis of starch. The majority of research on β-amylase is done in the product system.

β-Amylase has a softer impact with a more favorable outcome for maintaining product quality than α-amylase (Sarikaya et al. 2000). However, over-treatment with amylase can quickly result in excess hydrolysate, a bad surface, and a terrible flavor. It is important to create optimal circumstances for the enzyme to work well so that the product maintains its quality and shelf life. Experiment results are highly contradictory with one another because of the influence of different components, processing methods, and ambient variables in various starch-based products.

All in all, it can be said that treatment with β-amylase increases gelatinization temperature and decreases viscosity. It also inhibits the aging of wheat starch over the short and long terms, and processing time and effect correlate more strongly. At the same hydrolysis time, increasing β-amylase activity also has a larger inhibitory effect.

3.4.2.1 Application of β-Amylase Treated Modified Starch in Food

β-Amylase is a common enzyme utilized in the food industries because of its capacity to disassemble complex carbohydrates, particularly starches, into smaller sugar molecules. Brewing and baking businesses frequently use it (Shi et al. 2014). β-Amylase is used in baking to improve the volume and texture of baked goods (Yao et al. 2003). Maltose and other sugars are formed when starches are broken down, and yeast can then ferment these sugars to make carbon dioxide. The release of carbon dioxide helps in leavening the dough, leading to a lighter and fluffier final product. It's important for the brewing industry since it helps to produce beer. β-Amylase also can be utilized in the production of syrups, such as corn syrup and malt syrup. Ramie leaf β-amylase exhibits intermediate temperature stability, suitable for the baking industry at 65 °C (He et al. 2017).

3.4.3 Application of Debranching Enzymes in Starch Modification

Starch debranching enzymes (SDBEs) hydrolyze polysaccharides such as starch, amylopectin, pullulan, and glycogen α-1, 6 glycosidic links. SDBEs are also used to make resistant starch, sugar syrup, and cyclodextrin. Because of its potential to greatly improve production efficiency and raw material utilization in starch processing procedures like saccharification and modification, the synergistic catalysis of SDBEs and other starch-acting hydrolases has attracted a great deal of research attention in recent years. The two primary SDBEs, pullulanases and isoamylases, have highly different choice of substrates. Both isoamylases and pullulanases hydrolyze α-1, 6 bonds, with differing substrate requirements. Isoamylases require three parts of α-1, 4 bonded glucose, while pullulanases require only one or two units (Xia et al. 2021). Although pullulanase from *Klebsiella pneumoniae* hydrolyzes pullulan polysaccharides and branched dextrins well, it has been found that it lowered the activity when combined with amylopectins of high molecular weight (approximately 15% of the activity observed with pullulan and glycogen about only 1%) (Hii et al. 2012). On the other hand, isoamylases preferentially hydrolyze high-molecular-weight glycogen and amylopectin instead of pullulan, which they do not hydrolyze. Therefore, the lowest active substrate units for pullulanase and isoamylase are 6^2-α-maltosyl maltose and 6^3-α-maltosyl maltotriose, respectively. This crucial distinction serves as a guide for their use in various starch transformation processes that produce various products such as resistant starch (Li et al. 2018), glucose syrup (Xia et al. 2021), and cyclodextrins (Duan et al. 2013). Debranching starch (DBS) by SDBEs undergoes gelatinization when heated in water, melting the crystalline structure. Debranching modifications can increase the onset (T_o), peak (T_p), and end temperature (T_c) (Zeng et al. 2015). When DBS is synthesized by SDBEs and stored in a variety of ways, the resulting crystallites vary in degree of crystallinity and perfection, resulting in these enhanced gelatinizing properties (Hoover 2001). Crystal production is favored by chains with a length between DP 14 and DP 24, as opposed to chains that are either too long or too short (Xia et al. 2021).

Moreover, digestibility can be influenced by SDBEs. Amylolytic enzymes in the human digestive tract break down α-1, 4- and α-1, 6-glycosidic bonds during the complex process of starch digestion. Starch's incomplete digestion is hampered by the existence of α-1, 6-glycosidic bonds, which may cause gastrointestinal pain and a decreased glycemic response. SDBEs can increase the availability of glucose and increase the digestibility of starch by hydrolyzing these connections, which leads to a more favorable glycemic response.

3.4.3.1 Application of Debranching Enzyme Treated Modified Starch in Food

DBS by SDBEs can be used as fat replacer (Table 3.1). The epidemiological connection between dietary fat intake and health explains why high-fat foods are not only more calorically dense but also more likely to cause health issues. Carbohydrates and fat have different calorie contents. The prevention of calorie-related health risks is aided by replacing the fat in common foods with carbohydrates (Xia et al. 2021). Several crucial needs, including distinctive mouthfeel, flavor, viscosity, and other organoleptic and functional features, should be considered when using carbohydrates as fat substitutes in food. A fat-like carbohydrate called DBS that contains linear short chains has the potential to successfully substitute one or more dietary fats. For example, DBS can replace cream, fat, oil, water-in-oil or oil-in-water emulsions, as well as other lipid food components, in foods including ice cream, low-fat cheeses, low-fat spreads,, breaded meals, and baked goods (Table 3.1) (Kale et al. 2020). Due to the thermoreversible gels or high-strength gels offered by DBS dispersions, these fat substitutes have substantial advantages in food applications.

DBS by SDBEs can also be used as coating in ready-to-eat cereals (Table 3.1). Customers are drawn to ready-to-eat cereals because they can maintain their texture even after being dipped in milk. An essential factor is the amount of time a cereal may retain the correct crispness in milk (Chiu and Mason 1998). Sugar coatings are typically applied to ready-to-eat cereals to reduce water absorption, extend "bowl life," and add gloss. Excessive sugar consumption can lead to health issues like obesity, tooth decay, and hyperactivity. DBS by SDBEs is hydrophobic and may quickly create a gel network in an aqueous solution (Liu et al. 2015). As a result of the quick layer development, DBS exhibits good barrier properties. It may be applied as a coating for cereal (Table 3.1). DBS looks to be a suitable replacement for sugar. Ready-to-eat cereal's shelf life and dietary fiber content can increase by using DBS-based coatings (Luckett and Wang 2012) (Table 3.1).

3.4.4 Application of Cyclodextrin Glycosyltransferases in Starch Modification

Starch can only be transformed into cyclodextrins by cyclodextrin glycosyltransferases (CGTase, EC 2.4.1.19). Cyclodextrins, also known as cycloamylase dextrins (CD), are cyclic, non-reducible oligomers produced by cleaving and cycling α-D-glucopyranose, which consists of glucose units linked by α-(1,4) glycosidic bonds. Based on how many glucopyranose units a CD contains (6, 7, or 8 correspondingly), CDs can be classified as α, β, and γ (Crini et al. 2018). By modifying the reaction parameters, such as pH, temperature, and substrate concentration, the size of the CDs can be adjusted. The liquefaction of starch is accomplished by CGTases, which share an evolutionary origin with α-amylase.

3.4.4.1 Application of Cyclodextrin Glycosyltransferase Treated Modified Starch in Food

The food sector utilizes CGTase-modified starch extensively due to its numerous benefits and role in the creation of novel food products. It is possible to increase the stability of delicate ingredients, cover up unpleasant flavors, and improve the texture of food products by including CDs into food compositions. For example, starch is frequently employed in food recipes as a thickening and gelling ingredient. Native starches, however, frequently have texture, stability, and clarity issues. These characteristics can be improved by starch being subjected to CGTase modification and including CDs into food compositions, giving food items better texture and sensory qualities (Zhou et al. 2019). To allow for their regulated release during consuming, CDs can also be utilized to encapsulate tastes, perfumes, and nutraceuticals. For instance, CDs obtained by CGTase can help in preventing the deterioration and evaporation of volatile molecules, retaining their sensory properties (Gonzalez Pereira et al. 2021). Also, to increase the stability and bioavailability of delicate components like vitamins and nutraceuticals, CGTase-modified starch can be utilized as an encapsulant. For example, drinks, snacks, baked goods, and other food applications, encapsulated flavors obtained by CDs can be employed to stabilize flavor profiles (Table 3.1) (Gonzalez Pereira et al. 2021). These encapsulated substances can be added to fortified goods, nutritional supplements, and functional foods to give bioactive compounds with the controlled release (Gonzalez Pereira et al. 2021).

In addition, an accumulation of the enzymes cyclodextrinase (CDase) and cyclodextrin glycosyltransferase (CGTase) increases the retrogradation inhibition, which can be deduced from a drop-in enthalpy from 5.65 to 1.42 J/g. Adding 0–3.0% by weight of α-CD and γ-CD to wheat flour has some effects on the texture and aging of prebaked bread (Zhou et al. 2019). Prebaked bread has been observed to have less hardness, a longer shelf life, and a slower aging process. These effects may be the result of a decrease in enthalpy and a delay in starch retrogradation. Besides, CGTase-modified starch is used in the creation of low-calorie and low-fat food items. The viscosity and functionality of starch can be changed, allowing the creation of fat imitators and low-fat substitutes. While preserving the correct texture, mouthfeel, and sensory qualities, these modified starches can replace some of the fat in food compositions (Gonzalez Pereira et al. 2021).

3.5 Conclusion

In comparison to more traditional chemical and physical modification methods, enzymatic starch modification has a number of benefits. The absence of toxic chemicals and high temperatures makes this method more sustainable and kind to the environment. The modification process can be precisely controlled by enzymes, leading to the desired starch derivatives. The potential for enzymatic transformation of starch molecules to enhance its functioning and increase its widespread industrial applications appears exciting. Starch's solubility, stability, and viscosity can be enhanced by using enzymes with specific structural alterations. This enzymatic procedure makes it possible to create modified starches with unique properties, satisfying the needs of the food and pharmaceutical industries. Expanded commercial use of enzymatically modified starches remains a possibility, despite obstacles such as enzyme supply and process improvement. This long-term strategy for starch modification will support the creation of cutting-edge items across a range of sectors, boosting productivity and sustainability.

References

Alting AC, van de Velde F, Kanning MW, Burgering M, Mulleners L, Sein A et al (2009) Improved creaminess of low-fat yoghurt: the impact of amylomaltase-treated starch domains. Food Hydrocoll 23:980–987

Auh JH, Chae HY, Kim YR, Shim KH, Yoo SH, Park KH (2006) Modification of rice starch by selective degradation of amylose using alkalophilic Bacillus cyclomaltodextrinase. J Agric Food Chem 54:2314–2319

Ayudhaya PSN, Pongsawasdi P, Laohasongkram K, Chaiwanichsiri S (2016) Properties of cassava starch modified by amylomaltase from Corynebacterium glutamicum. J Food Sci 81(6):C1363–C1369
Bangar SP, Ashogbon AO, Singh A, Chaudhary V, Whiteside WS (2022) Enzymatic modification of starch: a green approach for starch applications. Carbohydr Polym 287:119265
Bijttebier A, Goesaert H, DElCouR JA (2008) Amylase action pattern on starch polymers. Biologia 63(6):989–999
Chen X, Zhang L, Li X, Qiao Y, Zhang Y, Zhao Y, Chen J, Ye X, Huang Y, Li Z, Cui Z (2019) Impact of maltogenic α-amylase on the structure of potato starch and its retrogradation properties. Int J Biol Macromol 45:325–331
Chiu CW, Mason WR (1998) U.S. patent no. 5,711,986. U.S. Patent and Trademark Office, Washington, DC
Cornejo-Ramírez YI, Martínez-Cruz O, Del Toro-Sánchez CL, Wong-Corral FJ, Borboa-Flores J, Cinco-Moroyoqui FJ (2018) The structural characteristics of starches and their functional properties. CYTA J Food 16(1):1003–1017
Crini G, Fourmentin S, Fenyvesi E, Torri G, Fourmentin M, Morin-Crini N (2018) Cyclodextrins, from molecules to applications. Environ Chem Lett 16:1361–1375
Duan X, Chen S, Chen J, Wu J (2013) Enhancing the cyclodextrin production by synchronous utilization of isoamylase and α-CGTase. Appl Microbiol Biotechnol 97:3467–3474
Gonzalez Pereira A, Carpena M, García Oliveira P, Mejuto JC, Prieto MA, Simal Gandara J (2021) Main applications of cyclodextrins in the food industry as the compounds of choice to form host-guest complexes. Int J Mol Sci 22(3):1339
Grewal N, Faubion J, Feng G, Kaufman RC, Wilson JD, Shi Y (2015) Structure of waxy maize starch hydrolyzed by maltogenic α-amylase in relation to its retrogradation. J Agric Food Chem 63(16):4196–4201
Guerra NP, Castro LP (2012) Modelling the effects of aging time of starch on the enzymatic activity of three amylolytic enzymes. Sci World J 2012:402439
He L, Park SH, Nguyen DHD, Duong HX, Duong TPC, Tran PL, Jong-Tae Park JT, Ni L, Park KH (2017) Characterization and thermal inactivation kinetics of highly thermostable ramie leaf β–amylase. Enzyme Microb Technol 101:17–23
Hii SL, Tan JS, Ling TC, Ariff AB (2012) Pullulanase: role in starch hydrolysis and potential industrial applications. Enzyme Res 2012:921362
Hoover R (2001) Composition, molecular structure, and physicochemical properties of tuber and root starches: a review. Carbohydr Polym 45:253–267
Ichikawa T (2017) Effects of beta-amylase on the properties of gluten-free bread prepared with finely powdered rice flour. J Jpn Soc Food Sci Technol-Nippon Shokuhin Kagaku Kogaku Kaishi 64(12):584–590
Iulek J, Franco OL, Silva M, Slivinski CT, Bloch C Jr, Rigden DJ, de Sá MFG (2000) Purification, biochemical characterisation and partial primary structure of a new α-amylase inhibitor from Secale cereale (rye). The international journal of biochemistry & cell biology 32(11-12):1195–1204
Kale RV, Sontakke MD, Raut GS, Chavan VR (2020) Use of enzyme modified sweet potato starch in formulation of ice cream. Int J Chem Stud 8(4):3002–3008
Kandra L (2003) α-Amylases of medical and industrial importance. J Mol Struct THEOCHEM 666:487–498
Kaur B, Arif F, Bhat R, Karim AA (2012) Progress in starch modification in the last decade. Food Hydrocoll 26:398–404
Keeratiburana T, Hansen AR, Soontaranon S, Blennow A, Tongta S (2020) Porous high amylose rice starch modified by amyloglucosidase and maltogenic α-amylase. Carbohydr Polym 230:115611
Kelly RM, Dijkhuizen L, Leemhuis H (2009) The evolution of cyclodextrin glucanotransferase product specificity. Appl Microbiol Biotechnol 84(1):119–133
Korompokis K, Deleu LJ, De Brier N, Delcour JA (2021) Use of amylomaltase to steer the functional and nutritional properties of wheat starch. Food Secur 10(2):303

Kumar P, Satyanarayana T (2009) Microbial glucoamylases: characteristics and applications. Crit Rev Biotechnol 29(3):225–255

Le Loan TK, Thuy NM, Le Tri Q, Sunghoon P (2021) Characterization of gluten-free rice bread prepared using a combination of potato tuber and ramie leaf enzymes. Food Sci Biotechnol 30:521–529

Lee H, Park I (2020) The influence of starch modification with amylosucrase treatment on morphological features. PRO 8(11):1409

Li D, Fu X, Mu S, Fei T, Zhao Y, Fu J, Li B-H, Ma Y, Zhao J, Hou J, Li X, Li Z (2021a) Potato starch modified by Streptococcus thermophilus GtfB enzyme has low viscoelastic and slowly digestible properties. Int J Biol Macromol 183:1248–1256

Li H, Cui B, Janaswamy S, Guo L (2019a) Structural and functional modifications of kudzu starch modified by branching enzyme. Int J Food Prop 22(1):952–966

Li J, Li L, Zhu J, Ai Y (2021b) Utilization of maltogenic α-amylase treatment to enhance the functional properties and reduce the digestibility of pulse starches. Food Hydrocoll 120:106932

Li J, Rashed MM, Deng L, Jin Z, Jiao A (2019b) Thermostable and mesophilic α-amylase: effects on wheat starch physicochemical properties and their applications in extruded noodles. J Cereal Sci 87:248–257

Li X, Wang Y, Lee B-H, Li D (2018) Reducing digestibility and viscoelasticity of oat starch after hydrolysis by pullulanase from *Bacillus acidopullulyticus*. Food Hydrocoll 75:88–94

Liu G, Hong Y, Gu Z, Li Z, Cheng L, Li C (2015) Preparation and characterization of pullulanase debranched starches and their properties for drug controlled-release. RSC Adv 5(117):97066–97075

Liu S, Yuan TZ, Wang X, Reimer M, Isaak C, Ai Y (2019) Behaviors of starches evaluated at high heating temperatures using a new model of rapid visco analyzer–rva 4800. Food Hydrocoll 94:217–228

Lucchiari S, Santoro D, Pagliarani S, Comi GP (2007) Clinical, biochemical and genetic features of glycogen debranching enzyme deficiency. Acta Myol 26(1):72–74

Luckett CR, Wang YJ (2012) Application of enzyme-treated corn starches in breakfast cereal coating. J Food Sci 77(8):C901–C906

Luo K, Kim NG, You SM, Kim YR (2019) Colorimetric determination of the activity of starch-debranching enzyme via modified tollens' reaction. Nano 9(9):1291

Menocci V, Goulart AJ, Adalberto PR, Tavano OL, Marques DP, Contiero J, Monti R (2008) Cyclodextrin glycosyltransferase production by new Bacillus sp. strains isolated from Brazilian soil. Braz J Microbiol 39(4):682–688

Miao M, Jiang B, Jin Z, BeMiller JN (2018) Microbial starch-converting enzymes: recent insights and perspectives. Compr Rev Food Sci Food Saf 17:1238–1260

Miao M, Li R, Huang C, Jiang B, Zhang T (2015) Impact of β-amylase degradation on properties of sugary maize soluble starch particles. Food Chem 177:1–7

Miao M, Xiong S, Ye F, Jiang B, Cui SW, Zhang T (2014b) Development of maize starch with a slow digestion property using maltogenic α-amylase. Carbohydr Polym 103:164–169

Nguyen DH, Tran PL, Ha HS, Lee JS, Hong WS, Le QT, Oh BC, Park SH (2015) Presence of b-amylase in ramie leaf and its antistaling effect on rice cake. Food Sci Biotechnol 24:37–40

Ojogbo E, Ogunsona EO, Mekonnen TH (2020) Chemical and physical modifications of starch for renewable polymeric materials. Mater Today Sustain 7–8:100028

Otache MA, Duru RU, Achugasim O, Abayeh OJ (2021) Advances in the modification of starch via esterification for enhanced properties. J Polym Environ 29(5):1365–1379

Park KH, Park JH, Lee S, Yoo SH, Kim JW (2008) Enzymatic modification of starch for food industry. In: Carbohydrate-active enzymes. Woodhead Publishing, pp 157–183

Pavezzi FC, Gomes E, da Silva R (2008) Production and characterization of glucoamylase from fungus Aspergillus awamori expressed in yeast Saccharomyces cerevisiae using different carbon sources. Braz J Microbiol 39(1):108–114

Potocki de Montalk G, Remaud-Simeon M, Willemot RM, Monsan P (2000) Characterisation of the activator effect of glycogen on amylosucrase from Neisseria polysaccharea. FEMS Microbiol Lett 186(1):103–108

Punia S, Omodunbi A, Singh A, Chaudhary V, Scott W (2022) Enzymatic modification of starch : a green approach for starch applications. Carbohydr Polym 287(December)
Rashed MM, Li J, Deng L, Jin Z, Jiao A (2019) Thermostable and mesophilic α-amylase: Effects on wheat starch physicochemical properties and their applications in extruded noodles. Journal of Cereal Science 87:248–257
Rashid N, Cornista J, Ezaki S, Fukui T, Atomi H, Imanaka T (2002) Characterization of an archaeal cyclodextrin glucanotransferase with a novel C-terminal domain. J Bacteriol 184(3):777–784
Ray RR (2004) β-Amylases from various fungal strains. Acta Microbiol Immunol Hung 51(1–2):85–95
Riaz M, Rashid MH, Sawyer L, Akhtar S, Javed MR, Nadeem H, Wear M (2012) Physiochemical properties and kinetics of glucoamylase produced from deoxy-d-glucose resistant mutant of *Aspergillus niger* for soluble starch hydrolysis. Food Chem 130(1):24–30
Sarikaya E, Higasa T, Adachi M, Mikami B (2000) Comparison of degradation abilities of α- and β-amylases on raw starch granules. Process Biochem 35(7):711–715
Seo DH, Yoo SH, Choi SJ, Kim YR, Park CS (2020) Versatile biotechnological applications of amylosucrase, a novel glucosyltransferase. Food science and biotechnology 29(1):1–16. https://doi.org/10.1007/s10068-019-00686-6
Shi M, Zhang Z, Yu S, Wang K, Gilbert RG, Gao Q (2014) Pea starch (Pisum sativum L.) with slow digestion property produced using β-amylase and transglucosidase. Food Chem 164:317–323
Van Der Maarel MJ, Capron I, Euverink GJW, Bos HT, Kaper T, Binnema DJ, Steeneken PA (2005) A novel thermoreversible gelling product made by enzymatic modification of starch. Starch-Stärke 57(10):465–472
Van der Maarel MJ, van der Veen B, Uitdehaag JC, Leemhuis H, Dijkhuizen L (2002) Properties and applications of starch-converting enzymes of the alpha-amylase family. J Biotechnol 94(2):137–155
Vitolo M (2020) Enzymatic modification of starch. World J Pharm Pharm Sci 9(4):1341–1358
Wang L, Wang W, Wang Y, Xiong G, Mei X, Wu W (2018) Effects of fatty acid chain length on properties of potato starch—fatty acid complexes under partially gelatinization. Int J Food Prop 21(1):2121–2134
Wangpaiboon K, Charoenwongpaiboon T, Klaewkla M, Field RA, Panpetch P (2023) Cassava pullulanase and its synergistic debranching action with isoamylase 3 in starch catabolism. Front Plant Sci 14(January):1–13
Woo SH, Kim JS, Jeong HM, Shin YJ, Hong JS, Choi HD, Shim JH (2021) Development of freeze-thaw stable starch through enzymatic modification. Foods 10(10):2269
Wu W, Jiao A, Xu E, Chen Y, Jin Z (2020) Effects of extrusion technology combined with enzymatic hydrolysis on the structural and physicochemical properties of porous corn starch. Food Bioprocess Technol 13:442–451
Xia W, Zhang K, Su L, Wu J (2021) Microbial starch debranching enzymes: developments and applications. Biotechnol Adv 50:107786
Yao Y, Zhang JM, Ding XL (2003) Partial beta-amylolysis retards starch retrogradation in rice products. J Agric Food Chem 51(14):4066–4071
Zeng F, Ma F, Kong F, Gao Q, Yu S (2015) Physicochemical properties and digestibility of hydrothermally treated waxy rice starch. Food Chem 172:92–98
Zhang R, Yip VLY, Withers SG (2010) Mechanisms of enzymatic glycosyl transfer. Compr Nat Prod II: Chem Biol 8:385–422
Zhou J, Ke Y, Barba FJ, Xiao S, Hu X, Qin X, Ding W, Lyu Q, Wang X, Liu G (2019) The addition of α-cyclodextrin and γ-cyclodextrin affects the quality of dough and prebaked bread during frozen storage. Food Secur 8(5):174
ZOU Jian, LIU Ya-wei (2011) Emulsification Properties of Alpha-Amylase Hydrolysates of Octenyl Succinic Anhydride Modified Maize Starch and Their Applications in Ice Cream[J]. FOOD SCIENCE, 32(14): 337–340.

Green Solvent Modification of Starch: Ionic Liquids (ILs) and Supercritical Carbon Dioxide ($ScCO_2$)

4

Nabilah Abdul Hadi and Md. Hafizur Rahman

Abstract

Supercritical carbon dioxide ($ScCO_2$) and ionic liquids (ILs) are two solvents that are safe for the environment and may be used in starch modification, which has the potential to improve its functionalities and broaden its uses. These environmentally friendly alternatives to conventional organic solvents are valued for their low environmental effect, non-toxicity, and potential for recycling and reuse. ILs provide the benefits of superior solubility, precise reaction control, and the ability to selectively target specific starch components, although they encounter obstacles related to their synthesis and high cost. In contrast, supercritical carbon dioxide $ScCO_2$, characterized as a non-polar solvent, achieves starch particle expansion, enhancing reactivity and acting as both a co-solvent and a carrier for indirect starch modifications. Notably, $ScCO_2$ is an ecologically beneficial alternative for starch modification due to its benefits including simplicity of accessibility, non-toxicity, and ability to be recycled. These environmentally friendly starch modifications are used in a variety of sectors, such as delivery systems, emulsifiers, adhesives, coatings, and bioplastics. These contribute to long-term sustainability goals by reducing emissions, waste output, and resource use. Addressing infrastructure, equipment, and scalability concerns, on the other hand, involves collaboration among regulators, corporations, and academics. This collaboration intends to enhance sustainable solvent synthesis, environmen-

N. A. Hadi (✉)
Faculty of Fisheries and Food Science, University Malaysia Terengganu, Kuala Nerus, Terengganu, Malaysia
e-mail: nabilah.abdhadi@umt.edu

M. H. Rahman
Department of Quality Control and Safety Management, Faculty of Food Sciences and Safety, Khulna Agricultural University, Khulna, Bangladesh

N. Mazumder, M. H. Rahman (eds.), *Advanced Research in Starch*,
https://doi.org/10.1007/978-981-99-9527-1_4

tally friendly methodologies, and commercial feasibility, opening the path for a more sustainable future in starch modification.

Keywords

Ionic liquids · Supercritical carbon dioxide · Starch modification · Green solvents · Nano-materials

4.1 Introduction

4.1.1 Overview of Starch Modification

Starch is one of the most widely used and sustainable biopolymers, with applications in the food, packaging, biomedical, and industrial industries. However, starch frequently only has a small range of capabilities and characteristics when it is in its natural state, which might not be sufficient to meet the demands in food, pharmaceutical, and cosmetic applications. Therefore, starch modification is necessary to enhance its properties and broaden the range of its applications (Fan and Picchioni 2020). In recent years, more focus has been placed on the use of environmentally friendly solvents for starch modification, such as ionic liquids (e.g., choline and amino acids) and supercritical carbon dioxide ($ScCO_2$) (Lukic et al. 2022; Zhang et al. 2023a). The main topics of this discussion are the significance of these green solvents and how they impact starch modification processes.

4.1.2 Importance of Green Solvents in Starch Modification

One of the key benefits of employing green solvents in starch modification is their environmental sustainability. Traditional solvents, such as organic solvents, i.e., dimethyl sulfoxide (DMSO) and pyridine, typically derived from fossil fuels, have significant levels of toxicity and high cost and pose environmental problems during manufacturing, use, and disposal (Fan and Picchioni 2020; Karić et al. 2021). Green solvents such as ionic liquids (ILs) and supercritical dioxide ($ScCO_2$) provide an alternative option which is eco-friendlier. For instance, ILs often include organic cations and inorganic or organic anions and have very low volatility and vapor pressure. They may be produced synthetically from renewable materials, and they are typically biodegradable, which might lessen their influence on the environment (Lu et al. 2013; Wu et al. 2009). On the other hand, $ScCO_2$ is a cheap, safe, inflammable, and easily recyclable solvent (Tabernero et al. 2022; Machado et al. 2022). Thus, procedures for modifying starches can reduce their environmental effect and help to a more sustainable future. In terms of the effectiveness of the starch modification process, green solvents provide several benefits over traditional solvents. Due to their distinct physicochemical characteristics, ILs have exceptional starch-solvating capacities, enabling effective dissolving and dispersion. This enables modifying processes and makes access to the starch granule structure simpler. Due to ILs' typically enhanced solubility for diverse reagents, it is furthermore simpler to introduce different modifying agents, such as crosslinkers or functional additives, into starch matrices (Zhang et al. 2023a; Chen et al. 2021; Devi and Das 2020). Similar to this, $ScCO_2$'s strong diffusivity and low viscosity enable it to readily enter the starch matrix, increasing its interactions with the starch molecules and speeding up the modification process. Studies from Araujo and Scopel (Araujo et al. 2023; De Marco et al. 2019) have discovered the capability of supercritical carbon dioxide in producing starch aerogels which can act as drug delivery and bioactive carriers such as polyphenols. Green solvents provide a wide range of possibilities for starch modification, enabling the creation of unique, property-rich starch-based products. It is simpler to produce modified starches with various structures and properties when using ILs, for instance, to produce modified starch and dissolve starch with a wide range of properties. Green solvents can also operate as reaction media for a variety of modification processes, including esterification, etherification, and grafting, providing fine control over the kind and degree of modification (Zhang et al. 2023a; Lu et al. 2020; Zhou et al. 2014). This is because to undergo starch modification, specifically for chemical modification, the process needs to be under a liquid environment. Thus, the ability of ionic liquids to function as nanomaterial dispersion agents enables the fabrication of starch-based nanocomposites with enhanced mechanical, barrier, or thermal characteristics. Super critical carbon dioxide also offers unique potential for starch modification, such as in enzymatic starch modification and the impregnation of starch matrices with bioactive compounds. The creation of starch-based materials with improved functions is made possible by the flexible modification capabilities offered by green solvents, hence extending the range of possible applications. The capacity to maintain or enhance the original characteristics of starch while introducing desired alterations is a vital component of starch modification. To achieve this equilibrium, the selection of the right green solvents is crucial. For instance, it has been demonstrated that ionic liquids have a minor effect on the crystalline structure of starch which effected the crystallinity of the starch structure and gelatinization temperature (Lu et al. 2020).

4.1.3 Rationale for the Use of Ionic Liquids (ILs) and Supercritical Carbon Dioxide ($ScCO_2$)

When compared to conventional organic solvents, the use of ILs and $ScCO_2$ as green solvents for starch modification offers substantial environmental advantages. The production of ILs frequently uses renewable resources, minimizing the need for fossil fuels. They have a small vapor pressure and low volatility, which reduces the amount of volatile organic compounds (VOCs) released into the environment. ILs may also be recycled and used again, which lowers waste production and encourages sustainability. $ScCO_2$, on the other hand, is a safer and more ecologically friendly substitute for traditional solvents since it is non-toxic, non-flammable, and readily recoverable. In starch modification methods, ILs and $ScCO_2$ show energy-efficient properties. Compared to conventional solvents, ILs demand less energy for reaction and dissolution, which results in reduced overall energy usage. $ScCO_2$'s benign working conditions make it possible for reactions to take place at lower temperatures and pressures, which further boosts energy efficiency. Additionally, it is simple to remove both ILs and $ScCO_2$ from the modified starch products and allows recycling and reducing energy-intensive purification procedures. Preserving or enhancing starch's natural qualities is a significant benefit of starch modification using green solvents. The crystalline structure of starch is not greatly changed by ILs because of their special solvation capabilities, which preserve crucial characteristics like gelatinization behavior. In situations where starch's ability to gelatinize is crucial, such as in food or medicinal formulations, this is especially advantageous. With its gentle working conditions, $ScCO_2$ helps starch maintain its integrity and usefulness by preventing heat degradation and unwanted structural changes (Araujo et al. 2023; Salimi et al. 2014; Vatansever et al. 2021).

4.2 Ionic Liquids (ILs)

4.2.1 Definition and Characteristics of Ionic Liquids

Ionic liquids (ILs) are an intriguing class of substances that have attracted considerable interest in numerous scientific and technological disciplines. Due to their exceptional properties, which set them apart from conventional molecular solvents, these exceptional substances have been the subject of extensive research. Ionic liquids consist of cations and anions, resulting in a salt-like structure with a low melting point, typically below 100 °C. The versatility of ILs has resulted in their use in numerous fields, including catalysis, electrochemistry, and material science. In this essay, we will examine the definition, functions, and characteristics of ionic liquids, concentrating on their application in starch modification as a case study.

4.2.2 Commonly Used Ionic Liquid (ILs) for Starch Modification

The most commonly used aprotic ILs consist of large cations such as imidazolium, pyridinium, and phosphonium and limited anions. 1-Butyl-3-methylimidazolium chloride (BMIM Cl) is an example of an ionic liquid used in starch modification (Li et al. 2022a; Roy Goswami et al. 2020). One of the first of these ILs was 1-butyl-3-methylimidazolium chloride, which was utilized in the chemically catalyzed synthesis of starch acetate using acetic anhydride (Biswas et al. 2006). Study has found that at 80 °C, starch and zein protein were soluble in BMIM Cl and 1-butyl-3-methylimidazolium dicyanamide (BMIMdca) at up to 10% (w/w) concentrations. BMIM Cl is a commonly investigated and utilized ionic liquid in a variety of chemical processes, including starch modification. It contains 1-butyl-3-methylimidazolium, a cation, and chloride, an anion. The combination of these cations and anions produces a solvent with unique properties that is suitable for starch modification reactions that is stable and modifiable. Typically, starch is dissolved in the ionic liquid to facilitate the reaction between the starch molecules and the modifying agents when starch is modified with BMIM Cl. The IL functions as both a solvent and a catalyst, allowing the modification reaction to take place under moderate conditions. Due to its advantageous solvation properties, BMIM Cl can effectively dissolve starch and the reagents used in the modification process. The synthesis of starch esters or ethers is a frequent application of BMIM Cl in starch modification. These modifications enhance the functional properties of starch, making it more adaptable for use in industries such as the food, pharmaceutical, and paper sectors.1-Ethyl-3-methylimidazolium acetate (EMIM Ac) is another ionic liquid that can be used in starch modification. EMIM Ac is a widely used ionic liquid for a variety of chemical processes, including biomass conversions and green chemistry applications. It is composed of 1-ethyl-3-methylimidazolium cation and acetate anion. This combination produces a solvent with unique properties that is stable and environmentally benign, making it suitable for starch modification reactions. In order to facilitate the modification reaction when starch is modified using EMIM Ac, the starch is dissolved in the ionic liquid. EMIM Ac, like BMIM Cl, functions as both a solvent and a catalyst, allowing the reaction to occur under comparatively benign conditions. Similar to BMIM Cl, EMIM Ac has been utilized in the synthesis of starch esters and ethers. These modifications strengthen the starch's properties, including its solubility, thermal stability, and film-forming ability, making it suitable for use in the food, textile, and pharmaceutical industries, among others (Domene-López et al. 2019; Amini et al. 2023). ILs with a short alkyl side chain and the combination of an imidazolium cation and an acetate anion had a low viscosity, resulting in the complete disruption of the crystalline structure of starch (Ren et al. 2022). A benefit of using EMIM Ac is its ability to dissolve starch efficiently, allowing for more effective interactions between starch molecules and modifying agents. In addition, EMIM Ac has a broad liquid range, remaining in the liquid state over a broad temperature range, allowing for reactions to occur at varying temperatures (Men et al. 2015) (Table 4.1).

4.2.3 Advantages and Limitations of Ionic Liquids (ILs) in Starch Modification

Despite the fact that many researchers have labeled ILs as environmentally friendly, their release into the environment presents a vast array of risks. It has been reported that certain imidazolium-based ionic liquids are more toxic than other classes of ionic liquids. Therefore, their use in specific applications, particularly those involving direct contact with humans or release into the environment, may necessitate a thorough evaluation and consideration of safety protocols (Thamke et al. 2022). Imidazolium-based ionic liquids may raise toxicity and environmental impact concerns, necessitating cautious evaluation and safety protocols for applications

Table 4.1 Comparison of organic solvents with ionic liquids (ILs)

Property	Organic solvents	Ionic liquids	Reference
Applicability	Wide range of applications	Versatile, but more specialized	Plechkova and Seddon (2008)
Catalytic properties	Limited	Often catalytic	Plechkova and Seddon (2008)
Vapor pressure	Moderate to high	Low	Ravula et al. (2019)
Flammability	Generally flammable	Non-flammable or low flammability	Chen et al. (2019)
Polarity	Variable	Highly polar	Hiraga et al. (2018)
Stability	Less thermally and chemically stable	More stable	Ngu (2022)
Water solubility	Some are water-soluble	Insoluble in water	Seddon et al. (2000)
Volatility	High volatility	Low volatility	Ngu (2022)
Solvation	Effective solvents	Excellent solvating capabilities	
Cost	Relatively low	Expensive	Faisal Elmobarak et al. (2023)
Toxicity	Some can be toxic	Generally low toxicity	Thamke et al. (2022)

Note: The characteristics of ionic liquids and organic solvents might differ based on particular kinds and compositions. In addition, these compounds' properties may vary over time due to study and improvements

involving direct human contact or environmental release. In addition, the synthesis and purification of imidazolium carboxylate ionic liquids can be more difficult and expensive, which may hinder their pervasive application in large-scale industrial processes (Ren et al. 2022). Therefore, it is suggested that the mixture of water and ILs was an effective solvent for starch modification which can reduce the cost of ILs usage.

The residual presence of ILs in the modified starch is an additional disadvantage. Even though efforts are made to recover and recycle ILs, traces may persist in the final product, potentially influencing its integrity and leading to unintended consequences. For instance, the ionic liquid 1-ethyl-3-methylimidazolium acetate (EMIM Ac) has been utilized in starch modification, but it may leave behind residues that could compromise the functionality of the modified starch. In addition, the high viscosity of some ILs, such as *N*-hexylpyridinium bis (trifluoromethylsulfonyl) imide (HPy TFSI), can hinder their penetration into the starch matrix, thereby reducing the modification process' efficacy. This diminished interaction between IL and starch may result in insufficient modifications or functionalization. In addition, some ionic liquids have demonstrated limited selectivity for particular starch functional groups, resulting in non-specific modifications that may not be optimal for certain applications. For instance, 1-butyl-1-methylpyrrolidinium bis (trifluoromethylsulfonyl) imide (BMPY TFSI) has been used in starch modification, but without precise control over the degree and distribution of substitutions, it may introduce various modifications.

4.2.4 Ionic Liquid (IL)-Based Methods for Starch Modification

4.2.4.1 IL-Based Dissolution of Starch

The term "IL-based dissolution of starch" describes the method of destroying the granular structure of starch and dissolving its two main constituents, amylose and amylopectin, using ILs as solvents. Starch dissolution was carried out by the combination of ILs and water mixture. The study by (Wang et al. 2019a) has demonstrated that starch dissolution was achieved with 1-ethyl-3-methylimidazolium acetate: water at ratio of 6:4. The distinctive solvation characteristics of ILs and their interactions with starch chains provide the basis of IL-based disintegration. High polarity of ILs enables them to create potent hydrogen bonds with the hydroxyl groups of starch, which causes the dissolution of starch granules and the solubilization of starch molecules (Wang et al. 2019a; Ren et al. 2019).

The selection of a suitable IL that satisfies the precise specifications of the starch modification process marks the start of the dissolving procedure. The selection of IL is influenced by properties such as polarity, viscosity, stability, and starch compatibility. ILs can be tailored by selecting specific cations and anions, allowing customization of their physicochemical properties to optimize starch dissolution.

The starch dissolution process typically involves the following steps:

1. Starch pretreatment: Granules of starch are semi-crystalline, which hinders them from fully integrating in ILs. Pretreatment procedures are used to increase starch solubility and starch granule disruption. When milling or high-pressure homogenization are used as physical preparation steps, mechanically shattering starch granules can open up more surface area for IL penetration (Zhang et al. 2023b). Enzymes (i.e., lipase B from *Candida antarctica*) (Zhao et al. 2008) or chemical pretreatments may also be used to modify starch granules and make them more responsive to IL solvation.
2. Starch-IL mixing: Following pretreatment, starch and ILs are combined under the proper temperature, IL concentration, and agitation conditions. ILs have great thermal stability and moderate volatility, making it possible to handle and manage the dissolution process safely. The starch granules are broken up, and the starch chains are solvated as a result of the mixing, which promotes the interaction between ILs and starch (Zdanowicz 2020).
3. Starch dissolution: Starch hydroxyl groups are solvated by ILs, which causes the starch chains to separate from the granule structure and enter the IL phase. IL concentration, temperature, and time are a few variables that can have an impact on the dissolving process. To accomplish effective starch dissolution, these factors must be optimized (Ptak et al. 2020).
4. Recovery of dissolved starch: The soluble starch components, especially amylose and amylopectin, are extracted from the IL solution after starch has been dissolved. To separate the dissolved starch from the IL, a number of methods can be used, including solvent extraction, precipitation, and membrane separation. To produce the necessary starch derivatives, the recovered starch might next go through further modifications or processing stages (Ren et al. 2022).

The ease of starch gelation induced by an aqueous ionic liquid (IL) is found to be unrelated to the amylose content but intricately controlled by the granule surface density. In particular, the facilitation of the aqueous IL in reaching and disrupting the most vulnerable regions of the starch granule is essential for initiating a cascading collapse of the entire granular structure and eventual gelation (Zhang et al. 2023b). The dissolving of starch with IL as opposed to conventional solvents has a number of benefits. The improved solubility and reactivity that ILs offer make it possible to dissolve starch with complicated structures and high molecular weights. Customizing the dissolving process to obtain optimum starch solubility and efficiency is made possible by the exact control over IL characteristics. ILs also provide chances for future adjustments since dissolved starch may be altered chemically, enzymatically, or physically to modify its characteristics for certain uses.

Compared to conventional solvents, the IL-based dissolving of starch has a number of benefits. It is possible to dissolve starch with high molecular weights and complicated structures because of ILs' improved solubility and reactivity. To obtain desired starch solubility and efficiency, the dissolving process may be tailored with the fine control over IL characteristics (Zhang et al. 2023b). Due to the fact that dissolved starch can go through chemical, enzymatic, or physical changes to modify its characteristics for certain purposes, ILs also present potential for later adjustments.

4.2.4.2 IL-Assisted Starch Derivatization

The technique of altering the structure or adding particular functional groups to the starch molecule while employing ILs as catalysts or reaction medium is known as "IL-assisted starch derivatization." The unique solvation characteristics of ILs and their interactions with the starch molecule serve as the basis for IL-assisted starch derivatization. ILs provide a number of benefits when used as reaction medium for starch derivatization. They have a high degree of polarity and can solvate both the starch chains and the derivatization reagents. This improves the efficiency and selectivity of the process by facilitating the transfer of chemicals to the starch molecule. The derivatization reaction may be precisely regulated with respect to reaction parameters including temperature, pressure, and reaction time in ILs as well. A study by (Shogren and Biswas 2010) discovered that the reaction between starch (maltodextrin) and vinyl acetate is influenced by the ionic liquid medium used, resulting in various substitution positions. Specifically, acetylation at C2 was preferred when [BMIM] dicyanamide (DCA) was used, whereas acetylation occurred more uniformly when [BMIM] acetate was used (Table 4.2).

Methods of IL-Assisted Starch Derivatization

1. Direct Derivatization: In IL-assisted starch derivatization, the choice of IL is essential since various ILs have distinct characteristics that might affect the results of the reaction. The capacity to solvate the starch molecule, stability under reaction conditions, and compatibility with the required reagent are just a few examples of the criteria that must be taken into consideration when choosing an optimal IL (Bakouri and Guemra 2019; Zarski et al. 2016).

Table 4.2 Comparison of ionic liquid-mediated starch modifications

Ionic liquid	Substrate	Reagent	Catalyst	Product	Reference
1-Butyl-3-methylimidazolium	Star-dri 1 corn maltodextrin	Vinyl acetate	No catalyst	Starch acetate	Shogren and Biswas (2010)
[Me(OEt)2-Al-Im]Ac and [Me(OEt)3-Al-Im]Ac	Corn starch	Methyl palmitate and methyl stearate	Novozyme 435	Fatty acid starch esters	Zhan et al. (2020)
1-Butyl-3-methylimidazolium chlorine/1-octyl-3-methylimidazolium nitrate	Starch	Octenyl succinic anhydride (OSA)	Novozyme 435	OSA starch	Li et al. (2016)
Choline chloride (Decaen et al. 2017)	Starch	Glycerol	Acid (e.g., hydrochloric acid)	Glycerol plasticized starch	Decaen et al. (2017); Crucean et al. (2021)
Imidazolium chloride, imidazolium carboxylate, and choline carboxylate	Maize starch	Acetic anhydride	No catalyst	Acetylated maize starch	Ren et al. (2022)
1-Butyl-3-methylimidazolium chloride ([BMIM]Cl) and 1-ethyl-3-methlyimidazolium acetate ([EMIM]AC)	Corn starch	Lauric acid, palmitic acid, and stearic acid	No catalyst	Fatty acid corn starch ester	Gao et al. (2012)
1-Butyl-3-methylimidazolium chloride [BMIM]Cl	*Zea mays* starch	Sodium monochloroacetate and phosphoryl chloride	No catalyst	Etherified and crosslinked starch	Bakouri and Guemra (2019)
1-Butyl-3-methylimidazolium chloride	Potato starch	Hydrophobic fatty acids	Lipase	Fatty acid starch esters	Zarski et al. (2019)

Note: Depending on the desired product and the particular ionic liquid, reagent, and catalyst, starch modification reactions can vary substantially. In addition, enzymatic approaches employing specific enzymes as catalysts are frequently used in starch modification reactions

2. Catalyst-Assisted Derivatization: For starch derivatization processes, ILs can potentially serve as catalysts. In this process, the IL serves as a mediator, promoting the interaction between the chemicals and the starch without really being involved in the reaction itself. The environment created by the IL makes it more likely for the reaction to take place, increasing its pace and selectivity. Derivatization with a catalyst helps to manage the reaction process better and under gentler circumstances (Roy Goswami et al. 2020).
3. Enzyme-Assisted Derivatization: For starch derivatization, ILs can be utilized in conjunction with enzymes (e.g., lipase/esterase, protease). By solvating both the starch and the enzymes, the ILs act as a reaction medium and create an ideal environment for the enzymatic process to occur. The advantages of enzyme-assisted derivatization include excellent specificity, gentle reaction conditions, and the potential for regioselective alterations. In order to effectively derivate starch, ILs are essential for stabilizing enzymes and boosting their activity (Shokri et al. 2022; Choi and Nidetzky 2022).

4.2.4.3 IL-Mediated Starch Nanocomposites

ILs are used into the production of starch-based nanocomposite materials to create IL-mediated starch nanocomposites. The high polarity, solvation capacity, and adjustable features of ILs, together with other distinctive traits, serve as the basis for IL-mediated starch nanocomposites. In the creation of starch-based nanocomposites, ILs function as efficient dispersants, plasticizers, or modifiers, which enhance material performance (Amini et al. 2023).

The addition of ILs to the starch matrix improves the interaction between it and reinforcing agents like nanoparticles or fibers, leading to better interfacial adhesion and dispersion (Menezes et al. 2021). Strong interfacial contacts are facilitated by the ILs, which serve as mediators between the starch and the reinforcing chemicals by enhancing their uniform dispersion. The resultant nanocomposite materials have improved mechanical, thermal, and barrier characteristics as a consequence. The study by Wang, Liang (Wang et al. 2021) has demonstrated that polymeric ionic liquid was successfully used as a plasticizer to plasticize corn starch, resulting in the formation of thermoplastic starch which makes them highly suitable for manufacturing a wide range of products (i.e., bottles, food packaging, or medical devices). Ionic liquid such as 1-ethyl-3-methylimidazolium acetate can be an excellent plasticizer for starch. It achieves this by converting the starch's crystalline structure to a flexible amorphous state, increasing its flexibility, and lowering the glass transition temperature. As the amount of 1-ethyl-3-methylimidazolium acetate increases, these plasticizing effects become more pronounced (Castro et al. 2023).

Methods of IL-Mediated Starch Nanocomposites

1. Solution Casting: In this technique, ILs are mixed with a starch solution before fibers or nanoparticles are added as reinforcement. The reinforcing agents' distribution and interfacial interactions are improved by the ILs' facilitation of reinforcing agent dispersion inside the starch matrix. The final IL-mediated starch nanocomposites are produced by casting the resultant solution into films or

molding it into the required forms, followed by drying or curing (Wang et al. 2021).
2. In Situ Synthesis: In situ synthesis techniques may also be used to create IL-mediated starch nanocomposites. In this method, ILs serve as templates or reaction media to create nanoparticles inside the starch matrix. The ILs regulate the size, shape, and distribution of nanoparticles inside the starch matrix by facilitating their nucleation and development. The nanocomposite structure and characteristics may be precisely controlled using this technology.
3. Melt blending: Melt blending is the process of directly combining ILs, starch, and reinforcing chemicals while they are still in the melt stage, followed by processing methods like extrusion or injection molding. The ILs function as plasticizers, boosting the starch matrix's processability and the reinforcing agents' ability to disperse more easily. For the purpose of creating IL-mediated starch nanocomposites, melt blending offers a scalable and economical technique (Castro et al. 2023; Yang et al. 2021).

4.3 Supercritical Carbon Dioxide ($ScCO_2$)

4.3.1 Introduction to Supercritical Fluids

Supercritical carbon dioxide ($ScCO_2$) is a state of carbon dioxide in which both the temperature and pressure are greater than its critical point. At this critical juncture, carbon dioxide exists in a state that is neither gas nor liquid, but possesses properties of both. Specifically, it behaves as a dense, highly compressible fluid with a density similar to that of a liquid but closer to that of a gas in terms of viscosity (Tabernero et al. 2022). The capacity of supercritical fluid technologies to improve the functionality and qualities of starch-based products has generated considerable attention in the field of starch modification. Substances with features of both gases and liquids are known as supercritical fluids (SCFs), which exist above their critical temperature and pressure (Tabernero et al. 2022). The use of supercritical fluids for starch modification is introduced in this section, with special emphasis placed on its distinctive qualities and possible advantages. In order to modify starch at the molecular level, supercritical fluids must be able to permeate the amorphous areas of starch granules, which is one of its key advantages. In order to facilitate the passage of reactants and modifying agents into the granule structure, the supercritical fluid has the ability to solubilize or plasticize the starch matrix. This permits the insertion of desirable alterations, such as elevated gelatinization temperature, higher thermal stability, increased digestibility, and modified rheological qualities (Salimi et al. 2014; Muljana et al. 2010). Supercritical fluids are ideal for starch modification because they possess a number of essential characteristics. The starch granules may be effectively wetted and penetrated by their low surface tension, which makes it easier for reactants and modifying agents to move through the matrix. Rapid mass transfer is made possible by the supercritical fluid's high diffusivity, which boosts the effectiveness of modification processes. Additionally, the density and polarity of supercritical fluids may be changed by adjusting the temperature (i.e., 35 and 45 °C) and pressure (8, 10, and 20 MPa) (Lukic et al. 2022). This characteristic makes it possible to tailor the processes involved in starch modification by allowing for the exact control of solubility, swelling, and mass transfer behavior. It is possible to make specific adjustments to satisfy required functional specifications by choosing the proper supercritical fluids and operating conditions.

4.3.2 Characteristics and Properties of Supercritical Carbon Dioxide

One of $ScCO_2$'s main benefits is its variable solvating power, which is influenced by temperature and pressure levels. $ScCO_2$ has a density that is comparable to a liquid in supercritical circumstances, which enables it to permeate the starch matrix and interact with its constituent parts, allowing the replacement of organic compounds typically employed in the synthesis of ureas in order to avoid the use of toxic reactants such as phosgene and isocyanates (Ingrosso and Ruiz-López 2017). The degree of swelling, dissolving, or contact with the starch structure may be controlled by adjusting the temperature and pressure of this solvating power. Another noteworthy quality that enables $ScCO_2$'s entry into the starch matrix is its low viscosity. Due to the effective mass transfer of reactants and modifying agents made possible by the low viscosity, the modification process is improved. To achieve desired modifications, such as raised gelatinization temperature, improved thermal stability, increased digestibility, or modified rheological properties, $ScCO_2$ must be able to enter the amorphous regions of starch granules and interact with its constituents at a molecular level. Choosing the right solvent is one of the most important factors in starch modification. Compared to conventional organic solvents, $ScCO_2$ has a number of benefits. First off, $ScCO_2$ is suitable for use in food and pharmaceutical applications since it is non-toxic and non-flammable (Klettenhammer et al. 2020). Because of this quality, $ScCO_2$ is a desirable replacement for organic solvents, which might be hazardous to one's health and safety. A cleaner and safer finished product is also ensured by the use of $ScCO_2$ as a solvent since it lessens the possibility of residual solvent contamination in the modified starch.

Additionally, $ScCO_2$ has a selective solvating ability, making it possible to remove particular constituents from starch without substantially altering its structure. This feature is very helpful in starch purification operations because it allows for the selective removal of undesirable components like lipids or proteins, producing a starch product of higher purity. $ScCO_2$'s selective solvating ability streamlines the modification process and lowers overall production costs by minimizing the need for further purification stages (Ingrosso and Ruiz-López 2017). The fact that $ScCO_2$ is ecologically friendly is another important benefit. $ScCO_2$ differs from conventional organic solvents in that it is plentiful, non-toxic, and inflammable. Using it as a solvent in starch modification procedures lessens the environmental effect of using and discarding solvents. $ScCO_2$'s sustainability is in line with the rising need for manufacturing methods that are more environmentally friendly and long-lasting

(Wu and Han 2019). Additionally, $ScCO_2$ has a large working range, making it flexible to change various kinds of starches and accomplish a wide variety of modifications. $ScCO_2$'s adaptability under different pressure and temperature circumstances offers the chance to customize the modification process to a given starch type and the required functional qualities. $ScCO_2$'s broad working range increases its usability and possible economic feasibility by making it appropriate for both industrial-scale production and lab-scale research. On top of that, downstream processing benefits from the modified starch's simplicity in removing $ScCO_2$. $ScCO_2$ is easily removed from the modified starch by simply releasing the pressure, leaving behind a residue with little residual solvent concentration. This streamlines the filtration and drying processes, cutting down on processing time and energy usage (De Marco et al. 2019). Supercritical carbon dioxide ($ScCO_2$) is used in the starch modification operations. The interaction of $ScCO_2$ with starch granules is simplified in the illustration. $ScCO_2$ has characteristics of both gases and liquids under supercritical circumstances, which enables it to effectively permeate the amorphous areas of the starch granules. Through this penetration, $ScCO_2$'s solvating ability can interact with the starch constituents and speed up the alteration process. Small circles depict the modification agents that $ScCO_2$ carries and transports into the starch matrix, where they can cause the desired modifications to the structure and characteristics of the starch.

4.3.3 Advantages and Limitations of $ScCO_2$ in Starch Modification

In terms of its influence on starch properties, supercritical carbon dioxide ($ScCO_2$) offers a number of benefits for starch modification. Because $ScCO_2$ operates at temperatures and pressures above its critical point, it is able to penetrate the starch matrix without causing significant thermal degradation due to the benign processing conditions it creates. Consequently, $ScCO_2$-based procedures preserve the native structure and properties of starch better than conventional high-temperature methods. By altering the pressure and temperature, scientists can optimize the dissolution and reaction of starch with modifying agents to achieve the desired substitution or functionalization levels (Ren et al. 2022; Ivanovic et al. 2016). This feature allows starch properties to be tailored to specific applications. An example of starch modification using $ScCO_2$ is the esterification of starch with fatty acids. In this process, $ScCO_2$ serves as both a solvent and a reactant carrier, dispersing fatty acids and promoting their reaction with starch molecules. The benign reaction conditions prevent extensive degradation of the starch structure and preserve the integrity of its functional groups, resulting in modified starch with enhanced properties, such as enhanced hydrophobicity and film-forming ability. Furthermore, $ScCO_2$ provides enhanced mass transfer kinetics, allowing for quicker and more efficient reactions compared to conventional solvent-based methodologies (Villegas et al. 2019). This advantage can considerably decrease processing time and increase productivity in starch modification processes.

Supercritical carbon dioxide ($ScCO_2$) can be used as a drying method aid in preserving the original porous structure and texture of a wet gel when it is transformed into a dried form. When traditional drying methods, such as evaporation or freeze-drying, are used to remove the liquid from a gel, the gel's structure often collapses or loses its original porosity and texture (Villegas et al. 2019; García-González et al. 2012). When supercritical carbon dioxide is used as a drying agent, it endures a unique phase transition to become a supercritical fluid, which gives it the properties of both a gas and a liquid. As a result, $ScCO_2$ is able to permeate the gel matrix more uniformly and efficiently, extracting the liquid while maintaining the open pores and texture of the gel. While supercritical carbon dioxide ($ScCO_2$) offers a number of benefits in starch modification, it also has a number of drawbacks, particularly in terms of its influence on starch properties. The relatively limited solubility of starch in $ScCO_2$ is a key limitation. Starch is a polar compound with a high molecular weight, making it difficult to achieve significant dissolution in $ScCO_2$. Consequently, the efficiency of modification may be limited, and achieving high levels of substitution or functionalization may necessitate prolonged reaction times or higher pressures.

The limited specificity of $ScCO_2$ toward starch functional groups is another disadvantage. Unlike other solvents that can target specific chemical bonds or functional groups in starch, the interactions between $ScCO_2$ and starch are primarily driven by dispersion forces. Therefore, attaining specific modifications or introducing novel functionalities in starch may be more challenging when compared to the use of solvents with more selective interactions. Additionally, the utilization of $ScCO_2$ can alter the crystallinity of starch. Carbon dioxide at the supercritical state can interact with starch molecules and disrupt the crystalline regions, resulting in altered starch structure and properties (Vatansever et al. 2021). This effect may be advantageous in some applications, but it can hinder the functionality of starch in others, such as the food industry, where specific starch properties are essential to product performance. In the synthesis of highly substituted starch esters, $ScCO_2$'s limitations in starch modification are illustrated. Due to its low solubility, achieving high levels of substitution requires extensive dissolution of starch in $ScCO_2$, which can be challenging. Consequently, the degree of substitution may be restricted in comparison to other solvent-based techniques.

4.3.4 $ScCO_2$-Based Methods for Starch Modification

Supercritical carbon dioxide ($ScCO_2$)-based methodologies have emerged as innovative and environmentally friendly approaches for the modification of starch, a widely used polysaccharide with numerous applications. Starch derived from renewable sources has inherent biodegradability and biocompatibility, but its functional properties are frequently enhanced by modification. Traditional starch modification techniques involve organic solvents (dimethyl sulfoxide (DMSO) and pyridine) (Ruamcharoen et al. 2020), which may pose environmental and health risks. $ScCO_2$ is an attractive alternative due to its unique properties, such as its tunable solvent power, low critical point, and eco-friendly nature. $ScCO_2$ exists as a supercritical fluid above its critical temperature (31.1 °C) and pressure (73.8 bar), exhibiting both liquid- and gas-like properties. $ScCO_2$ can penetrate the amorphous regions of starch granules when used as a solvent for starch modification, allowing for efficient modification.

The principle mechanisms underlying $ScCO_2$-based starch modification can be classified into physical and chemical modifications, each offering different advantages and opportunities. Physical modifications involve altering the structure and properties of starch through various phenomena such as gelatinization, expansion, and retrogradation inhibition. Gelatinization is a key process in starch modification, wherein the crystalline structure of starch granules is disrupted, leading to increased solubility, swelling, and improved functional properties (Fan and Picchioni 2020). During gelatinization, $ScCO_2$ effectively penetrates the amorphous regions of starch granules, causing them to absorb water and swell. This transformation enhances starch's ability to form gels, stabilize emulsions, and improve texture and mouthfeel in food applications.

Expansion is another physical modification technique facilitated by $ScCO_2$, wherein starch granules undergo physical expansion and the formation of a porous structure when subjected to high pressure and temperature conditions (De Marco et al. 2019). The expansion process not only increases the surface area of starch, but also alters its physical properties, such as particle size and porosity, which is called as aerogel starch. These changes provide starch with improved reactivity and functionality, making it suitable for applications like encapsulation of bioactive compounds, controlled release systems, and adsorbents for purification processes. Retrogradation inhibition is a significant mechanism achieved through $ScCO_2$-based modification, wherein the reassociation of amylose and amylopectin chains, which normally occurs during storage and cooling, is hindered. Retrogradation leads to the formation of insoluble starch polymers, resulting in undesirable texture changes and reduced shelf life. By utilizing $ScCO_2$, the amorphous regions of starch can be effectively protected, preventing the retrogradation process due to structural changes of inner part of starch granules. Vatansever and Whitney (Vatansever et al. 2021) have demonstrated that the retrogradation of pea starch is hindered by the incorporation of supercritical carbon dioxide and ethanol during starch isolation. This inhibition results in improved freeze-thaw stability and prolonged shelf life of starch-based products, particularly in the frozen food industry. Utilizing supercritical carbon dioxide ($ScCO_2$) as a reaction medium provides supplementary avenues for the customization of starch properties through chemical modifications. Esterification, crosslinking, and enzymatic reactions are frequently utilized chemical modification methodologies in the realm of $ScCO_2$-based starch modification. Esterification is a chemical process wherein starch undergoes a reaction with carboxylic acids or their derivatives, resulting in the incorporation of ester groups onto the starch molecules. The aforementioned modification has the potential to induce changes in the hydrophobic characteristics, solubility, and film-forming attributes of starch, thereby rendering it advantageous for various applications including the development of biodegradable packaging materials, drug delivery systems, and emulsifiers. The process of crosslinking starch in supercritical carbon dioxide ($ScCO_2$) entails the establishment of covalent bonds between starch molecules. This molecular transformation leads to enhanced mechanical characteristics, heightened thermal stability, and increased resistance against enzymatic degradation (Wang et al. 2019b). Crosslinked starch demonstrates an augmented capacity for gel strength, viscosity, and film-forming capability, rendering it well-suited for a diverse range of industrial implementations including adhesives, coatings, and bioplastics. Enzymatic reactions utilizing $ScCO_2$ as a reaction medium offer numerous benefits, including enhanced enzymatic activity, enhanced enzyme stability, and selective modification of starch. $ScCO_2$ can boost the activity and stability of enzymes involved in starch modification processes, including hydrolysis and transglycosylation. Enzymatic hydrolysis in $ScCO_2$ is capable of selectively breaking down starch into smaller molecules, such as oligosaccharides, which have a variety of functional properties and applications (Chorfa et al. 2022). Specifically, enzymatic hydrolysis of carbohydrate can generate prebiotic oligosaccharides that promote digestive health and serve as functional food constituents. Enzymatic transglycosylation in $ScCO_2$ can produce starch derivatives with tailored structures and properties. Transglycosylation reactions, for instance, can be used to produce starch-based prebiotics with enhanced resistance to digestion, which may have potential health benefits and applications in the food and pharmaceutical industries (Fu et al. 2022; Matsubara et al. 2021).

Recent developments in $ScCO_2$-based starch modification have demonstrated its utility in a variety of contexts. For instance, researchers have devised modified carbohydrates with enhanced properties for drug delivery systems that have been treated with $ScCO_2$ (Lukic et al. 2022; Machado et al. 2022; Villegas et al. 2020). By modifying starch with $ScCO_2$-based techniques, controlled release systems that permit the sustained release of pharmaceuticals and improve their therapeutic efficacy can be created. In addition, $ScCO_2$-assisted starch modification has been used to create biodegradable packaging materials. $ScCO_2$-treated starch films have improved barrier properties, mechanical strength, and biodegradability, making them environmentally preferable alternatives to petroleum-based packaging materials (Ivanovic et al. 2016). These modified starch coatings can be used to reduce the environmental impact of conventional food packaging materials. The modification of starch with $ScCO_2$ has also been investigated for the production of starch-based adhesives. Starch that has been crosslinked through $ScCO_2$-assisted reactions possesses enhanced adhesive properties, such as increased bond strength and water resistance (Patel et al. 2009). These adhesives derived from starch have applications in a variety of industries, including paper and packaging, wood composites, and textiles. In addition, starch has been modified using $ScCO_2$-based techniques for the production of starch-based nanomaterials. Nanoscale starch particles and nanocomposites have garnered considerable interest due to their unique properties, such as increased surface area, enhanced mechanical strength, and improved compatibility with other materials (Machado et al. 2022). There are prospective applications for these nanomaterials in drug delivery, tissue engineering, and the creation of advanced functional materials. In conclusion, $ScCO_2$-based methodologies provide versatile approaches for starch modification, allowing the modification of starch properties for a variety of applications. Through physical modifications such as gelatinization, expansion, and inhibition of retrogradation, as well as chemical modifications such as esterification, crosslinking, and enzymatic reactions, starch can be transformed into functional materials suitable for drug delivery systems, biodegradable packaging, adhesives, and nanomaterials, among other applications. These developments in $ScCO_2$-based starch modification contribute to environmentally benign and sustainable solutions in a variety of industries, fostering innovation and encouraging the use of renewable resources.

4.3.5 Supercritical Fluid-Assisted Starch Processing (SFASP)

SFASP is an innovative method for modifying and processing starch that employs supercritical fluids, such as carbon dioxide (CO_2). Starch, an abundantly occurring polysaccharide found in plants, is a versatile substance with numerous industrial applications. Nonetheless, native starch frequently demonstrates limitations in terms of solubility, reactivity, and processability. SFASP is a promising method for overcoming these limitations and improving the functionality of starch. SFASP is based on the unique properties of supercritical fluids, particularly supercritical carbon dioxide, which functions as both a solvent and a processing assist. In its supercritical state, CO_2 possesses both liquid- and gas-like properties, allowing it to permeate the amorphous regions of starch and facilitate various transformations (Villegas et al. 2020; Li et al. 2022b). The ability to modify starch without the use of toxic compounds or high temperatures is one of the primary benefits of SFASP, making it a more environmentally favorable and energy-efficient process.

Physical and chemical modifications of starch can be accomplished with SFASP. Physical changes include gelatinization, expansion, and nanostructure formation. Gelatinization disrupts the ordered structure of starch granules, resulting in enhanced solubility and functional properties. The presence of SFASP facilitates the penetration of supercritical CO_2 into the amorphous regions of starch, resulting in expansion and gel formation (Zhang et al. 2023c). The enhanced viscosity, stability, and film-forming properties of the gelatinized starch make it appropriate for use in the food, pharmaceutical, and cosmetic industries. By exposing starch granules to supercritical CO_2 at high pressure and temperature, the starch granules endure physical expansion and form a porous structure. This expanded starch has an increased surface area and enhanced reactivity, making it suitable for applications such as adsorbents, drug carriers, and composite material additives. Expanded starch's porous structure enables efficient encapsulation and release of active compounds, making it useful in controlled release systems and encapsulation-based applications. Additionally, SFASP permits the formation of nanostructures composed of starch. By manipulating the process conditions, such as pressure, temperature, and CO_2 flow rate, it is possible to produce starch nanoparticles, nanofibers, and nanocomposites (Ubeyitogullari and Ciftci 2017). These nanostructures possess unique properties, including a large surface area, enhanced mechanical strength, and increased compatibility with other substances. Alavi and Ciftci (2022) have demonstrated the production dual nano-/microporous starch bioaerogels with high porosity (80–99.8%) and surface area (156–190 m^2). They are used in areas such as drug delivery systems, tissue engineering, and the development of advanced functional materials.

SFASP can also be used to accomplish chemical modifications of starch. In the presence of carboxylic acids or their derivatives, for instance, starch can undergo esterification reactions by being exposed to supercritical CO_2. The supercritical CO_2 functions as a solvent, facilitating the reaction between the carboxylic acid and hydroxyl groups of starch, which results in the addition of ester groups to the starch molecules (Bai et al. 2017). Improved hydrophobicity, solubility, and film-forming ability are among the altered properties of esterified starch. This makes it useful for applications such as biodegradable packaging, drug delivery systems, and emulsifiers. SFASP has found numerous applications in a variety of industries. In the food industry, for instance, SFASP has been utilized to develop functional food constituents with enhanced properties, such as modified carbohydrates with improved fortifying and stabilizing properties. These modified carbohydrates are used in condiments, seasonings, and desserts, among other food items. In the pharmaceutical industry, SFASP has been used to develop starch-based drug delivery systems, such as nanoparticles and microparticles.

4.3.6 $ScCO_2$ as a Carrier for Starch Functionalization

$ScCO_2$ has garnered considerable attention as a versatile and environmentally benign solvent for various applications, including the functionalization of starch. The unique properties of starch, a naturally abundant complex carbohydrate, can be further enhanced through chemical modification. $ScCO_2$ is an excellent carrier for incorporating functional additives into starch, allowing for the production of starch-based materials with enhanced properties and expanded applications. $ScCO_2$'s unique properties as a supercritical fluid are the driving force behind its use as a carrier for starch functionalization. When carbon dioxide is brought to a supercritical state by increasing the temperature and pressure above its critical point, it possesses both liquid- and gas-like properties. $ScCO_2$ in this state has high diffusivity and low viscosity and can readily penetrate the amorphous regions of starch, which facilitates the incorporation of functional additives (Zhang et al. 2023c). Several methods, including impregnation, supercritical fluid deposition, and supercritical-assisted reactions, can be used to functionalize starch with $ScCO_2$. The impregnation process entails dispersing the functional additive in $ScCO_2$ and exposing the starch matrix to the supercritical fluid, which allows the additive to permeate and impregnate the starch matrix (Araujo et al. 2023; De Marco et al. 2019). This technique is frequently used to introduce bioactive compounds, antioxidants, antimicrobial agents, or flavors into starch, thereby augmenting its functionality for particular applications. $ScCO_2$ impregnation of starch with antimicrobial agents, for instance, can be used to produce antimicrobial food packaging materials or coatings, thereby extending the expiration life of perishable products. Another technique used for starch functionalization is supercritical fluid deposition. In this procedure, functional additives are dissolved in $ScCO_2$ to form a uniform solution. The solution is then deposited onto the surface of the starch, where the supercritical fluid rapidly dissipates, leaving the functional adjuvant evenly distributed on the starch particles (Villegas et al. 2020; Cerro et al. 2023). This technique is ideal for encapsulating starch particles with materials such as polymers, nanoparticles, or active substances to impart controlled release properties or specific functionalities. For example, the deposition of bioactive compounds onto starch particulates via $ScCO_2$-assisted supercritical fluid deposition can be utilized to develop starch-based drug delivery systems or functional dietary constituents.

Using $ScCO_2$ as a carrier, supercritical-assisted processes provide another method for functionalizing starch. $ScCO_2$ may be used to accomplish chemical processes, where the supercritical fluid serves as a solvent and improves the kinetics of the reaction. To add functional groups to the starch molecules, processes like esterification, etherification, or grafting may be used to chemically modify starch. As a result, starch characteristics like solubility, rheology, heat stability, or film-forming capacity be changed, providing customized functions for particular applications. For instance, starch-based emulsifiers or encapsulation materials with better emulsifying or encapsulating capabilities may be made using $ScCO_2$-assisted esterification of starch with fatty acids. There are various benefits to using $ScCO_2$ as a carrier for starch functionalization. First off, compared to traditional organic solvents, $ScCO_2$ is a safe and ecologically friendly choice since it is non-toxic, non-flammable, and easily accessible. Furthermore, $ScCO_2$ is simply eliminated from the process, leaving no trace of the solvent in the finished starch product. This is especially helpful in situations where solvent residues can compromise the final product's safety, quality, or compliance. Additionally, $ScCO_2$ has a high degree of tunability, enabling precise control of the functionalization process and the characteristics of the functionalized starch. This includes the ability to alter process variables including temperature, pressure, and residence duration. $ScCO_2$'s tunability also enables the selective extraction of certain substances or contaminants from starch, resulting in functionalized starch products that are purer and more refined. Numerous industries have identified uses for $ScCO_2$ as a carrier for starch functionalization. The functionalization of starch with $ScCO_2$ assistance may result in the creation of better food components with improved characteristics. When making emulsions, dressings, sauces, and other food items, for instance, starches that have been functionalized with emulsifiers or stabilizers may be used to enhance stability, texture, and mouthfeel. The texture, structure, and moisture retention of gluten-free goods may be enhanced by substituting functionalized starches for wheat flour.

$ScCO_2$-based starch functionalization offers prospects for the creation of drug delivery systems in the pharmaceutical sector. In order to obtain controlled release qualities, better therapeutic targeting, or increase drug stability, starch may be functionalized with polymers, nanoparticles, or bioactive substances. Utilizing functionalized starches allows for precise control of drug release kinetics and increases therapeutic effectiveness whether used in the creation of oral dosage forms, transdermal patches, or implantable devices. The agricultural and pharmaceutical industries are not the only ones that use $ScCO_2$ as a carrier for starch functionalization. Additionally, it has implications in the field of materials science. Functionalized starches can be incorporated into biodegradable plastics, films, and coatings in order to increase their mechanical strength, barrier properties, and environmental sustainability. Starch-based materials functionalized with antimicrobial agents can be employed in the production of antibacterial packaging, thereby reducing the need for synthetic preservatives and extending the expiration life of products. In addition, functionalized starches can be used to develop biocompatible materials for tissue engineering, such as scaffolds or hydrogels, which promote cell adhesion, proliferation, and tissue regeneration. Recent research has demonstrated the potential of $ScCO_2$ as a transporter for starch functionalization. For example, research has centered on the creation of novel functionalized starches with enhanced properties, such as enhanced thermal stability, increased solubility, or controlled release behavior. Several studies have examined the influence of process parameters, such as pressure, temperature, and exposure duration, on the functionalization of starch by $ScCO_2$ in an effort to optimize the process conditions for the desired functional outcomes (Soleimanpour et al. 2020). In addition, the combination of $ScCO_2$-based functionalization with other techniques, such as enzymatic or microwave-assisted reactions, has been investigated to accomplish synergistic effects and expand the functionalization options available. Hence, the use of supercritical carbon dioxide ($ScCO_2$) as a carrier for starch functionalization provides numerous benefits, including environmental tolerance, tunability, and the ability to accomplish physical and chemical modifications. $ScCO_2$ enables the incorporation of functional additives into starch, resulting in improved properties and expanded applications in numerous industries, including food, pharmaceuticals, and materials science. Current research in this field continues to investigate novel functionalization strategies, optimize process conditions, and broaden the spectrum of functionalized starch products. Utilizing $ScCO_2$ as a carrier for starch functionalization has great potential for the development of innovative and sustainable materials with tailored functionalities to satisfy the needs of various industries.

4.4 Comparison of Ionic Liquids and Supercritical Carbon Dioxide

4.4.1 Differences in Solvent Properties

Ionic liquids (ILs) and supercritical carbon dioxide ($ScCO_2$) have been extensively investigated as potential solvents for the purpose of starch modification. Nonetheless, these substances exhibit notable disparities in their solvent characteristics, thereby influencing their appropriateness and efficacy in the context of starch modification procedures.

4.4.1.1 Solubility

In starch modification procedures, solubility is crucial because it determines a solvent's ability to dissolve starch and provide access to its structure for modification. In this context, the solubility characteristics of ionic liquids (ILs) and supercritical carbon dioxide ($ScCO_2$) differ, which impacts their effectiveness in starch modification. The high solubility of ILs in a variety of substances, including starch, is well-known. This characteristic is a result of their ionic constitution and ability to form strong hydrogen bonding interactions. ILs are composed of large organic cations and tiny inorganic or organic anions, which contribute to their potent solubilizing properties. When ILs come into contact with starch, the charged components of the ILs can interact with the hydroxyl groups and other polar functional groups present in starch. These interactions disrupt the intermolecular hydrogen bonds between starch molecules, thereby causing starch to dissolve in ILs. Consequently, ILs facilitate access to the starch structure, allowing for efficient modification processes. Studies have demonstrated that various ILs can completely dissolve starch, generating homogenous solutions without the need for additional modifications or co-solvents. For example, ILs based on imidazolium, pyridinium, or phosphonium cations, combined with chloride, acetate, or other anions, have been investigated extensively for starch dissolution (Roy Goswami et al. 2020; Ren et al. 2019; Crucean et al. 2021). These ILs have exhibited the capacity to dissolve starch in significant quantities, enabling for efficient modification processes. The high solubility of starch in ILs has been utilized to investigate a variety of starch modification strategies. ILs have been used for chemical modifications, including esterification, etherification, and grafting reactions, where starch is dissolved in the IL and subjected to the appropriate reactants or catalysts. These reactions permit the incorporation of functional groups into the structure of starch, resulting in modified starches with tailored properties and improved functionalities. In contrast to ILs, $ScCO_2$ has limited starch solubility, particularly in its native form. The highly organized structure of starch consists of both crystalline and amorphous regions. Due to their densely packed structure and lack of polar sites that can interact with the supercritical fluid, the crystalline regions are resistant to solubilization in $ScCO_2$ (Falua et al. 2022). As a result, $ScCO_2$ alone is ineffective at dissolving starch. However, scientists have devised strategies to improve the solubility of starch in $ScCO_2$. The inclusion of co-solvents or modifiers to the $ScCO_2$ system is one approach. These co-solvents, including ethanol, methanol, and water, can form hydrogen bonding interactions with starch, thereby disrupting its crystalline structure and facilitating solubilization in $ScCO_2$. Recent research has investigated the influence of co-solvent type, concentration, and processing conditions on the solubility of starch in $ScCO_2$. Moreover, modifications to the structure of starch can enhance its solubility in $ScCO_2$. Gelatinization and pre-gelatinization alter the crystalline structure of starch, resulting in increased solubility in $ScCO_2$ (Villegas et al. 2020). In addition, chemical modifications that add hydrophilic functional groups to starch molecules can increase their affinity for $ScCO_2$. Due to the formation of hydrophilic ester groups, for instance, the esterification of starch with fatty acids can enhance its solubility in $ScCO_2$. Recent scientific findings have demonstrated $ScCO_2$'s potential for starch modification when combined with co-solvents or modified starches. Researchers have examined the effect of co-solvent type and concentration on the solubility of starch in $ScCO_2$ systems in an effort to optimize the conditions for effective starch modification. $ScCO_2$ has also been combined with modified starches, such as hydroxypropyl starch and carboxymethyl starch, to enhance their solubility and increase the efficacy of $ScCO_2$-based starch modification processes. In addition, the comprehension of the solubility behavior of starch in $ScCO_2$ has paved the way for the development of novel starch modification techniques. For example, scientists have investigated the use of $ScCO_2$ as a solvent for encapsulating bioactive compounds within starch matrices. The functionality and bioavailability of the bioactive compound can be increased by dissolving starch in $ScCO_2$ and incorporating the desired bioactive ingredient, followed by controlled release. Notably, the limited solubility of starch in $ScCO_2$ can be advantageous in some applications. $ScCO_2$ can be used, for instance, for fractionation or extraction, where specific components of starch, such as amylose or amylopectin, can be selectively dissolved or extracted, resulting in the isolation of starch fractions with the desired properties or functions. In conclusion, while ILs have a high solubility for starch, allowing for its complete dissolution and simple access to its structure, $ScCO_2$ has a low solubility for starch in its native form. However, the solubility of starch in $ScCO_2$ can be improved through the use of co-solvents, modifiers, and modified starches. Recent scientific findings have focused on improving the solubility of starch in $ScCO_2$ systems by optimizing these strategies. This knowledge offers up new opportunities for the efficient modification of starch with $ScCO_2$, such as encapsulation, fractionation, and extraction procedures. Considering its distinct solubility characteristics and the wide range of industrial applications of modified starch, additional research and development efforts are necessary to investigate and exploit the full potential of $ScCO_2$ as a solvent for starch functionalization.

4.4.1.2 Reaction Environment

The reaction environment is crucial to starch modification processes because it influences the types of chemical reactions that can occur and the control of reaction conditions. The vastly different reaction environments of ionic liquids (ILs) and supercritical carbon dioxide ($ScCO_2$) affect their suitability for starch modification. Due to their ionic character, ILs provide a polar reaction environment, whereas $ScCO_2$ is a non-polar solvent. Regarding their respective reaction environments, let's examine the comprehensive explanation and recent scientific findings.

ILs consist of large organic cations and tiny inorganic or organic anions, endowing them with an inherent ionic nature (Choi and Nidetzky 2022). This property enables ILs to generate a polar reaction environment in the presence of starch molecules. The charged components of ILs interact with the hydroxyl groups and other polar functional groups present in starch via electrostatic and hydrogen bonding interactions. These interactions facilitate the modification of starch via chemical reactions such as esterification, etherification, and grafting. Polarity of ILs permits precise control of reaction conditions. By selecting particular cations and anions, researchers can modify the composition of ILs to accomplish the desired properties or functionalities in modified starch. In addition, the viscosity, polarity, and acidity/basicity of the ILs can be modified to optimize the reaction conditions for starch modification. This high level of control over the reaction environment provided by ILs increases the efficacy and selectivity of chemical alterations. Recent scientific findings have highlighted the diverse amylase modification applications of ILs. Researchers have investigated numerous ILs, including imidazolium-, pyridinium-, and phosphonium-based ILs, combined with various anions such as chloride, acetate, and other organic or inorganic species. These ILs have been shown to facilitate a variety of chemical modifications on starch molecules, resulting in modified starches with enhanced functionalities and tailored properties. $ScCO_2$ is a non-polar solvent that lacks the inherent ionic characteristics of ILs. $ScCO_2$ cannot directly support certain types of chemical reactions, particularly those that require a polar environment or specific interactions with polar functional groups. $ScCO_2$ can still play an important role in starch modification processes as a co-solvent or carrier for compounds or catalysts. $ScCO_2$ can facilitate chemical modifications by solubilizing reactants, additives, and catalysts that are insoluble or incompatible with ILs or other conventional solvents. By dissolving these components in $ScCO_2$, they can be introduced into the starch matrix, thereby facilitating chemical reactions. Utilizing $ScCO_2$ as a co-solvent or carrier is advantageous due to its unique properties as a supercritical fluid. Near its critical point, $ScCO_2$ has exceptional diffusivity, low viscosity, and tunable properties, allowing for efficient mass transfer and reaction control. Recent scientific research has investigated the use of $ScCO_2$ in starch modification as a co-solvent or carrier. $ScCO_2$ has been used in the synthesis of starch-based nanoparticles, for instance, by dissolving starch and various additives in $ScCO_2$ and then utilizing controlled precipitation or polymerization. This strategy has demonstrated promise for the development of starch nanoparticles with tailored diameters and properties for applications such as drug delivery, food encapsulation, and nanocomposites. In addition, $ScCO_2$ has been utilized as a carrier for enzymes or catalysts in starch-based enzymatic reactions. The supercritical conditions increase the activity and stability of the enzymes, resulting in more efficient enzymatic starch modifications. This indirect role of $ScCO_2$ in facilitating enzymatic reactions provides numerous benefits, including increased reaction rates, enhanced enzyme stability, and selective modification of starch molecules. Study also demonstrated $ScCO_2$'s potential as a co-solvent or transport medium for starch modification. Numerous additives and catalysts that can be dissolved or dispersed in $ScCO_2$ to facilitate chemical reactions have been investigated by researchers. In order to catalyze specific reactions on starch, such as oxidation or crosslinking, transition metal complexes or nanoparticles have been incorporated into $ScCO_2$ systems. In terms of reaction efficacy, selectivity, and control, these $ScCO_2$-based catalytic systems have proven to be effective. In addition, the unique properties of $ScCO_2$ have been utilized to develop novel methods for starch modification. In the encapsulation of bioactive compounds within starch matrices, for instance, the use of $ScCO_2$ as a solvent and medium for controlled release has been explored. Dissolving starch in $ScCO_2$ enables the incorporation of bioactive constituents, followed by controlled precipitation or solidification to form encapsulated starch particles. This method permits the creation of starch-based delivery systems with improved stability, bioavailability, and controlled release properties. It is essential to observe that the choice between ILs and $ScCO_2$ as solvents for starch modification depends on the process' specific objectives and requirements. ILs provide a polar reaction environment and direct interactions with starch, allowing for precise control of modification reactions. They are ideal for chemically modifying starch with high selectivity and custom-tailored properties. $ScCO_2$ provides a non-polar solvent environment but can serve as a co-solvent or carrier for compounds or catalysts, enabling indirect modifications of starch and facilitating particular reactions or applications. In conclusion, the solvents employed in starch modification, whether ILs or $ScCO_2$, offer distinct benefits and methods. ILs provide a polar reaction environment and high solubility for starch, enabling precise control over direct chemical modifications. $ScCO_2$, on the other hand, is a non-polar solvent that can function as a co-solvent or carrier for additives or catalysts. Both ILs and $ScCO_2$ have demonstrated potential and versatility in starch modification processes, according to recent scientific findings. The choice of solvent depends on the intended modifications, reaction conditions, and specific applications, giving researchers valuable options for customizing starch properties and functionalities for a variety of industrial and biomedical applications. Further research and development are required to investigate the full potential of these solvents and expand their starch modification applications.

4.4.1.3 Selectivity

Selectivity is a crucial aspect of starch modification processes, as it permits the extraction or modification of specific components or functional groups within the starch matrix. The selectivity of ionic liquids (ILs) and supercritical carbon dioxide ($ScCO_2$) differs, which affects their suitability for the modification or purification of starch. High selectivity is exhibited by ILs, whereas the selectivity of $ScCO_2$ can be increased with the addition of co-solvents or modifiers. Let's examine a thorough explanation and recent scientific findings regarding the selectivity properties of these molecules. With their distinctive ionic nature and modifiable properties, ILs exhibit a high degree of selectivity for particular components or functional groups in starch. This selectivity is a result of the ability of ILs to establish specific interactions with various molecular species within the starch matrix. ILs can selectively solvate and interact with specific starch structural components, such as amylose or amylopectin, or functional groups, such as hydroxyl groups. This selectivity permits targeted modifications of starch by selectively reacting with or extracting particular components or functional groups. Recent scientific findings indicate that ILs are selective in starch modification. The use of ILs to dissolve or extract specific components or impurities from starch has been investigated by researchers (Khosousi et al. 2023). ILs have been used to extract amylose from starch, resulting in the isolation of amylose fractions that have been purified. This selectivity has significant implications for a variety of applications, including the development of specialty starches with specific functionalities and the production of high-value products, such as resistant starch and amylose-based materials. In addition,

ILs can selectively interact with impurities or by-products in starch, thereby facilitating the purification process. ILs can solubilize impurities such as lipids, proteins, and inorganic ions while maintaining the integrity of the starch matrix. This selective dissolution of impurities makes it possible to remove and purify the starch. In the food and pharmaceutical industries, where purity and quality are of the utmost importance, this feature is particularly advantageous. In contrast, $ScCO_2$ has a lower selectivity than ILs in general. $ScCO_2$ is a non-polar solvent, and its interactions with starch are governed primarily by dispersion forces. This non-specific interaction reduces its selectivity for particular components or functional groups within the starch matrix. However, co-solvents or modifiers can be used to increase the selectivity of $ScCO_2$. Recent research has centered on enhancing the selectivity of $ScCO_2$ in starch modification processes. The addition of co-solvents, such as ethanol or water, can modify the solvation properties and increase the selectivity of $ScCO_2$ for particular components or functional groups in starch. Adding water to $ScCO_2$ has been shown to increase the solubility and selectivity of polar components or functional groups in starch, for instance. In addition, the utilization of modified carbohydrates or functional additives can improve the selectivity of $ScCO_2$. Customized modified starches can have enhanced interactions with $ScCO_2$, resulting in greater selectivity for particular components or functional groups. In addition, the addition of functional additives or catalysts to the $ScCO_2$ system can modify the solvation behavior, allowing for selective starch modifications or targeted reactions.

The investigation of $ScCO_2$'s selective properties has been undertaken in diverse starch modification procedures in recent scientific studies. One example of the application of supercritical carbon dioxide ($ScCO_2$) is its utilization as a discerning solvent for the extraction of particular constituents, such as lipids or proteins, from starch. The investigation of $ScCO_2$'s selectivity has also been conducted in the context of forming composites or materials based on starch. This involves the selective incorporation of specific additives or fillers into the starch matrix. In brief, ionic liquids (ILs) demonstrate a notable degree of specificity toward particular components or functional groups within starch, thereby enabling precise modifications or purification procedures. Ionic liquids (ILs) possess the ability to selectively extract or modify particular constituents and interact with impurities, thereby facilitating the production of purified starch or specialized starches that exhibit customized functionalities. On the other hand, supercritical carbon dioxide ($ScCO_2$) typically exhibits reduced selectivity; however, the incorporation of co-solvents or modifiers can effectively enhance this selectivity. Recent scientific research has investigated various approaches to enhance the selectivity of supercritical carbon dioxide ($ScCO_2$). These approaches include the incorporation of co-solvents or the utilization of modified starches. The aforementioned advancements have broadened the scope of potential uses for supercritical carbon dioxide ($ScCO_2$) in the realm of starch modification procedures. This expansion allows for more precise and regulated modifications of starch, catering to a diverse range of industrial and biomedical applications.

4.4.1.4 Pressure and Temperature

In starch modification procedures, the pressure and temperature conditions required for solvent operation play a crucial role. Ionic liquids (ILs) and supercritical carbon dioxide ($ScCO_2$) have very different pressure and temperature requirements, which affect their availability, ease of handling, scalability, and associated costs. In general, ILs operate in ambient or moderate conditions, whereas $ScCO_2$ requires high pressure and temperature to attain its supercritical state. ILs are renowned for their adaptability in operating under ambient or moderate temperature and pressure conditions. Typically, the temperature range for ILs is between −20 and 200 °C, while the pressure range is atmospheric or slightly elevated. This makes ILs more accessible and manageable than $ScCO_2$ because they do not necessitate specialized apparatus or extensive safety precautions associated with high pressure and temperature. Studies have brought attention to the benefits of ionic liquids (ILs) in the modification of starch, particularly when conducted under moderate operating conditions. Scholars have conducted investigations into the application of ionic liquids (ILs) for various types of starch modifications, encompassing chemical modifications, enzymatic reactions, and physical transformations. The aforementioned studies have provided evidence regarding the efficacy of ionic liquids (ILs) in altering the characteristics of starch, all the while ensuring that the processes involved remain straightforward, productive, and secure. In contrast, supercritical carbon dioxide ($ScCO_2$) functions within an environment characterized by elevated pressure and temperature, enabling it to attain a supercritical state and manifest distinctive solvent characteristics. In order to attain supercritical conditions, it is generally necessary for $ScCO_2$ to be subjected to pressures exceeding 74 bar (1074 psi) and temperatures surpassing 31 °C. The elevated pressure and temperature requirements present challenges in terms of equipment design, operational safety, and cost. The industrial scalability of $ScCO_2$ can be limited, and the costs of equipment and operations can be increased due to its high-pressure requirement. The expenses related to the fabrication and upkeep of pressure vessels designed to endure the elevated pressures inherent in supercritical carbon dioxide ($ScCO_2$) can be substantial. Furthermore, it is imperative to establish specific safety measures and protocols to guarantee the secure management of supercritical carbon dioxide ($ScCO_2$) under high pressures. Several factors have the potential to impact the feasibility and commercial viability of large-scale industrial applications of starch modification processes utilizing supercritical carbon dioxide ($ScCO_2$). Recent scientific research, on the other hand, has centered on optimizing $ScCO_2$-based starch modification processes in order to surmount the challenges posed by high pressure and temperature. Researchers have investigated various process parameters, including pressure, temperature, and residence time, in an effort to improve the efficacy, selectivity, and yield of starch modifications using $ScCO_2$. In addition, improvements in apparatus design and engineering have increased the scalability and economic efficiency of $ScCO_2$-based processes.

In addition, novel approaches have been investigated to mitigate the constraints imposed by the need for high pressure in $ScCO_2$ systems. Utilizing co-solvents or modifiers that increase the solubility of starch in $ScCO_2$ at lower pressures and temperatures is one such strategy. By incorporating co-solvents or modifiers into the $ScCO_2$ system, the pressure and temperature requirements can be decreased, thereby making the process more accessible and cost-effective.

This shows that ILs operate under ambient or moderate temperature and pressure conditions, making starch modification processes accessible, simple, and easy to handle. Recent scientific research has shown that ILs are effective at modulating carbohydrate properties under these conditions. $ScCO_2$ requires high pressure and temperature to attain its supercritical state, which can present challenges for apparatus design, safety, scalability, and cost. Nonetheless, ongoing research seeks to optimize $ScCO_2$-based starch modification processes by investigating process parameters, apparatus design, and the use of co-solvents or modifiers to improve the efficiency, scalability, and cost-effectiveness of $ScCO_2$-based processes. These efforts further the development and commercialization of starch modification technologies based on $ScCO_2$.

4.4.1.5 Environmental Consideration

In selecting solvents for starch modification processes, environmental factors play a significant role. Compared to conventional organic solvents, both ionic liquids (ILs) and supercritical carbon dioxide ($ScCO_2$) provide environmental benefits. Due to their negligible volatility, potential for recycling, and reduced reliance on conventional organic solvents, ILs are frequently regarded as more environmentally friendly solvents. Nonetheless, ILs may be toxic and can be energy-intensive to produce. $ScCO_2$, on the other hand, is non-toxic, non-flammable, and readily available and has a minimal impact on the environment. In addition, the elimination of the need for conventional organic solvents makes the use of $ScCO_2$ consistent with the principles of green chemistry. Let's investigate these facets in depth and discuss the most recent scientific findings.

ILs are regarded as greener solvents due to a number of factors. Their limited volatility reduces VOC emissions. This property reduces air pollution and the risk of adverse health effects. In addition, ILs are recyclable, allowing for the recovery and utilization of the solvent, which contributes to sustainability and waste reduction. Recycling ILs reduces the demand for the production of new solvents and the environmental impact of the procedure as a whole. The use of ILs as alternative solvents in diverse starch modification processes, including chemical modifications, enzymatic reactions, and physical transformations, has been the subject of research. These studies have demonstrated the viability and efficacy of ILs for replacing conventional organic solvents, reducing environmental impacts, and enhancing the sustainability of starch modification processes.

However, potential toxicity concerns associated with ILs must be considered. Some ILs may possess toxic properties, posing environmental and health risks. The chemical composition of ILs influences their toxicity, and it is essential to select ILs with minimal toxicity profiles. Recent scientific research has concentrated on assessing the toxicity of various ILs and identifying safer alternatives for starch modification processes. Strategies such as designing ILs with lower toxicity, evaluating biodegradability, and researching potential environmental impacts are being actively pursued. $ScCO_2$, on the other hand, has various environmental advantages and is regarded as a green solvent. It is non-toxic, inert, and widely accessible. $ScCO_2$ reduces the requirement for conventional organic solvents, lowering the dangers associated with their usage, such as VOC emissions and flammability issues. $ScCO_2$ has a minimal environmental effect since it is readily recovered and reused, resulting in less waste formation. The capacity to recycle $ScCO_2$ helps to ensure the long-term viability of starch modification procedures. Recent research has emphasized the environmental benefits of $ScCO_2$ in starch modification. $ScCO_2$ has been studied as a solvent in a variety of applications, including starch extraction, fractionation, and encapsulation. $ScCO_2$'s non-toxic nature promotes worker safety and reduces the risk of environmental pollution. Furthermore, the capacity to collect and reuse $ScCO_2$ decreases solvent usage, conserves resources, and minimizes the total carbon footprint of starch modification procedures. It should be noted that both the manufacturing of ILs and the creation of supercritical conditions for $ScCO_2$ demand energy. The energy-intensive nature of IL synthesis, as well as the energy needs for sustaining high-pressure and high-temperature conditions for $ScCO_2$, may have an influence on these solvents' overall environmental performance. Recent scientific research, on the other hand, has concentrated on establishing energy-efficient techniques for IL generation and optimizing $ScCO_2$-based starch modification processes to reduce energy consumption and improve sustainability. Both ILs and $ScCO_2$ provide environmental advantages in starch modification procedures when compared to standard organic solvents. Because of their low volatility and recycling possibilities, ILs are considered greener solvents. However, the toxicity and energy-intensive nature of IL synthesis must be addressed. $ScCO_2$, on the other hand, is non-toxic, non-flammable, freely accessible, and easily recovered, according to green chemistry principles. Recent research has highlighted the environmental benefits of both ILs and $ScCO_2$ in starch modification procedures. To improve their environmental performance, ongoing research focuses on producing safer and more sustainable ILs, optimizing energy use in IL manufacturing, and enhancing the energy efficiency of $ScCO_2$-based starch modification procedures.

4.4.2 Suitability for Different Starch Modification Processes

When it comes to the suitability of ionic liquids (ILs) and supercritical carbon dioxide ($ScCO_2$) for various starch modification procedures, their distinct properties offer specific advantages. To elaborate, it is necessary to delve deeper into the characteristics and recent scientific findings associated with the use of ILs and $ScCO_2$ for starch modification. Due to their versatility and ability to dissolve a wide variety of substances, including starch, ILs have received considerable attention in the field of starch modification. Polarity created by the combination of large organic cations and small inorganic or organic anions in ILs facilitates specific interactions with starch molecules. This unique characteristic enables precise control over chemical reactions and selectivity in starch modification. ILs are used in reactions involving esterification, etherification, and grafting, where they serve as solvents and catalysts. Recent scientific research has focused on enhancing the efficacy and selectivity of IL-based modification processes. To accomplish desired modifications of starch, researchers have investigated various forms of ILs, including imidazolium-, pyridinium-, and phosphonium-based ILs, combined with various anions. ILs have been utilized effectively in the esterification of starch, resulting in the incorporation of ester groups onto starch molecules. This modification modifies the starch's hydrophobicity, solubility, and film-forming properties, making it suitable for use in biodegradable packaging materials, drug delivery systems, and emulsifiers. Recent scientific research has demonstrated the efficacy of ILs in modifying starch specifically, resulting in enhanced functionalities and tailored properties. In addition to chemical modifications, ILs have demonstrated promise in starch-related enzymatic reactions. ILs can boost enzymatic activity and stability, resulting in more efficient enzymatic starch modifications. This benefit has been ascribed to the unique solvation properties of ILs, which can enhance enzyme-substrate interactions and create a favorable reaction environment. Recent scientific research has investigated the use of ILs in enzymatic reactions such as hydrolysis and transglycosylation, revealing the potential of ILs as solvents for selective enzymatic modifications of starch (Table 4.3).

Table 4.3 Potential application of ionic liquids and supercritical carbon dioxide

Starch modification	Ionic liquids (ILs)	Supercritical carbon dioxide ($ScCO_2$)
Chemical modifications	– ILs offer a polar reaction environment that enables precise control over reactions and selectivity – Examples of IL-based chemical modifications include esterification, etherification, and grafting of starch (Ptak et al. 2020)	– $ScCO_2$ can serve as a co-solvent or carrier for additives or catalysts in chemical modifications – Parameters for $ScCO_2$-based chemical modifications may include pressure, temperature, and the addition of co-solvents or modifiers (Matsubara et al. 2021)
Physical modifications	– ILs have limited suitability for physical modifications of starch – Parameters for $ScCO_2$-based physical modifications may include pressure, temperature, and residence time (Chorfa et al. 2022)	– $ScCO_2$ is highly suitable for physical modifications of starch, such as gelatinization, expansion, and retrogradation inhibition – Examples of $ScCO_2$-based physical modifications include the expansion of starch granules and the inhibition of retrogradation (Ivanovic et al. 2016)
Enzymatic reactions	– ILs offer a favorable reaction environment for enzymatic reactions on starch – Examples of IL-based enzymatic reactions include hydrolysis and transglycosylation of starch (Lu et al. 2013)	– $ScCO_2$ has limited suitability for enzymatic reactions on starch – Parameters for $ScCO_2$-based enzymatic reactions may include the addition of co-solvents or the use of modified enzymes (Uwineza and Waśkiewicz 2020)

(continued)

Table 4.3 (continued)

Starch modification	Ionic liquids (ILs)	Supercritical carbon dioxide ($ScCO_2$)
Selective extraction	– ILs are highly suitable for the selective extraction of specific components or impurities from starch – Examples of IL-based selective extraction include the isolation of specific starch fractions, such as amylose (Morais et al. 2020)	– $ScCO_2$ has limited suitability for selective extraction processes on starch. – Parameters for $ScCO_2$-based extraction processes may include the addition of co-solvents or modifiers
Encapsulation	– ILs are suitable for encapsulating bioactive compounds within starch matrices – Parameters for IL-based encapsulation may include the type of IL used, concentration, and encapsulation techniques (Zhou et al. 2014)	– $ScCO_2$ has suitability for encapsulation processes involving starch – Examples of $ScCO_2$-based encapsulation include the incorporation of additives or fillers into starch matrices (Klettenhammer et al. 2020)
Purity enhancement	– ILs are suitable for enhancing the purity of starch through the selective dissolution of impurities – Examples of IL-based purity enhancement include the removal of lipids, proteins, or inorganic salts from starch	– $ScCO_2$ has limited suitability for purity enhancement processes involving starch – Parameters for $ScCO_2$-based purity enhancement processes may include the addition of co-solvents or modifiers (Sutil et al. 2022; Santos et al. 2012)
Controlled release systems	– ILs are suitable for the development of controlled release systems involving starch (Chatterjee et al. 2023)	– $ScCO_2$ has suitability for the development of controlled release systems with starch. Highest β-carotene loading (0.96 ± 0.07 mg β-carotene/g aerogel) – 80% of β-carotene released in simulated in vitro release (Dias et al. 2022)

$ScCO_2$ specializes in physical starch modifications and can serve as a co-solvent or carrier for compounds or catalysts in a variety of starch modification processes. $ScCO_2$ is an apolar solvent, and its interactions with starch are governed primarily by dispersion forces. This property restricts the direct chemical modifications that $ScCO_2$ can undergo. However, $ScCO_2$ offers unique benefits in starch physical transformations. Gelatinization is one of the most prominent physical modifications accomplished with $ScCO_2$. Gelatinization is the disruption of the crystalline structure of starch, which results in enhanced solubility, bloating, and functional properties. By subjecting starch to high-pressure and high-temperature conditions in $ScCO_2$, its crystalline structure is destabilized, resulting in increased solubility and altered physicochemical properties. Recent scientific research has investigated the influence of various process parameters, including pressure, temperature, and residence time, on the gelatinization behavior of starch using $ScCO_2$. In addition to

$ScCO_2$, starch granules can also be expanded with $ScCO_2$. Under supercritical conditions, starch granules physically swell, resulting in the formation of a porous structure with increased reactivity. This expanded starch provides enhanced functionalities and can be used in a variety of industrial applications, including dietary products, biodegradable materials, and pharmaceutical formulations. Recent research has centered on optimizing the expansion process parameters, such as pressure, temperature, and expansion time, to accomplish desired expansion levels and modify the properties of modified starch (Morais et al. 2020). Retrogradation inhibition is an additional physical modification that can be achieved with $ScCO_2$. Retrogradation refers to the reassembly of amylose and amylopectin chains in starch, which, upon chilling, results in the formation of insoluble structures. Retrogradation can be prevented by exposing starch to supercritical conditions in $ScCO_2$, resulting in enhanced freeze-thaw stability (Domene-López et al. 2019). This change has implications for the creation of preserved foods and other applications requiring increased stability. Recent scientific studies have investigated the effects of $ScCO_2$ treatment on the retrogradation behavior of starch, casting light on the mechanisms and factors that influence the inhibition of retrogradation.

In starch modification processes, $ScCO_2$ can also function as a co-solvent or carrier for additives or catalysts, in addition to physical modifications. The exceptional diffusivity, low viscosity, and modifiable properties of $ScCO_2$ near its critical point allow for effective mass transfer and reaction control. Recent research has centered on the use of $ScCO_2$ as a medium for the controlled release and encapsulation of bioactive compounds within starch matrices. This method entails dissolving starch in $ScCO_2$ and incorporating the desired bioactive constituent, followed by controlled precipitation or solidification to produce encapsulated starch particles. This method improves the bioactive compound's functionality, stability, and bioavailability, opening up opportunities in the food, pharmaceutical, and biomedical industries.

When evaluating the suitability of ILs and $ScCO_2$ for various starch modification processes, it is essential to evaluate the requirements of the desired modifications. ILs are especially advantageous for chemical modifications because they provide a polar reaction environment and enable precise control of reactions and selectivity. $ScCO_2$ specializes in physical modifications and can function as a co-solvent or transporter for additives or catalysts. The choice between ILs and $ScCO_2$ is determined by the modification's specific goals, the desired properties of the modified starch, and the availability of suitable apparatus and resources.

Therefore, ILs and $ScCO_2$ provide distinct benefits and applications for starch modification processes. ILs are ideally adapted for chemical modification because they permit precise control over reactions and selectivity. $ScCO_2$ is advantageous for physical modifications and can serve as a co-solvent or transport medium for additives or catalysts. Recent scientific discoveries have aided in the comprehension and optimization of IL- and $ScCO_2$-based starch modification processes, resulting in enhanced functionalities and tailored properties of modified starches. Ongoing research continues to investigate the potential applications of ILs and $ScCO_2$ in starch modification, providing opportunities for the development of sustainable and efficient starch-based materials for a variety of industries.

4.5 Environmental and Economic Considerations

4.5.1 Environmental and Economic Considerations of Ionic Liquids in Starch Modification

Ionic liquids (ILs) provide a number of benefits for the environment and the economy when used in starch modification procedures. Assessing ILs' effects on the environment and investigating their economic feasibility in industrial applications have been the main topics of recent study. ILs have low volatility, which reduces emissions during starch modification operations from an environmental standpoint. Comparing IL-based procedures to conventional solvent-based methods, studies have demonstrated that IL-based processes may dramatically reduce volatile organic compound (VOC) emissions. For instance, a comparison research by Salar-García and Ortiz-Martínez (Salar-García et al. 2017) found a stunning 75% decrease in VOC emissions using ILs. This decrease in emissions improves air quality and lowers the likelihood of health problems for nearby communities and employees. Their environmental sustainability is further increased by the possibility of recycling and reuse for ILs. The development of effective techniques for IL recovery and purification in starch modification has been the focus of recent study. To recover and purify ILs for reuse, methods including extraction, distillation, and membrane techniques have been investigated. The effective recovery and reuse of IL were proven by Zhou and Sui (Zhou et al. 2018) using a combination of extraction and membrane separation methods. The sustainability of the starch modification process is increased by the capacity to recycle ILs, which lowers resource consumption and trash production. Additionally, ILs show selectivity in the starch components they dissolve, improving extraction and separation procedures and decreasing waste production. According to studies, IL-based starch modification produces less by-products and waste streams than conventional techniques. For instance, Ptak and Zarski (Ptak et al. 2020) discovered that using IL-based starch hydrolysis produced less waste than using conventional acid hydrolysis techniques. This decrease aids in better resource management and lessens the process' total environmental effect. To thoroughly investigate the environmental effects of IL-based starch modification techniques, life cycle analyses (LCAs) have been carried out. These evaluations consider the ILs' energy use, water use, and greenhouse gas emissions. According to recent LCA results, ILs utilize less water and produce less greenhouse gas emissions than conventional solvent-based methods (Islam et al. 2022). This study emphasizes the possible environmental advantages of starch modification utilizing ILs. The cost analysis of IL-based starch modification techniques is a crucial factor from an economic standpoint. Recent studies have investigated the total cost-effectiveness of ILs by considering aspects including raw material costs, IL synthesis and recovery costs, equipment expenses, and possible savings from decreased waste creation, even if the cost of ILs may be greater than that of standard solvents. Currently, a study by Faisal Elmobarak and Almomani (Faisal Elmobarak et al. 2023) estimated that the lab-scale cost of ILs is approximately $1000 per kilogram, making them 100–1000 times more expensive than conventional solvents. Ptak and Zarski (Ptak et al. 2020) stated that, despite higher upfront costs, using ILs in large-scale manufacturing was justified due to the total cost reductions from less waste creation and greater product yields. Through market evaluations and case studies, the commercial potential and scalability of IL-based starch modification have been investigated. Consumer awareness of sustainable goods and governmental support for green practices are driving greater demand for sustainable starch products. Ren and Wang (Ren et al. 2020) predicted a sharp increase in the need for ILs in applications involving starch modification. The economic feasibility and scalability of IL-based processes are influenced by this expanding market potential. Recent developments in the application of ILs for starch modification include life cycle analyses (LCAs), sustainable IL design, and partnerships between academics and industry. Utilizing bio-based and renewable starting materials is a key component of sustainable IL design since it lowers the negative environmental effects of IL manufacturing. A more sustainable alternative to traditional ILs has been investigated by researchers via the synthesis of bio-based ILs from agricultural waste materials (Faisal Elmobarak et al. 2023). Greener synthesis approaches for ILs have been developed, with an emphasis on increasing energy efficiency, employing sustainable solvents, and creating greener separation methods (Ren et al. 2022; Thamke et al. 2022; Adak and Banerjee 2016). For IL-based starch modification to be economically viable and to promote innovation, industry and academic cooperation is essential. IL-based processes have become more effective and economical as a consequence of successful collaborations. In conclusion, using ILs in starch modification techniques has major positive effects on the environment and the economy. Recent research has shown the selectivity, low emissions, recycling potential, and decreased waste creation of ILs in starch modification. The cost-effectiveness and commercial potential of IL-based processes are shown by economic evaluations and market research. Recent developments include life cycle analyses, greener synthesis techniques, sustainable IL design, and industry-academia partnership. Researchers, business people, and legislators may benefit greatly from these discoveries and new trends as they work to improve the use of ILs in ecologically sound and financially rewarding starch modification techniques.

4.5.2 Environmental and Economic Considerations of Supercritical Carbon Dioxide in Starch Modification

The utilization of supercritical carbon dioxide ($ScCO_2$) as a solvent for starch modification has garnered considerable interest owing to its distinctive environmental and economic benefits. The utilization of supercritical carbon dioxide ($ScCO_2$) presents numerous advantages when viewed through an environmental lens. First and foremost, it is important to note that supercritical carbon dioxide ($ScCO_2$) possesses the advantageous characteristics of being non-toxic, non-flammable, and easily accessible. These attributes render it a more secure substitute for traditional organic solvents. The utilization of hazardous solvents in starch modification processes is effectively mitigated, thereby eliminating potential health risks. Furthermore, supercritical carbon dioxide ($ScCO_2$) exhibits a minimal environmental footprint due to its high recoverability and reusability. This practice results in a decrease in solvent usage and a reduction in waste production, thereby making a significant contribution toward achieving a more sustainable operational process. Current scholarly investigations have been primarily concerned with assessing the ecological consequences associated with starch modification processes utilizing supercritical carbon dioxide ($ScCO_2$). Life cycle assessments (LCAs) have been employed to assess the comprehensive environmental performance of supercritical carbon dioxide ($ScCO_2$) in comparison to conventional solvents. These assessments consider various factors, including energy consumption, water usage, greenhouse gas emissions, and waste generation. Research findings indicate that processes utilizing $ScCO_2$ (supercritical carbon dioxide) exhibit diminished environmental impacts in comparison to conventional solvent-based methods. These benefits encompass a decrease in carbon footprint and water consumption. An illustrative instance involves a recent life cycle assessment (LCA) carried out wherein the environmental ramifications of $ScCO_2$-based and conventional solvent-based starch modification procedures were juxtaposed. The study revealed that $ScCO_2$ demonstrated diminished greenhouse gas emissions and decreased water usage (De Marco et al. 2016, 2019). The present study emphasizes the prospective environmental advantages associated with the utilization of supercritical carbon dioxide ($ScCO_2$) in the process of starch modification. In conjunction with environmental factors, the economic implications associated with the utilization of supercritical carbon dioxide ($ScCO_2$) in the process of starch modification are equally noteworthy. The economic feasibility of processes utilizing $ScCO_2$ relies on various factors, including the expenses associated with equipment, operational activities, as well as the accessibility and pricing of carbon dioxide. Recent research has undertaken economic evaluations to analyze the cost-effectiveness of starch modification using supercritical carbon dioxide ($ScCO_2$) in comparison to conventional techniques. The aforementioned analyses have considered various factors, including the costs associated with raw materials, the costs of recovering and reusing solvents, the costs of managing waste, and the potential cost savings resulting from enhanced process efficiency. Recent trends in the use of $ScCO_2$ for starch modification have centered on improving its environmental and economic performance. An emerging trend is the optimization of process parameters to enhance the efficacy and selectivity of $ScCO_2$-based processes. Researchers are examining various combinations of pressure, temperature, and co-solvents to improve the solubility of starch and attain specific modification objectives. Additionally, the incorporation of $ScCO_2$-based processes with other green technologies, such as enzymatic reactions or microwave-assisted techniques, is being studied to increase process efficiency and reduce environmental impact. In addition, the development of novel catalysts and additives for $ScCO_2$-based starch modification is a current area of study. These catalysts and additives can increase the reactivity and selectivity of $ScCO_2$, allowing for more precise and efficient alterations (Salimi et al. 2014). Recent studies, for instance, have investigated the use of enzyme-coated nanoparticles as catalysts in $ScCO_2$-based starch modification, resulting in increased reaction rates and selectivity. In conclusion, the utilization of supercritical carbon dioxide ($ScCO_2$) in starch modification presents significant environmental and economic benefits. Its non-toxic nature, minimal environmental impact, and solvent recovery and reuse potential make it a desirable alternative to conventional solvents. Recent research indicates that $ScCO_2$-based procedures result in lower greenhouse gas emissions and lower water consumption compared to conventional methods. Considering factors such as raw material costs, solvent recovery, and waste management, economic analyses have demonstrated the affordability of $ScCO_2$-based processes. Ongoing research focuses on optimizing process parameters, developing novel catalysts and additives, and nurturing collaborations in an effort to further improve the environmental and economic performance of $ScCO_2$-based starch modification. These advancements contribute to the development of environmentally sustainable and economically viable starch modification practices.

4.6 Applications of Starch Modified by Green Solvents

The modification of starch with these eco-friendly solvents offers numerous opportunities to improve the starch's functional properties and expand its applications in a variety of industries. This discussion will examine the potential applications of starch modified by ILs and $ScCO_2$ and highlight recent developments and discoveries in this field. Food manufacturers rely heavily on starch for its thickening, gelling, and stabilizing abilities in a broad variety of products. The modification of starch with eco-friendly solvents creates new opportunities for enhancing its functionality and addressing particular food-related challenges. ILs can be used to improve the emulsifying properties of starch, facilitating its use as a natural emulsifier in culinary products, for instance. The addition of ILs during starch modification can enhance the emulsion stability and mouthfeel of food formulations. Additionally, IL-modified starch can be used to develop low-calorie and low-fat food products by replacing fat and enhancing sensory properties. In terms of food applications, supercritical carbon dioxide also offers benefits. $ScCO_2$-assisted starch modification can increase the digestibility and bioavailability of starch, making it appropriate for use in functional foods and dietary supplements. $ScCO_2$ can facilitate the encapsulation of bioactive compounds within starch matrices, leading to controlled release characteristics. This enables the creation of functional food products with improved nutritional profiles and health benefits. Additionally, $ScCO_2$ can be used to produce starch-based micro- and nanostructures that can be incorporated into food ingredients to enhance their texture, stability, and sensory qualities. The increasing demand for sustainable and eco-friendly packaging materials has led to the investigation of biodegradable alternatives derived from starch. The enhanced mechanical properties, barrier properties, and biodegradability of starch modified with eco-friendly solvents make it an attractive material for packaging applications. IL-modified starch can be used to create films and coatings with superior oxygen and moisture barrier properties. These films are biocompatible, renewable, and biodegradable, which makes them suitable for food packaging and other applications that require protective barriers.

The modification of starch with $ScCO_2$ can also improve its packaging properties. $ScCO_2$ can be used as a surfactant, allowing for the production of starch-based films with enhanced flexibility, elongation, and tensile strength. The incorporation of $ScCO_2$ into the film formation procedure can result in the formation of a more uniform and compact film structure, thereby enhancing the mechanical and barrier properties. $ScCO_2$-modified starch-based packaging materials offer an eco-friendly alternative to conventional petroleum-based plastics and help reduce the environmental impact of packaging detritus. Starch is vital to the pharmaceutical industry as an excipient, binder, disintegrant, and controlled release agent. By modifying starch with environmentally friendly solvents, it is possible to increase its functionality and expand its applications in pharmaceutical formulations. For controlled drug release applications, ILs can be used to modify starch. By incorporating pharmaceuticals into IL-modified starch matrices, sustained and targeted drug release profiles can be achieved, thereby enhancing therapeutic efficacy and patient adherence. In addition, ILs can enhance the solubility and stability of poorly soluble medications, facilitating their formulation into more potent pharmaceutical products. Pharmaceutical applications of $ScCO_2$-assisted starch modification are also promising. $ScCO_2$ can be utilized as a solvent or co-solvent to increase the dissolution and bioavailability of pharmaceuticals in starch-based formulations. $ScCO_2$ can increase drug loading efficiency, reduce the use of organic solvents, and reduce the degradation of heat-sensitive pharmaceuticals. In addition, $ScCO_2$ can facilitate the encapsulation of active pharmaceutical constituents within starch matrices, providing protection and controlled release characteristics. Starch is used in the personal care and makeup business because it can soak liquids, make things thicker, and keep things stable. Using ecological solvents to modify starch improves its functionality, thereby enhancing the efficacy of personal care and cosmetic products. ILs are capable of modifying starch for applications such as dry shampoo, talc-free granules, and facial preparations. IL-modified starch has enhanced oil absorption capacity, texture, and sensory properties, making it an excellent alternative to conventional ingredients.

Additionally, $ScCO_2$ can be utilized in the modification of starch for personal care products. $ScCO_2$-assisted starch modification can improve the gel strength and stability of starch-based gels, which are frequently found in hair styling products and skin care formulations. $ScCO_2$ can facilitate the formation of starch-based micro- and nanostructures, which can then be incorporated into cosmetic formulations to impart a desirable texture, enhanced stability, and controlled release properties. The use of $ScCO_2$ as a green solvent is consistent with the increasing consumer demand for eco-friendly and sustainable personal care products.

In the textile industry, modified starch is utilized as a sizing agent, thickener, and binder. Utilizing environmentally favorable solvents in starch modification enables the production of eco-friendly and sustainable textiles. ILs can be utilized to modify starch for textile applications including fabric sizing, printing, and finishing. IL-modified starch offers enhanced adhesion, film-forming properties, and resistance to mechanical stress, resulting in enhanced fabric performance and quality. In addition, ILs can be used as a dyeing solvent, reducing the need for traditional organic solvents and enhancing the environmental profile of textile dyeing.

In conclusion, the potential applications of starch modified by environmentally friendly solvents, namely, ILs and $ScCO_2$, are diverse and encompass numerous industries. The use of these eco-friendly solvents enables the improvement of the functional properties of starch and the expansion of its applications in the food industry, biodegradable packaging, pharmaceuticals, personal care and cosmetics, and the textile industry. Recent developments in the field of starch modification with environmentally friendly solvents have shed light on their potential applications and advantages.

4.7 Future Perspectives and Research Directions

Utilizing ionic liquids (ILs) and supercritical carbon dioxide ($ScCO_2$) as green solvents in starch modification has yielded optimistic results, offering numerous environmental and economic benefits. Several prospective perspectives and research directions can further improve the implementation of these environmentally friendly solvents in starch modification processes. The creation of more sustainable and environmentally benign ILs and $ScCO_2$ is an essential area of study. To reduce the environmental impact of IL production, researchers are investigating the synthesis of bio-based ILs from renewable feedstocks, such as agricultural refuse materials. The use of bio-based ILs can provide a more sustainable alternative for starch modification while maintaining the desired properties. Attempts are also being made to enhance the synthesis of $ScCO_2$, including the use of renewable energy sources and greener separation techniques, in order to reduce the carbon footprint associated with its production.

The optimization of IL- and $ScCO_2$-based processes to enhance their efficacy and selectivity in starch modification is an additional research direction. This involves analyzing the effects of different process parameters, such as temperature, pressure, and the inclusion of co-solvents or catalysts. By fine-tuning these parameters, scientists can accomplish more precise and controlled starch modifications, resulting in improved functional properties and product quality. Exploration of novel starch applications and modifications utilizing ILs and $ScCO_2$ will also be an important area of future research. Individual ILs can be engineered to have specific properties and functions, allowing for novel approaches to starch modification. For instance, researchers may investigate the use of ILs in the creation of starch-based materials with enhanced properties, such as enhanced mechanical strength, thermal stability, or controlled drug release. Similarly, the use of $ScCO_2$ as a carrier for active compounds or the encapsulation of bioactive substances within starch matrices provides promising avenues for the development of functional starch-based materials. In addition, the integration of ILs and $ScCO_2$ with other eco-friendly technologies for starch modification has enormous potential. Synergistic strategies, such as combining ILs with enzymatic reactions or employing $ScCO_2$-assisted microwave techniques, can improve the efficiency, selectivity, and overall sustainability of starch modification processes. This research can lead to the creation of environmentally sustainable, integrated processes that leverage the benefits of both ILs and $ScCO_2$. Collaboration between academia, industry, and government agencies will continue to play a crucial role in fostering innovation and facilitating the adoption of ILs and $ScCO_2$ as green solvents for starch modification. Collaborative research initiatives can help bridge the divide between fundamental research and industrial application by facilitating the scaling up of processes and resolving the economic viability of these environmentally friendly solvents. Additionally, partnerships with CO_2 suppliers, IL manufacturers, and starch processing industries can promote the ubiquitous use of ILs and $ScCO_2$ in starch modification by ensuring a sustainable supply chain and fostering knowledge exchange. Future perspectives and research directions for ILs and $ScCO_2$ as green solvents for starch modification are centered on the development of more sustainable solvents, optimization of processes, exploration of new applications and modifications, and integration with other green technologies. The application of ILs and $ScCO_2$ in starch modification can contribute to the development of environmentally friendly and economically viable processes, resulting in the production of sustainable starch-based materials with improved properties and functionalities.

4.8 Challenges and Opportunities in Commercialization

4.8.1 Scale-Up and Cost Considerations

The scaling-up process is one of the primary obstacles in commercializing modified starch using green solvents. Despite the fact that laboratory studies have demonstrated the efficacy of ILs and $ScCO_2$ in modulating starch, the transition to large-scale production presents unique obstacles. The process of scaling up calls for substantial investments in apparatus, infrastructure, and operational procedures. In addition, it can be difficult to ensure consistent quality, reproducibility, and efficiency at a larger scale. To overcome these obstacles, process optimization, equipment design, and scalability studies are required to facilitate a cost-effective scale-up and ensure seamless scaling. Recent discoveries in the field of starch modification with environmentally friendly solvents have shone light on scale-up obstacles and provided insight into their resolution. Understanding the impact of scale on the efficacy of modified starch products has been the focus of numerous studies, which have identified key parameters that must be optimized during the scale-up process. For instance, Ptak et al. (2020) investigated the scaling up of IL-based starch modification and found that certain process parameters, such as reaction temperature and residence time, needed to be carefully controlled to ensure product quality consistency. The study emphasized the significance of optimizing mixture conditions and reactor design in order to achieve homogenous dispersion and efficient reaction at larger scales. De Marco and Iannone (De Marco et al. 2018) investigated the scaling up of $ScCO_2$-assisted starch modification and emphasized the significance of process control and apparatus design. Adjusting the pressure, temperature, and discharge rate of $ScCO_2$ significantly affected the degree of modification and the functional properties of the modified starch, according to the researchers. This parameter optimization enabled consistent and reproducible results during scale-up. In addition, developments in process modeling and simulation have yielded valuable instruments for addressing scale-up obstacles. Large-scale production systems can be designed with the aid of computational models that foresee the behavior of starch modification processes under various operating conditions. These models aid in the optimization of process parameters, the reduction of trial-and-error experimentation, and the enhancement of scale-up processes' efficacy. In terms of cost consideration, the economic viability of commercializing modified starch with environmentally friendly solvents is a critical factor. Green solvents, such as ILs, are typically more expensive than conventional solvents, which can have an impact on the total cost of the modified starch. Cost considerations extend beyond the solvent itself to include factors such as the price of primary materials, purification processes, refuse management, and energy consumption. Techno-economic analyses and cost optimization strategies are required to identify cost drivers and develop cost-effective approaches to make the commercialization of modified starch utilizing green solvents economically viable. Several studies have conducted a techno-economic comparison of the production costs of modified starch using ILs and conventional solvents. The study found that even though the initial investment for IL-based production was higher, the reduced waste generation and enhanced product yields lead to long-term cost savings. The analysis highlighted the significance of contemplating the entire production process and evaluating the costs and benefits holistically. While maintaining product quality, cost optimization strategies aim to reduce total production costs. Recent research has investigated several avenues for reducing the cost of producing modified starch with environmentally friendly solvents. For instance, process intensification techniques, such as continuous flow systems and microreactor technology, have been studied in an effort to increase production efficiency and decrease operating expenses (Abouelela et al. 2020; Klein-Marcuschamer et al. 2011; Mahmood et al. 2017). There have also been investigations into the integration of process stages, waste minimization strategies, and the utilization of renewable resources to reduce the price of basic materials.

4.8.2 Regulatory Compliance

The commercialization of modified starch produced with green solvents must comply with all applicable regulatory requirements. Compliance with these regulatory frameworks is essential for market acceptance and product safety. Regulatory frameworks vary across regions and industries. It is crucial to understand the regulatory landscape, ensure compliance with safety and quality standards, and resolve any potential regulatory barriers. It is essential that academia, industry, and regulatory bodies work together to overcome these obstacles and establish clear guidelines for the commercialization of modified starch utilizing green solvents.

Conformity with regulations necessitates conducting exhaustive safety evaluations of modified starch products. These evaluations consist of analyzing the modified starch's toxicity, allergenicity, and potential health effects. Recent research has focused on evaluating the safety of modified carbohydrates made with ILs and $ScCO_2$. For instance, Thamke and Singh (Thamke et al. 2022) conducted a comprehensive safety evaluation of IL-modified starch and determined that it is safe for human consumption based on numerous toxicity studies. These safety evaluations contribute to regulatory compliance and consumer protection. Compliance with FDA regulations is essential for starch modification using green solvents, especially if the modified starch is intended for use in the food or pharmaceutical industries. The FDA provides guidelines for the use of food additives, including modified carbohydrates, in order to guarantee their safety and labeling requirements. Recent research has highlighted the importance of adhering to FDA regulations when developing and marketing modified starch products. Understanding the specific FDA requirements and guidelines ensures regulatory compliance and market acceptance. Environmental compliance is another important aspect of regulatory compliance in starch modification using green solvents. Due to their reduced environmental influence, green solvents such as ILs and $ScCO_2$ are chosen. Recent research has focused on assessing the environmental sustainability of starch modification processes employing ILs and $ScCO_2$. This includes evaluating factors like ecological footprint, water consumption, and refuse generation. Adhering to environmental regulations and sustainability standards ensures compliance with the applicable environmental requirements and advances the overall sustainability objectives. Emerging trend in the field of starch modification using green solvents is the harmonization of regulations across various regions and countries. International organizations, such as the International Union of Pure and Applied Chemistry (IUPAC) and the International Conference on Harmonization of Technical Requirements for Registration of Pharmaceuticals for Human Use (ICH), are making efforts to harmonize guidelines and standards pertaining to modified starches and green solvents. Compliance with these harmonized regulations simplifies the process of regulatory compliance and facilitates access to global markets. Regulatory compliance encompasses product labeling and openness. Providing accurate and comprehensible labeling information is essential for regulatory compliance and consumer trust. Recent studies have highlighted the significance of transparent labeling, including the identification of the green solvent used in the process of modification. Clear labeling facilitates consumer decision-making and ensures compliance with labeling regulations and guidelines.

4.8.3 Market Demand and Consumer Acceptance

The market demand for modified starch utilizing green solvents, such as ionic liquids (ILs) and supercritical carbon dioxide ($ScCO_2$), plays a significant role in these products' commercialization. Understanding market trends, consumer preferences, and the potential applications of modified starch utilizing environmentally friendly solvents is essential for satisfying market demands and ensuring consumer acceptance. Here are some important considerations regarding market demand and consumer acceptance.

Market research is necessary for identifying market trends, assessing consumer requirements, and determining the potential demand for modified starch products. This involves analyzing market size, growth rates, and the competitive environment. Recent market research indicates a growing demand for eco-friendly and sustainable products, such as modified carbohydrates (Kumar et al. 2023). Understanding market dynamics and recognizing market segments can aid in the development of targeted marketing strategies and product positioning. The market demand for modified starch products is greatly influenced by consumer preferences. Recent consumer surveys and studies indicate a growing preference among consumers for sustainable and environmentally favorable products. Consumers are increasingly cognizant of the environmental impact of the products they purchase and actively pursue alternatives that align with their values. The use of green solvents in modified starch products provides a sustainable and eco-friendly alternative that can appeal to environmentally conscious consumers.

Identifying potential applications for modified starch utilizing environmentally friendly solvents is essential for satisfying market demands. Recent studies have investigated a variety of modified starch applications, including food and beverages, pharmaceuticals, cosmetics, and packaging materials. These applications offer distinctive value propositions, including enhanced texture, stability, and biodegradability. Customizing product offerings and value propositions for specific market segments can increase consumer acceptance and stimulate market demand. Academics and educating consumers about the advantages of modified starch derived from environmentally friendly solvents is crucial for increasing consumer acceptance. Recent trends indicate a growing interest in eco-friendly and sustainable products, but consumers may be unfamiliar with modified starch and its benefits (Kumar et al. 2023; Hamaker 2021). Education and awareness campaigns can therefore play a vital role in highlighting the environmental benefits, performance characteristics, and potential applications of modified starch products. These campaigns may be conducted via various channels, such as social media, industry events, and partnerships with influencers or sustainability advocates.

4.9 Conclusion

The modification of starch with environmentally friendly solvents, such as ionic liquids (ILs) and supercritical carbon dioxide ($ScCO_2$), is a promising strategy for augmenting the properties and applications of starch in a sustainable manner. During the course of our discussion, several critical aspects regarding the use of green solvents for starch modification became apparent:

Green Solvents: Due to their minimal environmental impact, non-toxicity, and potential for recycling and reuse, ILs and $ScCO_2$ are regarded as environmentally friendly solvents. These solvents provide alternatives to conventional organic solvents, thereby decreasing the environmental impact of starch modification processes.

Principle Mechanisms: Physical and chemical mechanisms can be utilized by ILs and $ScCO_2$ to modify starch. Chemical modifications include esterification, crosslinking, and enzymatic reactions, while physical modifications include gelatinization, expansion, and inhibition of retrogradation. These mechanisms allow for the modification of starch properties and the introduction of new functions.

Advantages of ILs: The high solubility of starch provided by ILs enables simple access to the starch structure and precise control of reactions. They are selective for particular starch components and allow for targeted modifications. The synthesis, toxicity, and energy-intensive production of ILs may pose challenges.

Advantages of ScCO_2: ScCO_2 is a non-polar solvent that physically expands starch particles, resulting in enhanced reactivity and functional properties. It also functions as a co-solvent or carrier for additives or catalysts, thereby facilitating indirect chemical modifications. ScCO_2 is widely accessible, non-toxic, and recoverable for reuse. However, its limited solubility for native starch may necessitate modification or the addition of co-solvents.

Applications: Food and beverage, pharmaceutical, cosmetic, and packaging are just a few of the industries that can benefit from modified starch made with environmentally friendly solvents. Biodegradable packaging, drug delivery systems, emulsifiers, adhesives, coatings, and bioplastics are examples. The modified properties of modified starch increase its potential applications and contribute to the attainment of sustainable development objectives.

Environmental Considerations: By minimizing emissions, waste production, and resource consumption, starch modification with green solvents is consistent with environmental objectives. Recent research emphasizes the low volatility of ILs, the possibility of IL recycling, and the reduced refuse generation in IL-based processes. ScCO_2's non-toxicity and reusability contribute to its environmental advantages. A crucial factor is the commercial viability of modifying starch with eco-friendly solvents. Recent research has examined the cost implications, market potential, and scalability of IL-based procedures. Although green solvents may have higher up-front costs, savings can be realized through reduced waste production, increased product yields, and market demand for sustainable products.

Challenges and Opportunities: The use of green solvents to produce modified starch on a larger scale presents difficulties in terms of infrastructure, equipment, process optimization, and scalability. To solve these issues and provide clear rules, cooperation between regulatory agencies, business, and academics is crucial. The creation of sustainable ILs, the use of greener synthesis techniques, and partnerships aimed at maximizing economic viability are recent developments.

References

Abouelela AR et al (2020) Commercial aspects of biomass deconstruction with ionic liquids. In: Commercial applications of ionic liquids, pp 87–127

Adak S, Banerjee R (2016) A green approach for starch modification: esterification by lipase and novel imidazolium surfactant. Carbohydr Polym 150:359–368

Alavi F, Ciftci ON (2022) Developing dual nano/macroporous starch bioaerogels via emulsion templating and supercritical carbon dioxide drying. Carbohydr Polym 292:119607

Amini E, Valls C, Roncero MB (2023) Promising nanocomposites for food packaging based on cellulose—PCL films reinforced by using ZnO nanoparticles in an ionic liquid. Ind Crop Prod 193:116246

Araujo EJS et al (2023) Supercritical impregnation of polyphenols from passion fruit residue in corn starch aerogels: effect of operational parameters. J Food Eng 343:111394

Bai J et al (2017) Synthesis of octenylsuccinic-anhydride-modified cassava starch in supercritical carbon dioxide. Starch-Stärke 69(11–12):1700018

Bakouri H, Guemra K (2019) Etherification and cross-linking effect on physicochemical properties of Zea mays starch executed at different sequences in 1-butyl-3-methylimidazolium chloride [BMIM]Cl ionic liquid media. Int J Biol Macromol 125:1118–1127

Biswas A et al (2006) Ionic liquids as solvents for biopolymers: acylation of starch and zein protein. Carbohydr Polym 66(4):546–550

Castro JM et al (2023) Study of the plasticization effect of 1-ethyl-3-methylimidazolium acetate in TPS/PVA biodegradable blends produced by melt-mixing. Polymers 15(7):1788

Cerro D et al (2023) Nanoencapsulation of food-grade bioactive compounds using a supercritical fluid extraction of emulsions process: effect of operational variables on the properties of nanocapsules and new perspectives. LWT 184:115115

Chatterjee S et al (2023) Cationic starch: a functionalized polysaccharide based polymer for advancement of drug delivery and health care system—a review. Int J Biol Macromol 248:125757

Chen C-C et al (2019) Fire and explosion hazards of ionic liquid 1-methyl-1-propylpyrrolidinium bis(trifluoromethanesulfonyl)imide. J Loss Prev Process Ind 60:233–240

Chen J, Zeng X, Chen L (2021) Regulation nature of water-choline amino acid ionic liquid mixtures on the disaggregation behavior of starch. Carbohydr Polym 272:118474

Choi J, Nidetzky B (2022) Ionic liquid as dual-function catalyst and solvent for efficient synthesis of sucrose fatty acid esters. Mol Catal 526:112371

Chorfa N et al (2022) Physical and enzymatic hydrolysis modifications of potato starch granules. Polymers 14:2027. https://doi.org/10.3390/polym14102027

Crucean D et al (2021) Influence of the presence of choline chloride on the classical mechanism of "gelatinization" of starch. Polymers (Basel) 13(9):1509

De Marco I et al (2016) LCA of starch aerogels for biomedical applications. Chem Eng Trans 49:319–324

De Marco I et al (2018) An environmental study on starch aerogel for drug delivery applications: effect of plant scale-up. Int J Life Cycle Assess 23:1228–1239

De Marco I, Riemma S, Iannone R (2019) Life cycle assessment of supercritical impregnation: starch aerogel + α-tocopherol tablets. J Supercrit Fluids 143:305–312

Decaen P et al (2017) Choline chloride vs choline ionic liquids for starch thermoplasticization. Carbohydr Polym 177:424–432

Devi LS, Das AB (2020) Effect of ionic liquid on sol-gel phase transition, kinetics and rheological properties of high amylose starch. Int J Biol Macromol 162:685–692

Dias ALB et al (2022) Role of supercritical CO2 impregnation variables on β-carotene loading into corn starch aerogel particles. J CO2 Util 63:102125

Domene-López D et al (2019) Comparative study on properties of starch films obtained from potato, corn and wheat using 1-ethyl-3-methylimidazolium acetate as plasticizer. Int J Biol Macromol 135:845–854

Faisal Elmobarak W et al (2023) Current status of CO2 capture with ionic liquids: development and progress. Fuel 344:128102

Falua KJ et al (2022) Valorization of starch to biobased materials: a review. Polymers (Basel) 14(11):2215

Fan Y, Picchioni F (2020) Modification of starch: a review on the application of "green" solvents and controlled functionalization. Carbohydr Polym 241:116350

Fu Y, Jiang E, Yao Y (2022) New techniques in structural tailoring of starch functionality. Annu Rev Food Sci Technol 13:117–143

Gao J, Luo Z-G, Luo F-X (2012) Ionic liquids as solvents for dissolution of corn starch and homogeneous synthesis of fatty-acid starch esters without catalysts. Carbohydr Polym 89(4):1215–1221

García-González CA et al (2012) Supercritical drying of aerogels using CO2: effect of extraction time on the end material textural properties. J Supercrit Fluids 66:297–306
Hamaker BR (2021) Current and future challenges in starch research. Curr Opin Food Sci 40:46–50
Hiraga Y, Duereh A, Smith RL (2018) Aspects of solvent polarity and solvent properties in developing efficient systems for processing biomass with ionic liquid mixtures and supercritical CO2. J Supercrit Fluids 134:12–20
Ingrosso F, Ruiz-López MF (2017) Modeling solvation in supercritical CO2. ChemPhysChem 18(19):2560–2572
Islam N et al (2022) Screening of ionic liquids as sustainable greener solvents for the capture of greenhouse gases using COSMO-RS approach: computational study. Fuel 330:125540
Ivanovic J, Milovanovic S, Zizovic I (2016) Utilization of supercritical CO2 as a processing aid in setting functionality of starch-based materials. Starch-Stärke 68(9–10):821–833
Karić N et al (2021) A green approach to starch modification by solvent-free method with betaine hydrochloride. Int J Biol Macromol 193:1962–1971
Khosousi T et al (2023) Starch nanoparticles modified with ionic liquid: a novel and efficient catalyst for the synthesis of symmetrical azine derivatives at room temperature. Results Chem 6:101063
Klein-Marcuschamer D, Simmons BA, Blanch HW (2011) Techno-economic analysis of a lignocellulosic ethanol biorefinery with ionic liquid pre-treatment. Biofuels Bioprod Biorefin 5(5):562–569
Klettenhammer S et al (2020) Novel technologies based on supercritical fluids for the encapsulation of food grade bioactive compounds. Foods 9(10):1395
Kumar Y et al (2023) Starch: current production and consumption trends. In: Sharanagat VS et al (eds) Starch: advances in modifications, technologies and applications. Springer International Publishing, Cham, pp 1–10
Li D, Zhang X, Tian Y (2016) Ionic liquids as novel solvents for biosynthesis of octenyl succinic anhydride-modified waxy maize starch. Int J Biol Macromol 86:119–125
Li D et al (2022a) Intensifying the moderate electric field-induced modification of maize starch by 1-butyl-3-methylimidazolium chloride. Carbohydr Polym 292:119654
Li C-G et al (2022b) Study of the microstructure of chitosan aerogel beads prepared by supercritical CO2 drying and the effect of long-term storage. RSC Adv 12(33):21041–21049
Lu X et al (2013) Two-step method of enzymatic synthesis of starch laurate in ionic liquids. J Agric Food Chem 61(41):9882–9891
Lu L, Guo L, Cui B (2020) Comparative study on phase transition and morphology of starch from maize and potato in ionic liquid/water mixtures: effects of the different ratio. Int J Biol Macromol 147:911–920
Lukic I et al (2022) Supercritical CO2-assisted processes for development of added-value materials: optimization of starch aerogels preparation and hemp seed extracts impregnation. J CO2 Util 61:102036
Machado ND et al (2022) Supercritical CO2-assisted impregnation/deposition of polymeric materials with pharmaceutical, nutraceutical, and biomedical applications: a review (2015–2021). J Supercrit Fluids 191:105763
Mahmood H et al (2017) Ionic liquids pretreatment for fabrication of agro-residue/thermoplastic starch based composites: a comparative study with other pretreatment technologies. J Clean Prod 161:257–266
Matsubara M, Nakato Y, Kondo E (2021) Enhancing resistant starch content in brown rice using supercritical carbon dioxide processing. J Food Process Eng 44(2):e13617
Men Y et al (2015) Preparation of corn starch-g-polystyrene copolymer in ionic liquid: 1-Ethyl-3-methylimidazolium acetate. Carbohydr Polym 121:348–354
Menezes DB et al (2021) Starch-based biocomposite membrane reinforced by orange bagasse cellulose nanofibers extracted from ionic liquid treatment. Cellulose 28:4137–4149
Morais ES et al (2020) Use of ionic liquids and deep eutectic solvents in polysaccharides dissolution and extraction processes towards sustainable biomass valorization. Molecules 25:3652. https://doi.org/10.3390/molecules25163652

Muljana H et al (2010) Synthesis of fatty acid starch esters in supercritical carbon dioxide. Carbohydr Polym 82(2):346–354
Ngu LH (2022) Carbon capture technologies. In: Reference module in earth systems and environmental sciences. Elsevier
Patel S et al (2009) Development of cross-linked starch microcellular foam by solvent exchange and reactive supercritical fluid extrusion. J Appl Polym Sci 111(6):2917–2929
Plechkova NV, Seddon KR (2008) Applications of ionic liquids in the chemical industry. Chem Soc Rev 37(1):123–150
Ptak S, Zarski A, Kapusniak J (2020) The importance of ionic liquids in the modification of starch and processing of starch-based materials. Materials 13:4479. https://doi.org/10.3390/ma13204479
Ravula S et al (2019) Vapor pressure mapping of ionic liquids and low-volatility fluids using graded isothermal thermogravimetric analysis. ChemEngineering 3(2):42
Ren F et al (2019) Dissolution behavior of maize starch in aqueous ionic liquids: effect of anionic structure and water/ionic liquid ratio. ACS Omega 4(12):14981–14986
Ren F et al (2020) Applications of ionic liquids in starch chemistry: a review. Green Chem 22(7):2162–2183
Ren F et al (2022) Green synthesis of acetylated maize starch in different imidazolium carboxylate and choline carboxylate ionic liquids. Carbohydr Polym 288:119353
Roy Goswami S et al (2020) Catalyst-free esterification of high amylose starch with maleic anhydride in 1-butyl-3-methylimidazolium chloride: the effect of amylose content on the degree of MA substitution. Carbohydr Polym 234:115892
Ruamcharoen J, Munlee R, Ruamcharoen P (2020) Improvement of water vapor barrier and mechanical properties of sago starch-kaolinite nanocomposites. Polym Compos 41(1):201–209
Salar-García M et al (2017) Ionic liquid technology to recover volatile organic compounds (VOCs). J Hazard Mater 321:484–499
Salimi K et al (2014) Controlled graft copolymerization of lactic acid onto starch in a supercritical carbon dioxide medium. Carbohydr Polym 114:149–156
Santos DT et al (2012) Production of stabilized sub-micrometric particles of carotenoids using supercritical fluid extraction of emulsions. J Supercrit Fluids 61:167–174
Seddon KR, Stark A, Torres M-J (2000) Influence of chloride, water, and organic solvents on the physical properties of ionic liquids. Pure Appl Chem 72(12):2275–2287
Shogren RL, Biswas A (2010) Acetylation of starch with vinyl acetate in imidazolium ionic liquids and characterization of acetate distribution. Carbohydr Polym 81(1):149–151
Shokri Z et al (2022) Elucidating the impact of enzymatic modifications on the structure, properties, and applications of cellulose, chitosan, starch and their derivatives: a review. Mater Today Chem 24:100780
Soleimanpour M et al (2020) Fabrication of nanostructured mesoporous starch encapsulating soy-derived phytoestrogen (genistein) by well-tuned solvent exchange method. Int J Biol Macromol 159:1031–1047
Sutil GA et al (2022) Effects of incorporation of pure or multicomponent active agents in biopolymers for food packaging using supercritical CO2. Trends Food Sci Technol 120:349–362
Tabernero A et al (2022) Supercritical carbon dioxide and biomedicine: opening the doors towards biocompatibility. Chem Eng J 444:136615
Thamke V et al (2022) Current toxicological insights of ionic liquids on various environmental living forms. J Environ Chem Eng 10(2):107303
Ubeyitogullari A, Ciftci ON (2017) Generating phytosterol nanoparticles in nanoporous bioaerogels via supercritical carbon dioxide impregnation: effect of impregnation conditions. J Food Eng 207:99–107
Uwineza PA, Waśkiewicz A (2020) Recent advances in supercritical fluid extraction of natural bioactive compounds from natural plant materials. Molecules 25:3847. https://doi.org/10.3390/molecules25173847
Vatansever S et al (2021) Physicochemical and multi-scale structural alterations of pea starch induced by supercritical carbon dioxide + ethanol extraction. Food Chem 344:128699

Villegas M et al (2019) Development of an integrated one-pot process for the production and impregnation of starch aerogels in supercritical carbon dioxide. J Supercrit Fluids 154:104592
Villegas ME et al (2020) Commercial starch behavior when impregnated with food additives by moderate temperature supercritical CO2 processing. Starch-Stärke 72(11–12):1900231
Wang J et al (2019a) Toward a better understanding of different dissolution behavior of starches in aqueous ionic liquids at room temperature. ACS Omega 4(6):11312–11319
Wang D-W et al (2019b) Tapioca/polyvinyl alcohol thermoplastic starch materials processed with the aid of supercritical CO2. Food Packag Shelf Life 22:100425
Wang J et al (2021) Thermoplastic starch plasticized by polymeric ionic liquid. Eur Polym J 148:110367
Wu T, Han B (2019) Supercritical carbon dioxide (CO2) as green solvent. In: Han B, Wu T (eds) Green chemistry and chemical engineering. Springer, New York, pp 173–197
Wu R-L et al (2009) Green composite films prepared from cellulose, starch and lignin in room-temperature ionic liquid. Bioresour Technol 100(9):2569–2574
Yang Y et al (2021) Hydrophobic thermoplastic starch supramolecularly-induced by a functional sucrose based ionic liquid crystal. Carbohydr Polym 255:117363
Zarski A et al (2016) Esterification of potato starch by a biocatalysed reaction in an ionic liquid. Carbohydr Polym 137:657–663
Zarski A et al (2019) From high oleic vegetable oils to hydrophobic starch derivatives: I. Development and structural studies. Carbohydr Polym 214:124–130
Zdanowicz M (2020) Starch treatment with deep eutectic solvents, ionic liquids and glycerol. A comparative study. Carbohydr Polym 229:115574
Zhan J et al (2020) Designing lipase-compatible ionic liquids as novel solvents for starch ester biosynthesis. Starch-Stärke 72(1–2):1900120
Zhang X et al (2023a) Study on the effect of different concentrations of choline glycine ionic liquid-water mixtures on debranched starch butyrylation reaction. Carbohydr Polym 308:120680
Zhang B et al (2023b) For efficient treatment of starch using aqueous ionic liquid at room temperature. Sustain Mater Technol 36:e00592
Zhang P et al (2023c) Improved stability of β-carotene by encapsulation in SHMP-corn starch aerogels. Food Chem 406:135040
Zhao H et al (2008) Designing enzyme-compatible ionic liquids that can dissolve carbohydrates. Green Chem 10(6):696–705
Zhou G, Luo Z, Fu X (2014) Preparation and characterization of starch nanoparticles in ionic liquid-in-oil microemulsions system. Ind Crop Prod 52:105–110
Zhou J et al (2018) Recovery and purification of ionic liquids from solutions: a review. RSC Adv 8(57):32832–32864

Structural Studies of Starch: Chromatographic Methods

5

Luis A. Bello-Perez and Josue Moreno-Zaragoza

Abstract

The structural characterization of starch components and their chains is important to determine the best end uses and behavior (functionality, physicochemical properties, and digestibility) of starch. Structural studies on the polysaccharide with the aid of chromatographic techniques have been carried out for the past 45 years. The first works used gel-permeation chromatography after high-performance size-exclusion chromatography coupled with different detectors (multi-angle laser-light scattering, refractive index, intrinsic viscosity) and high-performance anion-exchange chromatography. Despite the advances in structural works, there are some gaps in the study of starch structure using chromatographic methods. Column type, use of field-flow fractionation channel, and type of mathematical fit require further research considering each type of starch to obtain precise information on starch structure. The structural characteristics of starch and their relationship with functionality and digestibility can be used to design novel starches and develop new and improved products.

Keywords

Structure · Molar mass · Gyration radius · Hydrodynamic radius · Chain-length distribution · Solubilization

L. A. Bello-Perez (✉) · J. Moreno-Zaragoza
Instituto Politécnico Nacional, CEPROBI, Yautepec, Morelos, Mexico
e-mail: labellop@ipn.mx

N. Mazumder, M. H. Rahman (eds.), *Advanced Research in Starch*,
https://doi.org/10.1007/978-981-99-9527-1_5

5.1 Introduction

Starch is the major biopolymer in food crops as cereals, pseudocereals, tubers, roots, legumes, and unripe fruits. It can be analyzed considering (1) its presence in foods, as cooked rice; fry, boiled, and oven cooked potato; boiled legumes (beans, lentils, pea); and meals including starchy food crops, and (2) its isolation from botanical sources and use as ingredient to produce functional and sensory characteristics. In end consumption of starchy food crops, the native (original) structure of starch is responsible for the physicochemical, rheological, functional, digestibility, and sensory features. Additionally, starch structure must be considered in food processing. Understanding structure-function, structure-rheology, and structure-digestibility is key to explain the characteristics of foods and choose the right starch or starchy source to obtain products with desirable qualities. For example, potato with reduced starch digestibility produces low-calorie products, while rice presents better cooking and texture characteristics, and maize and wheat show low retrogradation. There is an increasing interest in developing starchy crops with a specific starch structure of amylose and amylopectin at the right ratio since both are important to find the best end uses for crops as wheat, potato, maize, and rice, among others. The first attempts at doing so manipulated starch biosynthesis to reduce the component concentration given that amylopectin is responsible for viscosity and amylose for retrogradation and gel formation. In addition, the chain-length distribution (CLD) of amylopectin was modified because some isoforms of the enzymes that participate in starch biosynthesis can produce chains with different lengths (Nakamura et al. 2022). To find the end use of starchy food crops to create new products, tailor-made starches in those botanical sources have been investigated, and studies to find ideal structure-function ratios have also been carried out. Those studies analyze the CLD of the branched component (amylopectin) where the prevalence of short or long chains of amylopectin is associated with gelatinization and retrogradation (Zhang et al. 2008a; Miao et al. 2015). These works state that the prevalence of long chains in starch structure-digestion is responsible for hydrolysis resistance in digestive enzymes (Miao and Hamaker 2021).

Research relates the structure of starch components to the use of chromatographic methods. However, there are no systematic studies on starch structure showing more accurate and reproducible results. The controversies surrounding works on starch structure include solubilization, eluents, chromatographic columns, temperature, and detectors (Vilaplana and Gilbert 2010; Gilbert et al. 2010; Bello-Perez et al. 2019; Obadi et al. 2022).

This chapter describes the methods to solubilize starch, separation methods (size-exclusion chromatography columns, asymmetric field-flow fractionation), conditions (solvent type, temperature), detectors (MALLS, RI, intrinsic viscosity), and analysis of results. The advantage and disadvantages of each procedure should be considered to suggest the analysis procedure to obtain information of the starch structure and suggest applications.

5.2 Starch Structure

5.2.1 Amylose

Amylose is a glucose biopolymer with α-(1→4) glycosidic linkages and a molecular weight (Mw) of ~10^6 g/mol. It is an essentially linear component of starch granules with some branching points that do not affect physicochemical, functional, nor rheological properties. It is distributed in amylopectin amorphous areas (branching points) of the amylopectin cluster model; however, its location in the granule remains a topic of debate. This molecule is tangentially located to the radial orientation of amylopectin chains. Some works have proposed that amylose concentration is higher in the periphery of starch granules as compared to their cores (Li et al. 2019; Nakamura and Kainuma 2021). The linear structure of amylose produces a helix with a hydrophobic cavity that can bind to iodine and produce blue color, allowing to quantify amylose given the formation of an amylose-iodine complex. Additionally, that amylose cavity can produce a thermally stable complex with fatty acids, and the presence of internal starch lipids affects amylose concentration. The amylose-lipid and amylose-polyphenol complexes are presently of interest from a nutritional point of view since they are not substrate for digestive enzymes hydrolyzing starch and, therefore, glucose is not produced (Gutiérrez and Tovar 2021; Romero-Hernández et al. 2022). The thermal processing of foods containing starch and lipids produces V-type complexes, decreases blood glucose, and reduces lipid and polyphenol accumulation in the body.

Several methods have been used to determine the amylose content in starch, including iodine colorimetry where amylose produces a complex with iodine. Some additions have been proposed for this method. Colorimetry identifies the apparent amylose through the presence of some amylose-lipid complexes and the underestimation of biopolymer concentration. Therefore, defatted starch and butanol have been proposed to remove lipids and precisely quantify amylose levels. However, some authors consider that the amylose content remains as apparent amylose. Starch debranching in amylopectin produces linear chains, while amylose is intact and separated by size-exclusion chromatography (Hoyos-Leyva et al. 2017). A typical profile obtained by this method is shown in Fig. 5.1 where the first peak corresponds to amylose and peaks 2 and 3 to linear amylopectin chains obtained after debranching. This method separates and quantifies amylose with accuracy. The third method uses concanavalin A to precipitate amylopectin and separate amylose for quantification. In our experience at the laboratory, iodine colorimetry and concanavalin A work for specific starch sources; for example, iodine colorimetry produces results in agreement with amylose content in high-amylose starch.

The structural characteristics of amylose have been reported by Wang et al. (2019). Some structural studies on amylose report the degree of polymerization (DP). Amylose in cereals as barley and maize presents DPs of 5580 and 2500, respectively (Takeda et al. 1988, 1999). The amylose structure of tapioca has been analyzed by size-exclusion chromatography (SEC), and its DP values range between

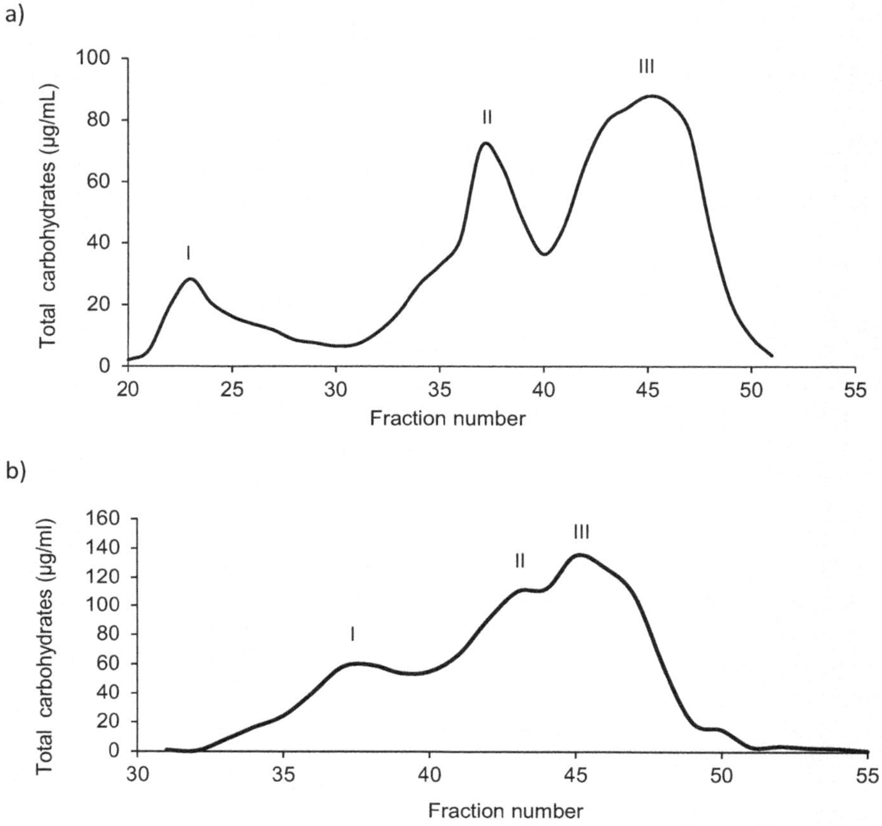

Fig. 5.1 Gel-permeation chromatography profile of (**a**) mango starch and (**b**) waxy maize starch. *I* amylose, *II* short chains of amylopectin, *III* long chains of amylopectin

580 and 22,400 (Shibanuma et al. 1994). According to reports, amylose is a mix of linear and branched molecules; the first account for 10–70%, while the second show a higher Mw (51, 52, 53). Amylose contains small but significant branching points with long chains. Two-dimensional correlation SEC (2D-CoSEC) is used to quantify the average number of branches in amylose as a function of Mw; for example, potato amylose shows two to four branches per molecule. The size independence of the average number of branches suggests that amylose molecules differ only in average chain length but not the number of branches. However, further analyses are necessary to reduce the probability that system artifacts be present. Based on these results, Wang et al. (2019) postulate that branching and chain elongation are two separated processes during amylose biosynthesis: Branching is carried out in early starch biosynthesis, and branches are elongated later on.

Recently, the nutritional importance of amylose in high-amylose starch has been reviewed due to its contribution to dietary fiber intake. Biotechnological approaches tailor the starch molecular structure to increase amylose content. It is well known that native high-amylose starch has reduced digestibility compared against regular starch. The molecular arrangement of starch components in granules with high amylose content is maintained even after thermal processing along with hydrolysis resistance by digestive enzymes.

However, we consider that resistance to hydrolysis by amylase depends not only on amylose but that amylose structure, chain length, branching, and Mw are also partially responsible for it.

5.2.2 Amylopectin

Amylopectin is a branched starch component which is more abundant than amylose in regular starches (~75%), and it constitutes ~99% of the starch granules in waxy varieties. Structurally, amylopectin is more complex than amylose, and its Mw is ~1×10^8 g/mol. It is responsible for viscosity in starch gels, and, because of its structure, it has been analyzed to identify relationships with physicochemical, rheological, and functional characteristics. Its structure has been studied to determine a structure-function relationship, and it is examined by debranching using enzymes as pullulanase and isoamylase to determine CLD. The following chains are proposed in amylopectin structure: (A) short chains with no branching points that are bound to other chains by the reducing end, (B) longer chains similarly attached but with one or more branches, and (C) single chain in each amylopectin molecule containing the reducing end and a greater number of branching points. The chains described are considered in the cluster model by Nikuni (1978) and French (1972) and have been refined by techniques as high-performance SEC (HPSEC). The study by Hizukuri (1986) reports that B chains can be subdivided into B2 (chain length of ~45) and B3 chains (chain length of ~70). The average DP in the major group of short chains was 11–15. More precise information on amylopectin CLD has been obtained using high-performance anion-exchange chromatography (HPAEC) with a pulsed amperometric detector (PAD).

Some years ago, the backbone model was implemented to explain the functionality of the complex amylopectin molecule. In it, amylopectin branches are apparently distributed within the clusters, and they are found in small groups known as building blocks (Bertoft 2007). The cluster model and the backbone model are presently accepted, but identifying which is correct would allow a better conceptualization of starch design, considering genetic, breeding, physical, enzymatic, and environmentally friendly chemical modifications for specific end uses (Hamaker 2021).

The average chain length of amylopectin from maize, wheat, rice, potato, and plantain is between 15 and 28. A relationship between CLD of amylopectin and crystalline arrangement has been proposed. Starches with A-type crystallinity (maize) possess a shorter chain length than B-type ones (potato), and this CLD is related to physicochemical, functional, and digestibility characteristics of starch. For example, amylopectin with a high content of short chains shows a lower tendency to retrogradation than amylopectin with long chains (Yuan et al. 1993). Amylopectin with long chains is more resistant to hydrolysis by digestive enzymes, while starches with a higher proportion of short chains are slowly hydrolyzed (Zhang et al. 2008b).

5.2.3 Size-Exclusion Chromatography

It is known that SEC can be divided in two methods known as gel-permeation chromatography (GPC) and HPSEC. The first has been more frequently used to debranch starch, and the second is commonly employed in branched and debranched samples. The SEC equipment used in branched starch is coupled with detectors (refractive index detector) that determine the concentration of the macromolecules. The multi-angle laser-light scattering (MALLS) and intrinsic viscosity determine the absolute Mw, gyration, and hydrodynamic radius. Significant advances in starch structure have been made after the development of those detectors that do not require the use of Mw standards.

5.2.4 Branched Starch

The study of branched and whole starch is key to elucidate the complete structure of amylose and amylopectin. It is also important to determine molecular characteristics related to starch structure as average Mw, Mw distribution, gyration, and hydrodynamic radius, which support starch properties in diverse applications. Starch solubilization is the first step to obtain more precise information on the molecular features of branched starch. Studies have reported solubilization methods as autoclaving, heating in boiling water bath, and microwaving using different solvents as water and DMSO. Fishman and Hoagland (1994) solubilized starch in water and heated it by microwaves in a pressure vessel with the objective to produce a complete starch solubilization before determining the structural features. Starch pretreatment and this technique showed that starches with different amylose/amylopectin ratio heated for 35 s exhibited a degree of solubilization of 87–96% without any significant degradation of the starch structure. The same technique has been used in yam starches, which were microwave heated for 38 s to obtain a solubilization recovery between 72% and 98%. The recovery in other botanical sources as maize, wheat, potato, and cassava was between 88% and 97% (Rolland-Sabaté et al. 2003). The solubilized starch was injected to an HPSEC system with TSK gel columns at 30 °C and coupled to MALLS and RI detectors. Maize starches with high amylose content showed lower Mw and gyration radius (2–4 and 3.5×10^7 g/mol, 163 and 177 nm) than waxy and regular maize (2.2 and 1.9×10^8 g/mol, 229 and 250 nm). Other botanical sources as potato and wheat present lower molecular characteristics than maize starches (5.4×10^7 g/mol, 149 nm and 7.1×10^7 g/mol,

178 nm, respectively) (Bello-Pérez et al. 1998a, b). The yam starches showed Mw between 1.88 and 3.27 × 10^8 g/mol and gyration radii of 258–396 nm (Rolland-Sabaté et al. 2003). The study of starches from different botanical sources applying the same chromatographic technique produces Mw values between 0.94 and 1.78 × 10^8 g/mol and gyration radii between 161 and 209 nm (Tetchi et al. 2007). The three different studies agree in that potato starch shows the lowest Mw (4.7 and 9.4 g/mol).

The use of different chromatography columns improves the separation of starch molecules. Table 5.1 shows column types, solvents, detectors, conditions, and molecular characteristics of starch components. Yoo and Jane (2002a, b) dispersed starches in DMSO based on X-ray diffraction patterns (A, B, and C types) and heated them in a boiling water bath for 1 h. They were precipitated with ethanol, and the pellet was redissolved in boiling water and heated for 30 min in a boiling water bath. Shodex columns (guard columns KB-G and KB-806 and analytical column KB-804) were used to separate starch components. Amylopectin of waxy starches showed a higher Mw than amylopectin of regular starches. Amylopectin of A-type starches showed a wider variation in Mw (0.7–56.8 × 10^8 g/mol) than that of B-type starches (1.7–3.4 × 108 g/mol), yet the gyration radii of A-type amylopectin were larger than those of B-type amylopectin with comparable Mw. A-type amylopectin showed highly compact spherical forms, and B-type presented differences in branch structures. This pioneer study of starch structure presents differences in the molecular features, depending on the crystalline arrangement of starch components in the granule. In our opinion, studies with solubilization procedure, HPSEC columns, and chromatographic conditions (flow rate, solvent, temperature) can help to bridge the gap between structural studies on branched starches.

GRAM columns have been used in recent years; the structural features of starch are more closely related to the physicochemical, functional, and digestibility properties of starch (Table 5.2). The amylose/amylopectin ratio can also be determined with this system. However, system choice (columns, detectors, solvent, temperature, etc.) should depend on the starch type (botanical source, amylose/amylopectin ratio) to find the true molecular characteristics of both starch components. It must be considered that there is no general chromatography system to structurally characterize starch. The use of series GRAM columns 30 and 3000 has been more frequently reported in structural studies of branched starches. The mobile phase is DMSO (99.5%) and LiBr (0.5%), and the columns are operated at 80 °C to decrease the shear stress of the solvent in the columns, which can produce depolymerization of starch components. As reported, the Mw of rice starch is lower than those in other columns (Table 5.1), but the Mw of potato amylose is within the range of amylose determined using other columns. The hydrodynamic radius of amylose is different in the botanical source and shows a high value of potato amylose. Still, contrasting values of waxy rice amylose (41.2 nm) and amylopectin (403.5 nm) are reported. The use of another GRAM column system (100 and 3000) does not show a different hydrodynamic radius than those reported with 30 and 3000.

Table 5.1 Molecular features of starch analyzed by high-performance size-exclusion chromatography

Sample	Solubilization method	Column	Detectors	AM/AP ratio	Mw amylose (g/mol) × 10^6	Rz amylose (nm)	Mw amylopectin (g/mol) × 10^8	Rz amylopectin (nm)	References
[a]Maize: Amylopectin Waxy Normal Hylon V Hylon VII	Microwave	TSK gel SWXL G4000–G300	MALLS-RI	NR	NR	NR	2.2 ± 2 2.0 ± 2 1.9 ± 3 0.35 ± 0.2 0.24 ± 0.2	229 ± 9 234 ± 9 250 ± 10 163 ± 5 177 ± 8	Bello-Pérez et al. (1998a)
Amaranth	Microwave	TSK gel SWXL G4000–G300	MALLS-RI	NR	NR	NR	2.7 ± 2	176 ± 12	Bello-Pérez et al. (1998b)
Waxy: Maize Potato Wheat – Normal: Maize Potato Wheat Rice Taro	Boiling bath water	SHODEX KB-806 and KB-804	MALLS-RI	NR	NR	NR	8.3 ± 0.2 2.0 ± 0.2 5.2 ± 0.4 – 4.9 ± 0.8 1.7 ± 0.2 3.1 ± 0.3 26.8 ± 2.9 12.6 ± 3.6	372 ± 11 344 ± 37 328 ± 6 – 312 ± 23 356 ± 36 302 ± 3 581 ± 41 560 ± 15	Yoo and Jane (2002a, b)
Maize (Hylon VII)	Boiling bath water	SHODEX KB-806 and KB-804	MALLS-RI	70:30	0.28	43	1.7	389 ± 57	Yoo and Jane (2002a, b)

Sample	Solubilization method	Column	Detectors	AM/AP ratio	Mw amylose (g/mol) × 10^6	Rz amylose (nm)	Mw amylopectin (g/mol) × 10^8	Rz amylopectin (nm)	References
Mandioca: M. esculenta Potato: S. tuberosum Wheat: T. aestivum Waxy maize: Zea maïs	Microwave	TSK gel SWXL G2000	MALLS-RI	17: 83 23: 77 26: 74 0: 100	26.17 ± 1.10 4.91 ± 0.40 4.97 ± 0.25	129 ± 3 64 ± 12 133 ± 4	1.72 ± 0.17 0.90 ± 0.08 1.17 ± 0.11 2.46 ± 0.50	169 ± 6 127 ± 10 149 ± 7 196 ± 16	Rolland-Sabaté et al. (2003)
Ginger: Zingiber officinale Rice normal: Oryza sativa wheat: Triticum aestivum	Microwave	TSK gel SWXL G2000	MALLS-RI	NR	NR	NR	1.46 ± 0.04 1.43 ± 0.04 2.54 ± 0.08	169 ± 2 165 ± 2 210 ± 3	Rolland-Sabaté et al. (2007)
Oats Wheat Durum	Boiling bath water	Waters Ultrahydrogel 6–13 µm, 7.8 × 300 mm	RI	20.5:79.5 23.25:76.75	0.168; 0.136	NR	0.136 0.0727	NR	Simsek et al. (2013)
Maize: Normal Hylon V Waxy	Microwave	SHODEX KB-807 and KB-804	MALLS-RI	25:75 55:45 0: 100	12 ± 0.1 16 ± 0.3	131 ± 1.4 97 ± 0.9	2.3 ± 0.1 1 ± 0.1 2.5 ± 0.2	229 ± 2.9 219 ± 3.6 232 ± 3.4	Bello-Perez et al. (2019)

Note: NR not reported, *Mw* molecular weight, *Rz* radii of gyration, *MALLS* multi-angle laser-light scattering, *RI* refractive index

[a]Apparent weight-average molar mass

Table 5.2 Structural studies of branched starch using HPSEC-GRAM columns

Column type[a]	Solvent and mobile phase	Detectors	Starch	AM/AP ratio	Mw (g/mol)	R_{hAM} (nm)	R_{hAP} (nm)	References
GRAM 30–3000 PSS	DMSO 99.5%/ LiBr 0.5%	MALLS-RI	Rice	27:73	7.49×10^7	31.62	300	Vilaplana and Gilbert (2011)
GRAM 30–3000 PSS	DMSO 99.5%/ LiBr 0.5%	RI	Waxy rice	1.4:98.6	NR	41.2 ± 2.2	403.5 ± 16.1	Li et al. (2016)
GRAM 30–3000 PSS	DMSO 99.5%/ LiBr 0.5%	RI	Maize waxy potato maize Hylon VII	NR	NR	– 21.74 ± 1.05 13.32 ± 0.12 10.40 ± 0.40	130.75 ± 0.36 94.13 ± 3.79123.23 ± 2.9111.32 ± 1.95	Martínez, Li et al. (2018)
GRAM 30–3000 PSS	DMSO 99.5%/ LiBr 0.5%	MALLS-RI	Amylose potato	NR	2.87×10^6	27.29	NR	Wang et al. (2019)
GRAM 100–3000 PSS	DMSO 99.5%/ LiBr 0.5%	RI	Rice chickpea	29:71 33:67	NR	10.0 ± 0.8 17.4 ± 0.4	89.9 ± 0.3 77.0 ± 0.6	Zhao et al. (2020)

R_{hAM} hydrodynamic radius amylose, R_{hAP} hydrodynamic radius amylopectin, *nm* nanometer, *Am* amylose, *Ap* amylopectin, *MALLS* multi-angle laser-light scattering, *RI* refractive index, *M.P.* mobile phase, *I.V.* injection volume, *O.T.* oven temperature, *NR* not reported, *Mw* average molecular weight

[a]GmbH, Mainz, Germany

5.2.5 Debranched Starch

Knowledge of debranched enzymes as isoamylase and pullulanase led to progress in the complex structure of amylopectin (Manners and Matheson 1981). Those studies were based on separating the linear chains of amylopectin bound to the main molecule by α-1–6 linkages and determined the DP to piece the puzzle of starch structure. The first studies used gel-permeation chromatography with a stationary phase (BioGel, Sephadex, Sephacryl, and Sepharose, among others) and different pore size to determine the average DP of each chain population. Two major linear chain populations of amylopectin were identified: One peak eluted at shorter time involving the long chains (B-), and the second peak is the short chains (A-) of amylopectin cluster model (Biliaderis et al. 1979, 1981; Jane and Chen 1992; Paredes-López et al. 1994).

In HPSEC, three columns are connected in sequence, and TSK G3000SW and G2000SW are used to separate linear chains after debranching (Table 5.3). Hizukuri (1986) describes a polymodal distribution: two peaks of short A and B1 chains and three peaks with long B2, B3, and B4 chains. Rice amylopectin shows the highest peak of A chains. However, the author reports that this polymodal distribution of amylopectin chains is consistent with the cluster model.

Amylopectin structure in waxy maize genotypes has been analyzed using HPSEC with a guard column (Pelliguard LC-18, Supelco), two 30-cm SEC columns (Zorbax PSM 60S, Du Pont), and a refractive index detector. A calibration curve has also been constructed following pullulan standards (Yuan et al. 1993).

Wheat starches with different amylose content (0.2, 21.5, 26.2, and 26.6% amylose) have been debranched using isoamylase and linear chains separated by HPSEC (Shodex OHpak SB-803 HQ) with an RI detector. The short chains (DP = 6–12) in the four wheat starches range between 24.2% and 27.2%, and the main fraction shows a DP of 13–24 (44.2% and 46.1%); around 20% of the total linear chains show DP ≥ 37. Regular wheat starches display extra-long chains (peak, ~DP 770) that are not found in waxy wheat starch nor starches from a different botanical origin (Yoo and Jane 2002a, b). A similar system is applied by Wang et al. using two Shodex OHpak columns (KB-802 and KB-804) for native (unmodified) and oxidized barley and corn starches. Three fractions are identified: Fraction I elutes at the shortest time and is assigned non-hydrolyzed amylose that maintains its molecular size, fraction II includes long B amylopectin chains, and fraction III shows short amylopectin chains (A). Oxidized starch shows a decrease in fraction I content since amylose is degraded to amylose molecules with smaller molar mass, yet fraction II increases along with oxidation (Chávez-Murillo et al. 2008).

Debranched starch has also been analyzed using GRAM columns as shown in Table 5.4. Studies report two fractions corresponding to short and long amylopectin chains and another to amylose DP. The CLD mainly depends on the length of amylose chain and the starch type. The DP of amylopectin short chains ranges between 14.4 and 19.4, while that of long chains is 39.6–47.5, depending on the botanical origin related to starch biosynthesis. Potato and Hylon VII present longer amylopectin chains than waxy and regular maize (DP ~39); potato amylose showed the

Table 5.3 Structural parameters of starches debranched by HPSEC

Starch	Column	System	FrI Mass fraction (%)	FrI DP_n	FrI DP_w	FrII Mass fraction (%)	FrII DP_n	FrII DP_w	FrIII Mass fraction (%)	FrIII DP_n	FrIII DP_w	References
Cassava	TSK-gel G5000–G400 PW	MALLS-RI	58.1 ± 2.4	NR	11.8 ± 2.2	24.9 ± 2.2	NR	24.6 ± 1.2	17.0 ± 2.2	NR	62.0 ± 0.2	Charles et al. (2005)
Normal maize	TSK-gel G3000–G2500PWXL	MALLS-RI	23.69 ± 0.14	900.0 ± 66.3	1913.0 ± 49.8	20.59 ± 0.12	50.5 ± 1.1	54.2 ± 1.0	55.71 ± 0.32	18.0 ± 0.2	19.6 ± 0.4	Lin et al. (2011)
Native (rice; maize; potato)	Shodex OHpak KB-804 (8 mm ID × 30 cm)	RI	15.1; 18.5; 18.6	NR	NR	19.6; 18.1; 28.6	NR	NR	65.2; 63.3; 52.8	NR	NR	Palma-Rodríguez et al. (2012)
Cassava	MZ-GEL SD Plus LS 5506–5606	MALLS-RI	19.30 ± 0.46;	1626.5 ± 116.5	3776.9 ± 152.7	24.03 ± 0.31;	38.4 ± 6.9	44.6 ± 9.4	56.67 ± 0.40;	15.2 ± 3.2	17.0 ± 3.8	Lin et al. (2023)

Note: FrI AM; FrII (long B2 and B3 chains of AP); FrIII (A chains and short B1 chains of AP); DP_w and DP_n, weight- and number-average degree of polymerization; *MALLS* multi-angle laser-light scattering, *RI* refractive index, *NR* not reported

Table 5.4 Molecular size distribution of debranched starches using HPSEC-GRAM

Column type	Solvent and mobile phase	Detectors	Conditions	Starch	AM/AP ratio	Zone AP1 Log (Rh/nm)$_{de}$/DP$_X$	Zone AP2 Log (Rh/nm)$_{de}$/DP$_X$	Zone AM3 Log (Rh/nm)$_{de}$/DP$_X$	References
GRAM 10–1000	DMSO 99.5%/LiBr 0.5%	MALLS-RI	C = 1–2 mg/mL I.V. = 0.1 mL flow: 0.6 mL/min O.T. = 80 °C	Rice	27:73	16	76	7200	Vilaplana and Gilbert (2010)
GRAM 100–1000	DMSO 99.5%/LiBr 0.5%	RI	C = 4 mg/mL I.V. = NR flow: 0.6 mL/min O.T. = 80 °C	Rice	13.34:86.66	20.6 ± 0.1	43.3 ± 0.3	738.0 ± 63.8	Li et al. (2016)
GRAM 100–1000	DMSO 99.5%/LiBr 0.5%	RI	C = 8 mg/mL I.V. = 0.1 mL flow: 0.6 mL/min O.T. = 80 °C	Maize waxy, potato, maize, maize Hylon VII	NR	14.4 ± 0.4; 18.5 ± 0.0; 14.6 ± 0.2; 19.4 ± 2.2	39.6 ± 0.7; 43.8 ± 1.4; 39.8 ± 0.9; 47.5 ± 0.7	NR; 1560.9 ± 694.5; 839.2 ± 95.0; 466.5 ± 268.4	Martínez et al. (2018)
GRAM 100–1000	DMSO 99.5%/LiBr 0.5%	RI	C = 4 mg/mL I.V. = 0.1 mL flow: 0.6 mL/min O.T. = 80 °C	Normal maize, Hylon VIII maize, potato, tapioca, wheat	18.8: 81.2; 83:17; 30.5:69.5; 20.7:79.3; 24:76	23.0 ± 0.4; 63.8 ± 02; 32.9 ± 0.3; 25.9 ± 0.2; 23.8 ± 0.3	NR	1265.2 ± 23.5; 1072.9 ± 5.6; 11055.4 ± 2801.2; 2576.1 ± 58.3; 1848.7 ± 30.6	Cao et al. (2022)
GRAM 100–1000	DMSO 99.5%/LiBr 0.5%	RI	C = NR I.V. = NR Flow: 0.6 mL/min O.T. = 70 °C	Tef (white, brown)	34.5: 65.5; 35.8: 64.2	11.5; 11.7	33.5; 32.2	1087; 1170	Vela et al. (2023)

Note: MALLS multi-angle laser-light scattering, *RI* refractive index, *NR* not reported longest DP (1560) as compared to regular and Hylon VII maize starches (839 and 466, respectively) (Martinez et al. 2018). The structural features of starch play an important role in the physicochemical, functional, and digestibility characteristics of the polysaccharide.

Tef flour is treated with ultrasound at different times to analyze the impact on starch structure. The DP of amylopectin chains after debranching tef starch has been reported. The control samples show similar DPs in both amylopectin fractions, while the average DP of short chains is 11.2 and 11.5, that of long chains is 32.7 and 33.9, and amylose presents a DP of 996 and 1090. The ultrasound treatment of starch does not produce significant changes in DP (Vela et al. 2023).

5.3 Asymmetric Flow Field-Flow Fractionation

Asymmetric flow field-flow fractionation (AF4) is a technique to separate macromolecules (Table 5.5) where a channel coupled to MALLS and refractive index detectors is suggested for the routine molecular characterization of starch (Chiaramonte et al. 2012). The technique is conducted in a trapezoidal channel with a permeable membrane of regenerated cellulose (Mw cut-off of 10 kDa). The sequence characteristics for sample elution into the AF4 channel should be determined according to variables as detector flow, focus flow, cross flow, and inject and elution time. These have been analyzed with the profile of macromolecules separated in the channel (Bello-Pérez et al. 2017). Waxy starches and amylopectin isolated from regular starches have been examined through AF4 coupled with MALLS, obtaining high mass recoveries (81–100%) and molar mass ranging between 1.05×10^8 (amylose-free potato variety) and 3.18×10^8 g/mol (waxy maize starch). The low values of molar mass and gyration radius can be related with a lower starch solubilization (Rolland-Sabaté et al. 2007). The effect of amylopectin dissolution has been analyzed to identify conditions where the biopolymer does not degrade; amylopectin is characterized afterward using an AF4-MALLS-RI system. The solvent used in the dissolution is key since the water at 140 °C and DMSO at 100 °C is used without any visible degradation of the biopolymer. Amylopectin dissolved in water shows a molar mass of 2.17×10^8 g/mol and gyration radii of 233 nm, while that in DMSO presents a molar mass of 2.5×10^8 g/mol and gyration radii of 238 nm. The authors conclude that poor reproducible results are obtained when the sample is solubilized in microwaves (Perez-Rea et al. 2015). The molar mass and gyration radius of both starch components are analyzed in maize starches with different amylose/amylopectin ratios (Chiaramonte et al. 2012). The molar mass of amylose of 2.0×10^5 g/mol and 4.2×10^5 g/mol increases along with the amylopectin content in the starch. The molar mass of amylopectin is between 0.9×10^8 g/mol in starch with the highest amylose content (Hylon VII) and 4.1×10^8 g/mol in regular maize starch, which is higher than that in amylopectin from waxy maize (3.2×10^8 g/mol). The gyration radii of amylopectin range between 145 and 316 nm. The authors conclude that the system used for starch characterization is a practical analytical platform to

Table 5.5 Structural parameters of branched starches by asymmetric flow field-flow fractionation (AF4)

Sample	Solubilization method	Detectors	AM/AP ratio	Mw amylose (g/mol) × 10^6	Rz amylose (nm)	Mw amylopectin (g/mol) × 10^8	Rz amylopectin (nm)	References
Normal maize amylopectin	DMSO-microwave	MALLS-RI	0:100	NR	NR	2.42 ± 0.10	207.0 ± 2.1	Rolland-Sabaté et al. (2007)
Normal maize	Microwave	MALLS-RI	32:68	1.5	54	0.42	140	Juna and Huber (2012)
Maize Hylon (V, VII)	DMSO-microwave	MALLS-RI	55:45; 75:25	0.3672; 0.2040	NR	2.51; 0.943	251.1; 94.3	Chiaramonte et al. (2012)
Waxy maize	Autoclave heating	MALLS-RI	0:100	NR	NR	2.17 ± 13	233 ± 3	Pérez-Rea et al. (2015)
Taro Colocasia esculenta var. Esculenta; Malanga: Xanthosoma	DMSO-microwave	MALLS-RI	7.8:92.2 10.2:89.8	25 ± 0.1 27 ± 0.1	253 261	8.0 ± 0.1 13 ± 0.2	354 311	Hoyos-Leyva et al. (2017)
Normal maize Potato Barley Waxy; Hylon V: (maize)	DMSO-microwave	MALLS-RI	32:68 24:76 26:74 0:100; 45:55	0.27 0.64 0.37 NR; 0.58	154 194 156 NR; 160	1.2 2.4 2.0 6.7; 2.1	211 251 239 331; 216	Bello-Pérez et al. (2017)
Chickpea	DMSO-microwave	MALLS-RI	23.3:76.7	0.28 ± 2.5	165 ± 0.6	2.94 ± 0.9	260 ± 1.8	Yniestra-Marure et al. (2019)

Notes: *NR* not reported, *Mw* molecular weight, *Rz* radii of gyration, *nm* nanometers, *MALLS* multi-angle laser-light scattering, *RI* refractive index, *Am* amylose, *Ap* amylopectin identify the structure-function relationship in starch. When comparing separation techniques of starch components, HPSEC-MALLS-RI and AF4-MALLS-RI are used in starches with different amylose/amylopectin ratios solubilized under similar conditions. The structural features of starch (molar mass, gyration radius, and apparent particle density) in regular maize starch are superior when using HPSEC-MALLS-RI than AF4-MALLS-RI but lower for waxy Hylon V starch. The structural parameters obtained with HPSEC-MALLS-RI are within the values reported in the recent literature (Bello-Perez et al. 2019). More comparative studies between both systems are necessary to obtain more precise results on the structural characteristics of starch.

5.4 High-Performance Anion-Exchange Chromatography (HPAEC) and Fluorophore-Assisted Carbohydrate Electrophoresis (FACE)

Debranched starch has been analyzed by HPAEC and FACE to determine the CLD of amylopectin. The first method separates the individual amylopectin chains based on length (Fig. 5.2): Glucose elutes first, followed by maltose and trioses. It is worth noting that HPAEC cannot distinguish between isomers with a similar number of monomers. The early studies to determine CLD used one of the two debranched enzymes (pullulanase and isoamylase), although isoamylase and a mix of the two debranched enzymes (pullulanase and isoamylase) have been recently proposed to complete amylopectin debranching (Table 5.6). The HPAEC profiles detected in chains with DP of 85 lead to grouping those chains in four populations: DP 6–12, DP 13–24, DP 25–36, and DP $\geq$ 37 (McPherson and Jane 1999). This classification is based on the cluster model where chains with DP 6–12 are classified as "A chains (short chains)," those with DP 13–24 are named B1 chains, those with DP 25–36 are

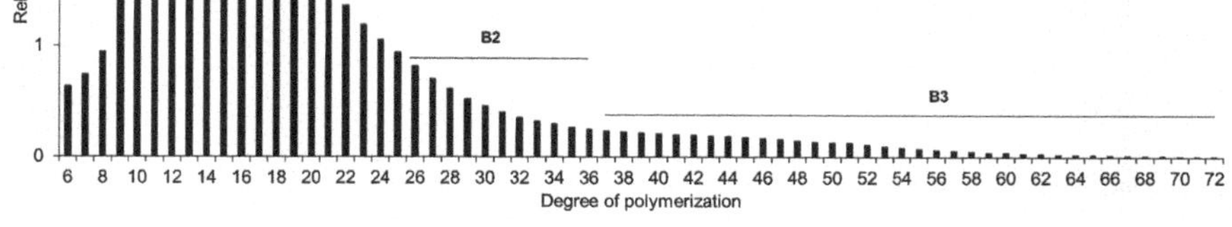

Fig. 5.2 Chain-length distribution of amylopectin of waxy maize starch analyzed by high-performance anion-exchange chromatography and pulsed amperometric detector. A, short chains; B1, B2, and B3 long chains

Table 5.6 Chain-length distribution of isoamylase debranched amylopectin analyzed by high-performance anion-exchange chromatography (HPAEC) with pulsed amperometric detector (PAD) and fluorophore-assisted capillary electrophoresis (FACE) HPAEC-PAD y FACE

Starch	Average chain length (DP)	% Chain-length distribution				References
		A chain (DP 6–12)	B1 chain (DP 13–24)	B2 chain (DP 25–36)	B31 chain (DP ≥ 37)	
Waxy rice	18.9	29	50	11	9	Hanashiro et al. (1996)[a]
Wheat	20.1	27	49	14	10	
Potato	23.6	18	48	15	18	
Waxy wheat	20.4 ± 0.2	21.5 ± 0.2	45.0 ± 0.2	14.8 ± 0.0	18.7 ± 0.7	Yoo and Jane (2002a, b)[a]
Barley	19.86 ± 0.0	27.03 ± 0.8	48.34 ± 0.5	16.04 ± 0.1	8.56 ± 0.5	Chávez-Murillo et al. (2008)[a]
Corn	20.48 ± 0.3	23.55 ± 0.9	50.81 ± 0.6	15.49 ± 0.1	10.14 ± 0.4	
Waxy starch	NR	43.6	46.9	7.9	1.6	Blazek et al. (2009)[b]
High-amylose maize (HM) native	26.9 ± 0.6	18.5 ± 0.4	45.3 ± 0.6	14.4 ± 0.1	21.8 ± 1.0	Lin et al. (2016)[b]
Native potato	NR	23.5 ± 0.3	51.2 ± 0.3	14.4 ± 0.2	9.2 ± 0.3	Na et al. (2020)[a]

DP degree of polymerization
[a]HPAEC-PAD
[b]FACE

B2 chains, and B3+ chains show DP ≥ 37, as was suggested by Hizukuri (1986). The HPAEC method has been used in oxidized maize and barley starch, finding increased A and B1 chains in modified starches and a reduced concentration of long chains (Chávez-Murillo et al. 2008). Additionally, HPAEC is combined with gel-permeation chromatography to analyze CLD. The fractions corresponding to short and long debranched amylopectin chains are pooled and injected into the HPAEC system. Three separate groups are identified: A (short chains), BS (short B chains), and BL (long B chains), averaging DPs of 16, 26, and 45, respectively (Hoyos-Leyva et al. 2017). The combination of the two chromatographic methods is complementary and can be used for a more precise characterization of the starch structure.

The CLD of amylopectin can be analyzed through capillary electrophoresis, a method used to separate oligosaccharides (Morell et al. 1998) and adapted to identify the CLD of amylopectin. The amylopectin chains are separated individually and grouped as in HPAEC. The highest detectable DP ranges between 80 and 89, depending on the starch source (Espinosa-Solís et al. 2009), and the results are comparable with those from HPAEC.

5.5 Conclusions and Future Trends

The chromatographic methods used in the structural characterization of starch can yield more accurate information if the solubilization method, instruments (columns, detectors, AF4), and conditions (solvent, temperature, flow rate) are appropriately chosen, without breaking the structure and obtaining an efficient separation. The methods for the structural characterization of starch are complementary, and more information can be obtained when they are combined. This information is key to identify the structure-function and structure-digestibility relationships and determine the more appropriate starch for specific end uses and the design of tailor-made starch with a structure that produces specific functionality and digestibility.

The advances in instruments and analytical methods, as columns, can support the development of techniques for the structural characterization of starch. Still, the best conditions to separate starch components and their chains remain to be found. Further works are necessary to create a method that provides more precise information on the structural parameters of starch to explain its functionality and digestibility.

References

Bello-Pérez LA, Colonna P, Roger P, Paredes-López O (1998a) Macromolecular features of amaranth starch. Cereal Chem J 75(4):395–402

Bello-Pérez LA, Roger P, Baud B, Colonna P (1998b) Macromolecular features of starches determined by aqueous high-performance size exclusion chromatography. J Cereal Sci 27(3):267–278

Bello-Pérez LA, Rodriguez-Ambriz SL, Lozano-Grande MA (2017) Molecular characterization of starches by AF4-MALS-RI: an alternative procedure. J Cereal Sci 75:132–134

Bello-Perez LA, Agama-Acevedo E, Lopez-Silva M, Alvarez-Ramirez J (2019) Molecular characterization of corn starches by HPSEC-MALS-RI: a comparison with AF4-MALS-RI system. Food Hydrocoll 96:373–376
Bertoft E (2007) Composition of building blocks in clusters from potato amylopectin. Carbohydr Polym 70(1):123–136
Biliaderis CG, Grant DR, Vose JR (1979) Molecular weight distribution of legume starches by gel chromatography. Cereal Chem 56:475–480
Biliaderis CG, Grant DR, Vose JR (1981) Structural characterization of legume starches. I. Studies on amylose, amylopectin and beta-limit dextrins. Cereal Chem 58:496–502
Blazek J, Salman H, López-Rubio A, Gilbert E, Hanley T, Copeland L (2009) Structural characterization of wheat starch granules differing in amylose content and functional characteristics. Carbohydr Polym 75:705–711
Cao P, Wu G, Yao Z, Wang Z, Li E, Yu S, Li S (2022) Effects of amylose and amylopectin molecular structures on starch electrospinning. Carbohydr Polym 296:119959
Charles AL, Chang YH, Ko WC, Sriroth K, Huang TC (2005) Influence of amylopectin structure and amylose content on the gelling properties of five cultivars of cassava starches. J Agric Food Chem 53(7):2717–2725
Chávez-Murillo CE, Wang YJ, Bello-Pérez LA (2008) Morphological, physicochemical and structural characteristics of oxidized barley and corn starches. Starch/Stärke 60:634–645
Chiaramonte E, Rhazi L, Aussenac T, White DR (2012) Amylose and amylopectin in starch by asymmetric flow field-flow fractionation with multi-angle light scattering and refractive index detection (AF4–MALS–RI). J Cereal Sci 56(2):457–463
Espinosa-Solís V, Jane JL, Bello-Pérez LA (2009) Physicochemical characteristics of starches from unripe fruits of mango and banana. Starch/Stärke 61:291–299
Fishman ML, Hoagland PD (1994) Characterisation of starches dissolved in water by microwave heating in a high pressure vessel. Carbohydr Polym 23:175–183
French D (1972) Fine structure of starch and its relationship to the organization of starch granules. J Jpn Soc Starch Sci 19:8–25
Gilbert RG, Gidley MJ, Hill S, Kilz P, Rolland-Sabaté A, Stevenson DG, Cave RA (2010) Characterizing the size and molecular weight distribution of starch: why it is important and why it is hard. Cereal Foods World 55(3):139
Gutiérrez TJ, Tovar J (2021) Update of the concept of type 5 resistant starch (RS5): self-assembled starch V-type complexes V. Trends Food Sci Technol 109:711–724
Hamaker BR (2021) Current and future challenges in starch research. Curr Opin Food Sci 40:46–50
Hanashiro I, Abe JI, Hizukuri S (1996) A periodic distribution of the chain length of amylopectin as revealed by high-performance anion-exchange chromatography. Carbohydr Res 283:151–159
Hizukuri S (1986) Polymodal distribution of the chain lengths of amylopectins, and its significance. Carbohydr Res 147(2):342–347
Hoyos-Leyva JD, Bello-Pérez LA, Alvarez-Ramirez J, Agama-Acevedo E (2017) Structural characterization of aroid starches by means of chromatographic techniques. Food Hydrocoll 69:97–102
Jane J, Chen JF (1992) Effects of amylose molecular size and amylopectin branch chain length on paste properties of starch. Cereal Chem 69:60–65
Juna S, Huber A (2012) Characterisation of normal corn starch using asymmetrical flow field-flow fractionation. Starch-Stärke 64:18–26
Li H, Prakash S, Nicholson TM, Fitzgerald MA, Gilbert RG (2016) The importance of amylose and amylopectin fine structure for textural properties of cooked rice grains. Food Chem 196:702–711
Li H, Yu W, Dhital S, Gidley MJ, Gilbert RG (2019) Starch branching enzymes contributing to amylose and amylopectin fine structure in wheat. Carbohydr Polym 224:115185
Lin AHM, Chang YH, Chou WB, Lu TJ (2011) Interference prevention in size-exclusion chromatographic analysis of debranched starch glucans by aqueous system. J Agric Food Chem 59(11):5890–5898

Lin L, Cai C, Gilbert RG, Li E, Wang J, Wei C (2016) Relationships between amylopectin molecular structures and functional properties of different-sized fractions of normal and high-amylose maize starches. Food Hydrocoll 52:359–368
Lin HT, Yeh AI, Ho THD, Pan MH, Chen SH (2023) Molecular weight, chain length distribution and long-term retrogradation of cassava starch modified by amylomaltase. Food Hydrocoll 134:108027
Manners DJ, Matheson NK (1981) The fine structure of amylopectin. Carbohydr Res 90:99–110
Martinez MM, Li C, Okoniewska M, Mukherjee I, Vellucci D, Hamaker BR (2018) Slowly digestible starch in fully gelatinized material is structurally driven by molecular size and A and B1 chain lengths. Carbohydr Polym 197:531–539
McPherson AE, Jane JL (1999) Comparison of waxy potato with other root and tuber starches. Carbohydr Polym 40(1):57–70
Miao M, Hamaker BR (2021) Food matrix effects for modulating starch bioavailability. Annu Rev Food Sci Technol 12:169–191
Miao M, Jiang B, Cui SW, Zhang T, Jin Z (2015) Slowly digestible starch—a review. Crit Rev Food Sci Nutr 55(12):1642–1657
Morell MK, Samuel MS, O'Shea MG (1998) Analysis of starch structure using fluorophore-assisted carbohydrate electrophoresis. Electrophoresis 19(15):2603–2611
Na JH, Kim HK, Lee JS, Park HJ, Moon TW, Lee CJ (2020) Structural characteristics of low-digestible sweet potato starch prepared by heat-moisture treatment. Int J Biol Macromol 151:1049–1057
Nakamura Y, Kainuma K (2021) On the cluster structure of amylopectin. Plant Mol Biol 1:1–16
Nakamura Y, Kubo A, Ono M, Yashiro K, Matsuba G, Wang Y, Kainuma K (2022) Changes in fine structure of amylopectin and internal structures of starch granules in developing endosperms and culms caused by starch branching enzyme mutations of japonica rice. Plant Mol Biol 108(4–5):481–496
Nikuni Z (1978) Studies on starch granules. Starch/Stärke 30:105–111
Obadi M, Qi Y, Xu B (2022) High-amylose maize starch: structure, properties, modifications and industrial applications. Carbohydr Polym 1:120185
Palma-Rodriguez H, Agama-Acevedo E, Mendez-Montealvo G, Gonzalez-Soto RA, Vernon-Carter EJ, Bello-Pérez LA (2012) Effect of acid treatment on the physicochemical and structural characteristics of starches from different botanical sources. Starch/Stärke 64:115–125
Paredes-López O, Bello-Pérez LA, López MG (1994) Amylopectin: structural, gelatinisation and retrogradation studies. Food Chem 50:411–417
Perez-Rea D, Bergenståhl B, Nilsson L (2015) Development and evaluation of methods for starch dissolution using asymmetrical flow field-flow fractionation. Part I: dissolution of amylopectin. Anal Bioanal Chem 407:4315–4326
Rolland-Sabaté A, Georges-Amani NG, Dufour D, Guilois S, Colonna P (2003) Macromolecular characteristics of ten yam (*Dioscorea spp*) starches. J Agric Food Chem 83:927–936
Rolland-Sabaté A, Colonna P, Mendez-Montealvo MG, Planchot V (2007) Branching features of amylopectins and glycogen determined by asymmetrical flow field flow fractionation coupled with multiangle laser light scattering. Biomacromolecules 8(8):2520–2532
Romero-Hernández HA, Gutiérrez TJ, Bello-Pérez LA (2022) Can starch-polyphenol V-type complexes be considered as resistant starch? Food Hydrocoll 124:107226
Shibanuma K, Takeda Y, Hizukuri S, Shibata S (1994) Molecular structures of some wheat starches. Carbohydr Polym 25:111–116
Simsek S, Whitney K, Ohm JB (2013) Analysis of cereal starches by high-performance size exclusion chromatography. Food Anal Methods 6:181–190
Takeda Y, Shitaozono T, Hizukuri S (1988) Molecular structure of corn starch. Starch/Stärke 40:51–54
Takeda Y, Takeda C, Mizukami H, Hanashiro I (1999) Structures of large, medium and small starch granules of barley grain. Carbohydr Polym 38:109–114
Tetchi FA, Rolland-Sabaté A, Amani GNG, Colonna P (2007) Molecular and physicochemical characterisation of starches from yam, cocoyam, cassava, sweet potato and ginger produced in the Ivory coast. J Sci Food Agric 87(10):1906–1916

Vela AJ, Villanueva M, Li C, Hamaker BR, Ronda F (2023) Ultrasound treatments of tef [*Eragrostis tef* (Zucc.) trotter] flour rupture starch α-(1,4) bonds and fragment amylose with modification of gelatinization properties. LWT 174:114463

Vilaplana F, Gilbert RG (2010) Two-dimensional size/branch length distributions of a branched polymer. Macromolecules 43(17):7321–7329

Vilaplana F, Gilbert RG (2011) Analytical methodology for multidimensional size/branch-length distributions for branched glucose polymers using off-line 2-dimensional size-exclusion chromatography and enzymatic treatment. J Chromatogr A 1218(28):4434–4444

Wang K, Vilaplana F, Wu A, Hasjim J, Gilbert RG (2019) The size dependence of the average number of branches in amylose. Carbohydr Polym 223:115134

Yniestra-Marure LM, Núñez-Santiago MC, Agama-Acevedo E, Bello-Perez LA (2019) Starch characterization of improved chickpea varieties grown in Mexico. Starch 71(3–4):1800139

Yoo SH, Jane JL (2002a) Molecular weights and gyration radii of amylopectins determined by high-performance size-exclusion chromatography equipped with multi-angle laser-light scattering and refractive index detectors. Carbohydr Polym 49:307–314

Yoo SH, Jane JL (2002b) Structural and physical characteristics of waxy and other wheat starches. Carbohydr Polym 49(3):297–305

Yuan RC, Thompson DB, Boyer CD (1993) Fine structure of amylopectin in relation to gelatinization and retrogradation behavior of maize starches from three wx containing genotypes in two inbred lines. Cereal Chem 70:81–89

Zhang G, Ao Z, Hamaker BR (2008a) Nutritional property of endosperm starches from maize mutants: a parabolic relationship between slowly digestible starch and amylopectin fine structure. J Agric Food Chem 56(12):4686–4694

Zhang G, Sofyan M, Hamaker BR (2008b) Slowly digestible state of starch: mechanism of slow digestion property of gelatinized maize starch. J Agric Food Chem 56(12):4695–4702

Zhao Y, Tan X, Wu G, Gilbert RG (2020) Using molecular fine structure to identify optimal methods of extracting starch. Starch-Stärke 72:1900214

6 Revealing Ultrastructural Analysis of Starch Using Scanning Electron Microscopy

Anusha, Baishali Das, and Nirmal Mazumder

Abstract

Starch serves as the main source of energy and carbohydrates in human nutrition. It has a big part in plant foods and offers a lot of desirable technological qualities, notably those connected to nutrition and texturizing abilities. Starch granules are composed of 20–30% of amylose and 70–80% of amylopectin. The ratio of amylose to amylopectin and their molecular weight distributions are two factors that affect many of the physicochemical characteristics of starches that determine their appropriateness for different end uses. The microstructure of starch granules cannot be fully understood by optical microscopy. Better resolution photos are produced using SEM. With the aid of SEM images, the surface morphology of starch granules may be more easily observed, and the size shape of the granules can also be determined. SEM makes it possible to see the hydrolyzed starch granules' breakdown pattern clearly. A scanning electron microscope (SEM) is employed to examine the microstructure of starch granules generated through the action of α-amylase, offering valuable insights into the enzymatic degradation patterns. This review primarily centers on providing comprehensive information and characterizing starch derived from various sources, including corn, rice, maize, and millet.

Keywords

Starch · Rice · Corn · Millet · Potato · Scanning electron microscope

Anusha · B. Das · N. Mazumder (✉)
Department of Biophysics, Manipal School of Life Sciences, Manipal Academy of Higher Education, Manipal, Karnataka, India
e-mail: nirmal.mazumder@manipal.edu

N. Mazumder, M. H. Rahman (eds.), *Advanced Research in Starch*,
https://doi.org/10.1007/978-981-99-9527-1_6

6.1 Introduction

The primary source of energy and carbohydrate in human nutrition is starch. It plays a significant role in plant foods and offers a number of attractive technological properties, especially those related to its texturizing ability (Lehmann and Robin 2007). It frequently appears in leaves, unripe fruits, unripe seeds, roots, and tubers. Starch is a polysaccharide, and the native form of starch is in the form of granules. The polysaccharide is chemically composed of amylose and amylopectin which are polymers of glucose. Amylopectin is highly branched, comprising chains of D-glucose units linked by both one to four and one to six linkages. Amylose is predominantly linear and has α-1-4-linked glucosyl units, and it can have ten branches or more. The semicrystalline structure of amylose and amylopectin has amorphous and crystalline zones. The crystalline zones in starch granules are caused due to the linear parts of the amylopectin chains (Magallanes-Cruz et al. 2017).

The primary enzymes responsible for starch synthesis are starch synthase and starch-branching enzyme (Božić et al. 2017). The starch-branching enzyme is responsible for rupturing the linear of (one to four) linkages and constructing (one to six) linkages, whereas starch synthase is responsible for the addition of glucose molecules and the production of (one to four) linkages as shown in Fig. 6.1. In turn, these processes cause the synthesis of amylose and amylopectin. Typically, the amylopectin content of starch molecules ranges from 70% to 80% to 20% to 30%, while the amylose level is much lower (Kowsik and Mazumder 2018).

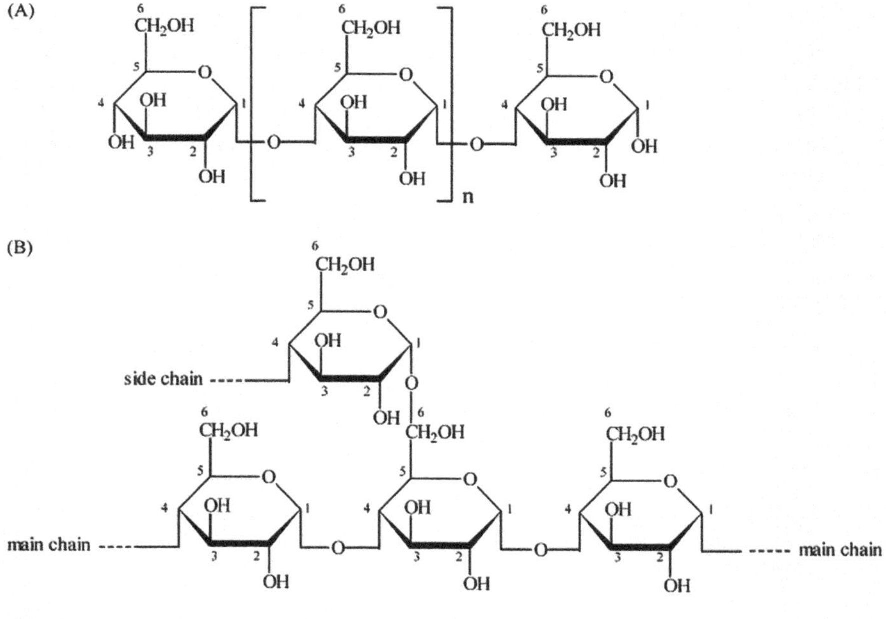

Fig. 6.1 Two components of starch—amylose (**a**) and amylopectin (**b**) showing glycosidic linkages. (The figure is reproduced from Herrero-Martínez et al. (2004))

The size of the starch granules ranges between 2 μm and 100 μm, which are found to have different shapes like polygonal, oval, and spherical (Dhital et al. 2010). When observed under optical microscopy, starch granules consist of semicrystalline and amorphous growth rings that originate from helium. These rings are made of amylopectin and amylose (Buléon et al. 1998). According to Božić et al. (2017), amylose makes up 20–30% of the starch granules, while amylopectin makes up 70–80%. The suitability of starches for different applications is determined by their physicochemical characteristics, which are heavily influenced by the balance between amylose and amylopectin content as well as their distribution of molecular weights. More specifically, the ratio of these two polysaccharides significantly impacts attributes like how easily they dissolve, their thickness or viscosity, the ability to form gels, and the temperature at which they turn into a gel-like state (gelatinization or solubilization). Consequently, the amount of amylose and the ratio of amylose to amylopectin are crucial factors when it comes to the quality of starches during their processing (Herrero-Martínez et al. 2004).

Hydrolases work on starch granules to break them down or degrade them. These enzymes can be isolated from a number of sources and are employed in industry. The rate at which the enzymes hydrolyze the starch granules enables humans to determine the type of food that is best to eat and its nutritional value (Kowsik and Mazumder 2018). The nutritional value of starch is markedly influenced by its processing and its state or condition. The two main physiological characteristics of starch are the rate of digestion and the release of glucose as a source of energy for the body. It is possible for the digestibility of a substance to alter in the human small intestine from being easily absorbed, as is the case with starch hydrolysis products, to becoming indigestible, as is the case with resistant starch. The amylose/amylopectin ratio of starch was shown to be one of the primary factors controlling both its physiological response and its ability to be digested. Figure 6.2 shows the structural changes that occur in modified starch. Native maize starch, occurring naturally in plants, undergoes a slow digestion process, providing the advantage of postponing the elevation of blood sugar levels following meals and sustaining them at a steady rate over an extended duration. This differs from rapidly digestible starch, which is quickly metabolized within the body (Lehmann and Robin 2007). Conversion of starch into by-products such as maltose, dextrin, glucose, etc., is majorly done by α-amylase. α-Amylases are produced by various microorganisms (El-Fallal et al. 2012).

Unmodified starches, which are naturally found in crops like maize, rice, potatoes, and wheat, do not possess all the desirable physicochemical characteristics due to their vulnerability to shear forces, acidic conditions, and a tendency to retrograde (Fashi et al. 2023). These unmodified starches exhibit suboptimal processing attributes, including poor solubility in cold water, a lack of emulsifying capabilities, and resistance to mechanical manipulation. However, their suitability for a broader range of industrial applications, including those in the food, plastic, textile, and various other industries, can be enhanced through modifications (Wang et al. 2022).

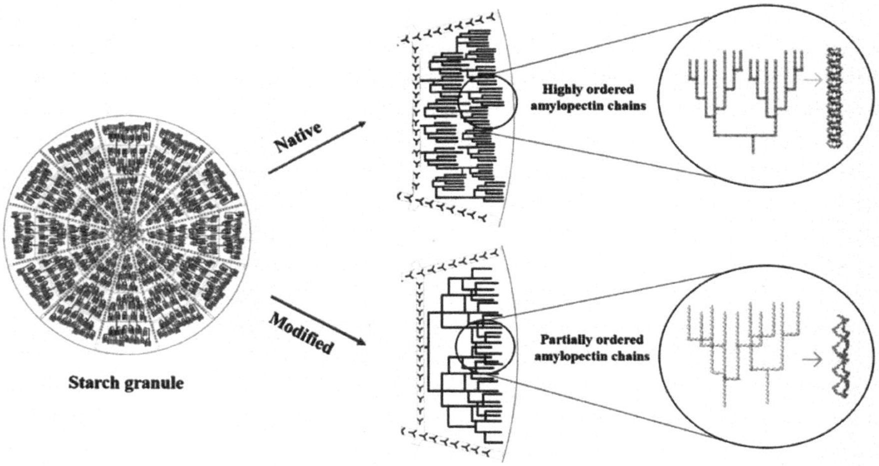

Fig. 6.2 Native and modified starch structure. (The figure is reproduced from Magallanes-Cruz et al. (2017))

6.2 Sources of Starch

6.2.1 Rice Starch

Among the cereal starches, rice starch stands out as a frequently encountered variety that possesses unique and innovative functional attributes. Amylose concentration in rice starches ranges from 40% to 80%. Rice starch is nonallergenic, tasteless, white, and smooth in texture and has no discernible flavor. Rice consists of 6–8% protein, 2% fats, and 73–75% carbs. Compared to other cereal grains, rice has the most carbohydrates. In water, salt, ethanol, and alkali, rice protein dissolves into 5% albumin, 12% globulin, 3% prolamin, and 80% glutelin. Since these proteins are soluble in an alkaline solution, rice starch has been isolated using an alkaline approach utilizing 0.5% NaOH (Wani et al. 2012). The food industry accounts for the majority of the industrial uses for rice starch. It can also be used to make bioethanol, a biofuel that can replace petroleum, as well as products for the laundry industry, which stiffens clothes, as a cosmetic dusting powder, as custard powder, and as baby food in some European nations. Rice starch can also be used to make rice wine, also known as yellow wine, which is well known in China, Japan, and other Asian nations (Akoh et al. 2008). Microscopy and particle size studies reveal that the rice starch granules are polyhedral and irregular in shape, with sizes ranging from 2 to 7 μm. Non-waxy and long-grain rice starches exhibit a more significant range of granule size variation when compared to waxy rice starches. In comparison to starches from different cereals, rice starch granules are generally smaller, and waxy and long-grain rice starches have amylose content spanning from nearly zero to approximately 35% (Wani et al. 2012).

6.2.2 Potato Starch

In general, potatoes are a great source of starch, which helps to give many foods their texture. In culinary and other industrial uses, potato starch can be used to thicken and stabilize colloidal suspensions. The distinctive composition and structure of potato starch sets it apart from cereal starches, resulting in variations in how they function. Potato starch granules typically range in size, with tiny granules falling within the 1–20 μm range and larger granules measuring between 20 and 110 μm. The amylose concentration of giant potato starch granules is higher than that of small granules (Singh et al. 2009). B-type crystalline structure can be seen in potato starch (Wang et al. 2022).

6.2.3 Maize Starch

Maize, a widely grown crop, finds its primary applications in the culinary, chemical, paper, and textile sectors. Typically, maize starch exhibits an A-type crystalline structure and contains approximately 25% amylose. When distinguishing between A- and B-type starches, their diverse chemical structures and crystal orientations are the key factors. A-type starch is characterized by a denser crystal structure, while B-type starch has a higher ratio of long-chain to short-chain amylopectin and contains more amylose (Wang et al. 2022). This distinction in crystal structure leads to a larger crystalline region in A-type starch and reduced digestibility. Due to its functional properties, which include high viscosity, a propensity for syneresis and retrogradation, and limited water solubility, maize starch cannot be used in food or medication. Several changes, including sonication, electric field, cross-linking, etherification, esterification, oxidation, and enzymatic, genetic, and pre-gelatinization genetic approaches, have been developed to address these shortcomings (Fashi et al. 2023).

6.2.4 Millet Starch

In addition to their nutritional advantages, millets are more productive on marginal lands than main cereal crops because they use less water during cultivation, are highly adaptable to unfavorable climatic conditions, and have a shorter growing season (Mahajan et al. 2021). Millets are high in dietary fiber (12–20%), protein (6–19%), and carbs (60–70%). They also contain 2–4% minerals. These are also an excellent source of carbohydrates, vitamin B, lipids, dietary fiber, polyphenols, and minerals, depending on the variety. Millet starch has traditionally been separated using wet-milling methods. To facilitate the separation of starch from other components, such as whole millet kernels or millet flour, they are soaked in an aqueous solution for several hours before the grinding process, as described in a study by Zhu (2014). The production percentage and chemical composition of millet starch exhibit significant variations in different research studies. Starch yields can range from 52% to 68.2%, while amylose content can vary from 6.8% to 38.6%. Lipid content falls within the range of 0.16% to 2.9%, protein content spans from 0.4% to 4.3%, and ash content spans from 0.02% to 1.4% (Annor et al. 2014).

6.3 Microscopic Analysis of Starch

The structure of starch is studied using microscopy techniques including scanning electron microscopy (SEM) and transmission electron microscopy (TEM). In order to obtain precise information on starch granules, optical microscopy is insufficient. Higher resolution pictures of the starch granules are obtained using scanning electron microscopy (Argast and Tennis 2004). SEM is used to get high-resolution pictures in order to better comprehend the starch granules' morphology. The sample's surface morphology, the starch's structure, and the morphology before and after enzymatic hydrolysis can all be studied with the use of SEM. Information on the microstructure of different extruded starch samples and samples that have been treated with amylase is provided by SEM (Ye et al. 2018). SEM helps in imaging specimens with a higher resolution. It uses electron beam instead of light beam as in case of optical microscopes. Electrons that are generated from cathode are maintained as a single beam with the help of vacuum chambers through a series of lenses. Primary, secondary, and auger electrons are the three different types of electrons that are produced. The backscatter electrons that are produced by the materials are used for imaging in SEM along with the secondary electrons. These electrons touch the specimen's surface, where they are detected by a detector. They are then displayed on a screen, creating sharp images (Inkson 2016).

6.3.1 Rice Starch

Samples are mounted on a metal stub and sputtered with gold to add conductivity. SEM photos are taken while the voltage increased and at different magnifications (Han et al. 2020). Ye et al. (2016) performed SEM analysis by attaching freeze-dried cut surface samples onto double-sided sticky tape that was affixed to a circular specimen stub to investigate and compare the morphology change of improved extrusion cooking technology (IECT)-modified starch with those of native starch. Examination through SEM revealed that the structure of rice starch was modified by IECT. As the number of freeze-thaw cycles increased, the honeycomb pattern of the IECT-altered starch gels gradually became more pronounced. IECT-modified starch gels, however, had holes that were smaller than those of the native starch. Yang et al. (2016) performed SEM analysis to show morphological difference between hydroxypropylated, phosphorylated, and hydroxypropyl-phosphorylated rice starch by placing on double-sided adhesive tape mounted on an aluminum stub, and a thin layer of platinum was applied to coat them under vacuum. Polyhedral granules with smooth surfaces and no cracks were visible using SEM. Hydroxypropylated starch showed less smoothness and a less noticeable cracking than native starch. On the other hand, both untreated native starch and phosphorylated starch exhibited similar morphological characteristics. To investigate how ultrasound affects the starch's morphology, SEM was employed. The samples were initially dehydrated in an oven to eliminate moisture interference. After drying, these samples were affixed to a stub using double-sided adhesive tape, and a layer of gold was applied through vacuum evaporation. In the case of native rice starch, the granules were observed to have a polyhedral structure with flat edges and smooth surfaces. Even when subjected to low-intensity ultrasonic treatment, surface fissures and visible pores were still present on the starch granules. As the ultrasonic power was increased, the outer layer of the starch granules gradually separated from the central portion. Hu et al. (2014) used SEM to examine the morphological changes that occur during the hydrolysis of native rice starch by 1-butanol-HCl. Samples were prepared by adhering native and 1-butanol-HCl-hydrolyzed normal rice starches to a specimen holder using a silver plate and covered with a thin layer of gold in a vacuum evaporator. SEM results showed that granules of native rice starch had smooth surfaces and an irregular polygonal shape. Most of the starch granules retained their original polygonal forms after 1-butanol-HCl hydrolysis, but some of them showed minute surface fractures. Chen et al. (2023) used SEM for the characterization of Japonica waxy rice starch by coating the samples with gold in a vacuum evaporator. Images were taken using a scanning electron microscope with a 3 kV accelerating voltage. All of the forms of the Japanese waxy rice starch were polyhedral and irregular, according to the results of the SEM study. The waxy rice starches displayed surfaces that combined both concave and convex features. The convex side appeared relatively smooth, while the concave side had a notably rough texture. This textural variation might be attributed to the presence of multiple starch granules within rice endosperm, with each amyloplast potentially containing several dozen granules. These granules could protrude and contribute to the formation of the uneven and rough concave surfaces. Govindaraju et al. (2022) used SEM in order to see the shape of the rice starch granules more clearly. Starch samples that had been lyophilized were applied on carbon tape, mounted on an aluminum stub, and then treated to silver sputtering. Images were captured at magnifications of 5000x and 10,000x to enable high-resolution structural investigation of both native and hydrolyzed rice starches. When analyzed through SEM (scanning electron microscopy), natural rice starch granules exhibit a surface structure characterized by distinct edges and an irregular shape, often appearing as polygons or polyhedrons. The natural starch granules had smooth, multidimensional surfaces and no splits or fissures of any kind. The native starch granules' average size ranged from 4.8 to 6.7 μm. Despite having the same botanical origin as the starch granules, hydrolyzed starch revealed diverse structural deformations which is due to the porosity of the surface, the ratio of the pore area to the area of the starch granules, or the surface area of the starch granules.

The effects of various hydrostatic pressures on the physicochemical properties of rice starch with a high water content were examined by Li et al. (2017) (as shown in Fig. 6.3). A starch sample was affixed to an SEM stub using double-sided adhesive tape and subsequently coated with a layer of gold before capturing images. In the

Fig. 6.3 SEM images of native (**a**) hydroxypropylated (24 h), (**b**) phosphorylated (150 min), (**c**) hydroxypropyl-phosphorylated (150 min), and (**d**) glutinous rice starch granules. (The figure is reproduced from Yang et al. (2016))

SEM images, the granules of native rice starch appeared polyhedral and had irregular shapes, with diameters ranging from 2 to 6 mm. Because rice starch granules have an amorphous nature, any alterations in granule size, shape, and surface resulting from high hydrostatic pressure (HHP) treatments between 120 and 480 MPa were not immediately evident. However, following HHP treatment at 600 MPa, the starch granules exhibited an irreversible change as they lost their particle-like structure and instead took on a viscous, gel-like appearance.

6.3.2 Potato Starch

After being subjected to ultrasonic treatment, potato starch underwent structural analysis using SEM by Zhu et al. in 2012. The samples were dried after receiving ultrasonic treatment. The samples were viewed under a microscope after being covered in a thin coating of gold film. The natural starch granules' surfaces were smooth, with the exception of a few depressions and salients that were probably caused by swelling and drying, according to SEM analysis. After being treated with ultrasound, starch samples acquired a notch and groove on their surface. Zarski et al. (2016) performed SEM analysis to determine morphology of native starch and starch after ionic liquid esterification with unsaturated fatty acids. The samples were not coated with any conductivity material, and it was observed that the majority of potato starch granules had an oval to elliptical shape. The starch lost some of its granular structure during the initial pre-gelatinization procedure, and some of the granules became more amorphous. The starch oleate had a markedly different morphology from the original and pre-gelatinized starches, in which the starch granules were entirely disintegrated and lost their distinctiveness and smoothness. Li et al. (2017) performed SEM analysis to characterize the structure of acid-hydrolyzed potato starch in granular form when exposed to induced electric field (IEF). Samples were attached to an aluminum specimen holder with double-sided adhesive tape for observation, and then they were coated with gold using vacuum evaporation, and SEM analysis showed the native potato starch granules had smooth surfaces and an oval or spherical form. Most starch granules preserved their form after IEF-HCl hydrolysis; however, some surface damage or roughness was seen. IEF-HCl may have disrupted the inside of the granules, according to SEM. Altayan et al. (2017) used SEM to characterize the morphology of thermoplastic potato starch (TPPS). The potato starch granules were either physically or cryogenically broken up, and all the materials were subsequently coated with a small layer of graphite to prevent electrical charge under the electron beam. SEM showed that potato starch granules were molten or physically broken up as compared to native starch granules; the TPPS's morphology revealed a homogenous, clean, and smooth surface devoid of any pores or fissures. Velásquez Herrera et al. (2017) evaluated the morphology of native potato starch using SEM; samples were subjected to analyses at 20 kV accelerated electron voltage and 12–20 Pa chamber pressure with high vacuum and high voltage mode using secondary electron signal (as shown in Fig. 6.4).

6.3.3 Maize Starch

Du et al. (2019) studied the impact of persimmon tannin (PT) on the surface structure and retrogradation properties of starch gel after 14 days of storage using scanning electron microscopy. Utilizing double-sided glue, the samples were placed on an aluminum (silver) specimen holder before being sputtered with gold in a vacuum evaporator. SEM showed that in comparison to the control, more bulges and cavities were visible on the surface of retrograded starch when PT was present, indicating, in the presence of PT, the retrograded starch structure is less compact. Chen et al. (2020) performed SEM analysis to understand about pullulan's (PUL) morphological impact on the amount of oil that native maize starch (NMS) granules absorb after being fried. Using double-sided tape and a coating of gold-palladium, the defatted samples were spread out and mounted on the specimen holder. SEM analysis was utilized to examine the structures of fried NMS (native maize starch) and a mixture of NMS with 5% PUL (pullulan). In the case of fried NMS, most of the starch granules were observed to be distorted, deformed, and fused together into large conglomerates, with only a limited number of recognizable granules remaining within. In contrast, when analyzing the NMS + 5% PUL sample, the majority of

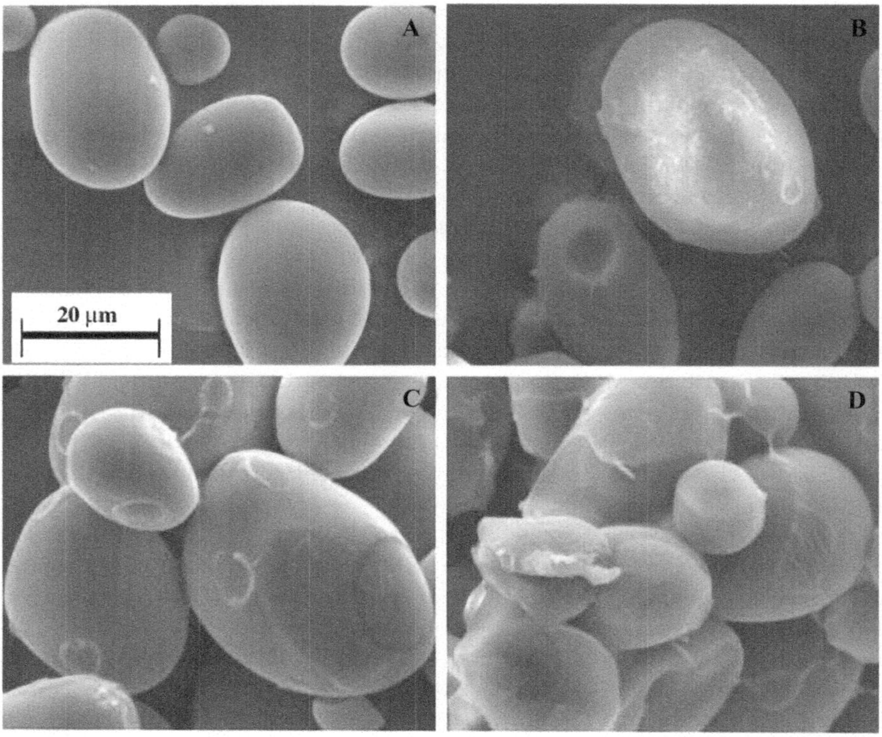

Fig. 6.4 SEM micrographs of potato starches treated at different electric field strengths: (**a**) native, (**b**) 30 kV cm^{-1}, (**c**) 40 kV cm^{-1}, and (**d**) 50 kV cm^{-1}. (The figure is reproduced from Han et al. (2009))

the starch granules maintained their original granular shape after frying, indicating that their swelling and gelatinization were effectively prevented. Furthermore, SEM revealed the presence of some irregular fragments on the surfaces of these granules, which could potentially be pullulan or amylose molecules that had leached and attached themselves to the starch granule surfaces. In Wang et al. 2021, Wang et al. employed scanning electron microscopy (SEM) to investigate the impact of proanthocyanidins (PA) on the structural arrangement of retrograded starch during storage. They used a portable microscope to examine how retrograded starch and starch mixed with 2% PA underwent morphological changes over time. The lyophilized samples were gold-coated on their cross sections, and images were captured at a magnification of 500x. According to SEM analysis, the well-defined porosity network structure that was previously present in the lyophilized retrograded starch was primarily created by water evaporation after the starch molecules were re-aggregated, re-ordered, and re-crystallized during cooling-storage. During storage, this pure starch network became uneven and thick, and it eventually vanished totally to be replaced with coarse-surfaced irregular fractures. SEM characterizations and photographic observations both supported the finding that PA had a propensity to obstruct the reorganization of starch chains and prevent maize starch retrogradation. Liu et al. (2021) performed SEM analysis to determine how a maize starch (MS)/stearic acid (SA)/carboxymethyl cellulose (CMC) composite film's physicochemical characteristics are affected by ultrasonic treatment. The samples that had been ultrasonically treated and those that had not were adhered to the testing surface using conductive carbon tape. Each sample was photographed after being coated with a layer of gold. SEM showed that the surface of the native starch sample lacked SA and CMC. This native sample's rough surface is likely the result of agglomerations of granules of insoluble starch and incompletely gelatinized starch on the surface of the polymer matrices. After the film-forming suspension received ultrasonic treatment, the surface of the polymer film improved in smoothness and homogeneity compared to the untreated material. This result can be attributed to the efficient blending of the hydrophobic and hydrophilic components made possible by ultrasonication.

6.3.4 Millet Starch

Gautam et al. (2020) used SEM with a 15 kV acceleration potential to examine the structure of isolated finger millet granules. SEM pictures of the extracted starch from the finger millets revealed granules of finger millet starch from three different varieties, including VL-315, VL-324, and VL-347, exhibited smooth-surfaced polygonal forms. The isolated starch granules' structure and shape were studied. The presence of unbroken starch granules was also visible in all starch granules. The starch isolation purity is indicated by the extracted finger millet starch granules' smooth, unbroken surfaces. Adebowale et al. (2005) used SEM to characterize the native finger millet starch. Using a vacuum evaporator and the secondary electron technique with an accelerating voltage of 10 kV, samples were coated with Au/Pd. SEM was utilized to examine the surface traits and form of the starches at a 3000x magnification. Most of the native finger millet starch consisted of granules that were irregular and polygonal in shape, ranging in size from 5 to 15 mm. When seen via an SEM, the surfaces seem smooth. Cavities are present in less than 1% of the starch granules, especially the smaller ones. Siroha and Sandhu (2018) used SEM to study the effect of cross-linking agent on the morphology of pearl millet starch to make a 1% suspension; starch samples were suspended in ethanol. Using double-sided adhesive tape, one drop of the starch-ethanol solution was applied to an aluminum stub. Micrography was performed at a 5 kV accelerating potential. The SEM examination indicated that the starch cross-linking process led to minimal changes in its granule structure when contrasted with unmodified native starch. Native starch granules were polygonal or spherical, possessing distinct edges. In contrast, the cross-linked starch granules displayed a somewhat rougher surface and the presence of grooves, suggesting minor fragmentation. Sun et al. (2022) reported the effects of cold plasma and extremely high pressure on the structural characteristics of proso millet starch. Cold plasma (CP) and ultrahigh pressure (UHP) treatment at pressures of 0, 200, 400, and 600 MPa were applied to pearl millet starch, and analyses of samples' morphological traits were conducted using SEM. A magnification of 8000x was used to capture photographs of the CP, UHP200, UHP400, UHP200-CP, and UHP400-CP samples, while a magnification of 3000x was used to capture images of the UHP600 and UHP600-CP samples. SEM showed that the natural starch of proso millet has a smooth surface and an uneven, polygonal shape. After CP treatment, the starch granules' surface had a small modification, but the granules' integrity had been preserved. No differences in morphology were seen between UHP200 and UHP400 starch granules and native starches suggesting that the native starch's granule structure was not affected by these pressure treatments. In contrast, the shape of the starch granules in UHP600 samples was completely disturbed, demonstrating that the starches were gelatinized at 600 MPa. SEM results indicated that as plasma treatment continued, the degree of damage increased, demonstrating that the dual modification treatment's impact is greater than that of the single modification treatment.

6.4 Conclusion

The way images are created while using transmission microscopical techniques significantly reduces the amount of information that can be obtained. Starch granules are translucent when viewed via a light microscope, and the assemblage of specimen structure at several levels on the optical axis could result in uncertainty. Also, the depth of high magnifications causes interference due to field limits by contrasting out-of-focus images with those that are in focus; optical microscopy does not provide detailed information on microstructure of starch granule. Scanning electron microscopy, by removing the drawbacks associated with transmission techniques, is a valuable tool for investigating the structure of modified starch granules. It offers supplementary information that complements what is obtained through transmission microscope studies. SEM is used to produce images with a better resolution. SEM images can also assist in assessing the size and shape of starch granules, enhancing the visualization of their surface structure. The use of SEM allows researchers to investigate the morphology both before and after enzymatic hydrolysis and to obtain high-resolution pictures that help them better grasp the starch granules' form. Using SEM, it is possible to see the breakdown pattern of hydrolyzed starch granules and learn more about the microstructure of different extruded starch samples and samples that have been treated with amylase. In order to have a thorough understanding of the patterns of enzymatic attack, it is helpful to observe the microstructure of starch granules caused by the action of amylase. Due the ability of SEM to provide detailed information on starch extracted from different sources such as corn, rice, maize, and millet, it is widely used for the their characterization.

Acknowledgments We thank Science and Engineering Research Board (SERB), Department of Science and Technology (DST), Government of India, India (Project Number-SERB/MTR/2020/000058). We also thank Manipal School of Life Sciences, Manipal Academy of Higher Education, Manipal, Karnataka, India, for providing the infrastructure needed to perform the project.

References

Adebowale KO, Olu-Owolabi BI, Olawumi E, kehinde, & Lawal, O. S. (2005) Functional properties of native, physically and chemically modified breadfruit (Artocarpus artilis) starch. Industrial Crops and Products 21(3):343–351. https://doi.org/10.1016/j.indcrop.2004.05.002

Akoh CC, Chang S-W, Lee G-C, Shaw J-F (2008) Biocatalysis for the production of industrial products and functional foods from rice and other agricultural produce. Journal of Agricultural and Food Chemistry 56(22):10445–10451. https://doi.org/10.1021/jf801928e

Altayan MM, Al Darouich T, Karabet F (2017) On the plasticization process of potato starch: preparation and characterization. Food Biophys 12(4):397–403. https://doi.org/10.1007/s11483-017-9495-2

Annor GA, Marcone M, Bertoft E, Seetharaman K (2014) Physical and molecular characterization of millet starches. Cereal Chem J 91(3):286–292. https://doi.org/10.1094/cchem-08-13-0155-r

Argast A, Tennis CF (2004) A web resource for the study of alkali feldspars and Perthitic textures using light microscopy, scanning electron microscopy and energy dispersive X-ray spectroscopy. J Geosci Educ 52(3):213–217. https://doi.org/10.5408/1089-9995-52.3.213

Božić N, Lončar N, Slavić MŠ, Vujčić Z (2017) Raw starch degrading α-amylases: an unsolved riddle. Amylase 1(1). https://doi.org/10.1515/amylase-2017-0002

Buléon A, Colonna P, Planchot V, Ball S (1998) Starch granules: structure and biosynthesis. Int J Biol Macromol 23(2):85–112. https://doi.org/10.1016/s0141-8130(98)00040-3

Chen L, McClements DJ, Zhang Z, Zhang R, Bian X, Jin Z, Tian Y (2020) Effect of pullulan on oil absorption and structural organization of native maize starch during frying. Food Chem 309:125681. https://doi.org/10.1016/j.foodchem.2019.125681

Chen R, Ma M, Zhao J, Fang J, Danino D, Sui Z, Corke H (2023) Characterization of multi-scale structure and physicochemical properties of starch from diverse japonica waxy rice cultivars. J Cereal Sci 109:103592. https://doi.org/10.1016/j.jcs.2022.103592

Dhital S, Shrestha AK, Gidley MJ (2010) Relationship between granule size and in vitro digestibility of maize and potato starches. Carbohydr Polym 82(2):480–488. https://doi.org/10.1016/j.carbpol.2010.05.018

Du X, Zhang Y, Pan X, Meng F, You J, Wang Z (2019) Preparation and properties of modified porous starch/carbon black/natural rubber composites. Composites Part B: Engineering 156:1–7. https://doi.org/10.1016/j.compositesb.2018.08.033

El-Fallal A, Abou M, El-Sayed A, Omar N (2012) Starch and microbial α-amylases: from concepts to biotechnological applications. In: Carbohydrates-comprehensive studies on glycobiology and glycotechnology. https://doi.org/10.5772/51571

Fashi A, Delavar AF, Zamani A, Noshiranzadeh N (2023) Lactic acid esterification of maize starch in solid state: assistant effect of pullulanase modification on digestibility and physicochemical properties. J Cereal Sci 109:103617. https://doi.org/10.1016/j.jcs.2022.103617

Gautam N, Garg S, Yadav S (2020) Underutilized finger millet crop for starch extraction, characterization, and utilization in the development of flexible thin film. J Food Sci Technol 58(11):4411–4419. https://doi.org/10.1007/s13197-020-04926-0

Govindaraju I, Pallen S, Umashankar S, Mal SS, Kaniyala Melanthota S, Mahato DR, Zhuo GY, Mahato KK, Mazumder N (2020) Microscopic and spectroscopic characterization of rice and corn starch. Microsc Res Tech 83(5):490–498. https://doi.org/10.1002/jemt.23437

Govindaraju I, Zhuo G-Y, Chakraborty I, Melanthota SK, Mal SS, Sarmah B, Baruah VJ, Mahato KK, Mazumder N (2022) Investigation of structural and physico-chemical properties of rice

starch with varied amylose content: a combined microscopy, spectroscopy, and thermal study. Food Hydrocoll 122:107093. https://doi.org/10.1016/j.foodhyd.2021.107093
Han Z, Zeng XA, Yu SJ, Zhang BS, Chen XD (2009) Effects of pulsed electric fields (PEF) treatment on physicochemical properties of potato starch. Innovative Food Sci Emerg Technol 10(4):481–485
Han X, Zhang M, Zhang R, Huang L, Jia X, Huang F, Liu L (2020) Physicochemical interactions between rice starch and different polyphenols and structural characterization of their complexes. LWT Food Sci Technol 125:109227. https://doi.org/10.1016/j.lwt.2020.109227
Herrero-Martínez JM, Schoenmakers PJ, Kok WT (2004) Determination of the amylose–amylopectin ratio of starches by iodine-affinity capillary electrophoresis. J Chromatogr A 1053(1–2):227–234. https://doi.org/10.1016/j.chroma.2004.06.048
Hu X, Li H, Wei B, Xu X, Jin Z, Tian Y (2014) Hydrolysis process of normal rice starch by 1-butanol–hydrochloric acid. Food Hydrocoll 41:27–32. https://doi.org/10.1016/j.foodhyd.2014.03.027
Inkson BJ (2016) Scanning electron microscopy (SEM) and transmission electron microscopy (TEM) for materials characterization. In: Materials characterization using nondestructive evaluation (NDE) methods, pp 17–43. https://doi.org/10.1016/b978-0-08-100040-3.00002-x
Kowsik PV, Mazumder N (2018) Structural and chemical characterization of rice and potato starch granules using microscopy and spectroscopy. Microsc Res Tech 81(12):1533–1540. https://doi.org/10.1002/jemt.23160
Lehmann U, Robin F (2007) Slowly digestible starch—its structure and health implications: a review. Trends Food Sci Technol 18(7):346–355. https://doi.org/10.1016/j.tifs.2007.02.009
Li W, Bai Y, Mousaa SA, Zhang Q, Shen Q (2011) Effect of high hydrostatic pressure on physicochemical and structural properties of rice starch. Food Bioprocess Technol 5(6):2233–2241. https://doi.org/10.1007/s11947-011-0542-6
Li D, Yang N, Zhou X, Jin Y, Guo L, Xie Z, Jin Z, Xu X (2017) Characterization of acid hydrolysis of granular potato starch under induced electric field. Food Hydrocoll 71:198–206. https://doi.org/10.1016/j.foodhyd.2017.05.012
Liu P, Gao W, Zhang X, Wang B, Zou F, Yu B, Lu L, Fang Y, Wu Z, Yuan C, Cui B (2021) Effects of ultrasonication on the properties of maize starch/stearic acid/sodium carboxymethyl cellulose composite film. Ultrason Sonochem 72:105447. https://doi.org/10.1016/j.ultsonch.2020.105447
Magallanes-Cruz PA, Flores-Silva PC, Bello-Perez LA (2017) Starch structure influences its digestibility: a review. J Food Sci 82(9):2016–2023. https://doi.org/10.1111/1750-3841.13809
Mahajan P, Bera MB, Panesar PS, Chauhan A (2021) Millet starch: a review. Int J Biol Macromol 180:61–79. https://doi.org/10.1016/j.ijbiomac.2021.03.063
Singh J, Kaur L, McCarthy OJ (2009) Potato starch and its modification. Adv Potato Chem Technol 1:273–318. https://doi.org/10.1016/b978-0-12-374349-7.00010-6
Siroha AK, Sandhu KS (2018) Physicochemical, rheological, morphological, and *in vitro* digestibility properties of cross-linked starch from pearl millet cultivars. Int J Food Prop 21(1):1371–1385. https://doi.org/10.1080/10942912.2018.1489841
Sun X, Saleh ASM, Lu Y, Sun Z, Zhang X, Ge X, Shen H, Yu X, Li W (2022) Effects of ultra-high pressure combined with cold plasma on structural, physicochemical, and digestive properties of proso millet starch. Int J Biol Macromol 212:146–154. https://doi.org/10.1016/j.ijbiomac.2022.05.128
Velásquez Herrera JD, Lucas Aguirre JC, Quintero Castaño VD (2017) Physical-chemical characteristics determination of potato (solanum phureja juz. & bukasov) starch. Acta Agronóm 66(3):323–330. https://doi.org/10.15446/acag.v66n3.52419
Wang M, Chen J, Chen S, Ye X, Liu D (2021) Inhibition effect of three common proanthocyanidins from grape seeds, peanut skins and pine barks on maize starch retrogradation. Carbohydr Polym 252:117172. https://doi.org/10.1016/j.carbpol.2020.117172
Wang L, Wang M, Zhou Y, Wu Y, Ouyang J (2022) Influence of ultrasound and microwave treatments on the structural and thermal properties of normal maize starch and potato starch: a comparative study. Food Chem 377:131990. https://doi.org/10.1016/j.foodchem.2021.131990

Wani AA, Singh P, Shah MA, Schweiggert-Weisz U, Gul K, Wani IA (2012) Rice starch diversity: effects on structural, morphological, thermal, and physicochemical properties-a review. Compr Rev Food Sci Food Saf 11(5):417–436. https://doi.org/10.1111/j.1541-4337.2012.00193.x

Yang L, Zhou Y, Wu Y, Meng X, Jiang Y, Zhang H, Wang H (2016) Preparation and physicochemical properties of three types of modified glutinous rice starches. Carbohydr Polym 137:305–313. https://doi.org/10.1016/j.carbpol.2015.10.065

Ye J, Hu X, Zhang F, Fang C, Liu C, Luo S (2016) Freeze-thaw stability of rice starch modified by improved extrusion cooking technology. Carbohydr Polym 151:113–118. https://doi.org/10.1016/j.carbpol.2016.05.026

Ye J, Liu C, Luo S, Hu X, McClements DJ (2018) Modification of the digestibility of extruded rice starch by enzyme treatment (β-amylolysis): an in vitro study. Food Res Int 111:590–596. https://doi.org/10.1016/j.foodres.2018.06.002

Zarski A, Ptak S, Siemion P, Kapusniak J (2016) Esterification of potato starch by a biocatalysed reaction in an ionic liquid. Carbohydr Polym 137:657–663. https://doi.org/10.1016/j.carbpol.2015.11.029

Zhu F (2014) Structure, physicochemical properties, and uses of millet starch. Food Res Int 64:200–211. https://doi.org/10.1016/j.foodres.2014.06.026

Zhu J, Li L, Chen L, Li X (2012) Study on supramolecular structural changes of ultrasonic treated potato starch granules. Food Hydrocoll 29(1):116–122. https://doi.org/10.1016/j.foodhyd.2012.02.004

Outlining the Role of Vibrational Spectroscopy in Characterizing Starch

7

Rajib Biswas

Abstract

Vibrational spectroscopy plays a very vital role in the characterization of materials. It is one of the first pillars of spectroscopic tools that enable researchers to investigate the primary configuration of the object under investigation. Be it a monatomic, diatomic, or polyatomic molecule, vital rudimentary information can be accrued with the help of this. Starch is considered as one of the pivotal food elements of humans. Being complex in nature owing to the large constituent chains involved, the effective characterization of it is very important. In this regard, vibrational spectroscopy assays give structural information of starch to a large extent. Although all constituents cannot be assessed via vibrational spectroscopy; however, it helps to build a preliminary database which in turn enables investigators to implement more sophisticated analytical tools to deeply study therein. Accordingly, this chapter outlines the role of vibrational spectroscopy in characterizing starch. The scopes and challenges are also fairly dealt with herein.

Keywords

Vibrational spectroscopy · Starch · Characterization · Morphology · Structure

7.1 Introduction

The primary source of energy in the human diet is the starch that is stored in the seeds and tubers of numerous agricultural crops, including maize, wheat, rice, barley, potato, and cassava (Zhang et al. 2021; Yu et al. 2018; Teng et al. 2021; Kumar

R. Biswas (✉)
Department of Physics, Tezpur University, Tezpur, Assam, India
e-mail: rajib@tezu.ernet.in

N. Mazumder, M. H. Rahman (eds.), *Advanced Research in Starch*,
https://doi.org/10.1007/978-981-99-9527-1_7

et al. 2018). The primary component of cereal grains is starch, and changes in its biophysical and biochemical characteristics are related to the amount and ratio of amylose and amylopectin, which influence and affect properties like viscosity and gelatinization and determine its end-use characteristics (e.g., bread, malt, beer, and polymers) (Zhang et al. 2013a, 2021; Kumar et al. 2018; Jadhav et al. 2010; Kannan 2018).

Polysaccharide macronutrients are crucial for human nutrition, and among them, starch holds significant importance. Typically, starch is subjected to processing before the breakdown of other dietary components, as per existing literature (Zhang et al. 2021). Cereal scientists have consistently demonstrated interest in exploring various analytical methods to examine the correlation between starch structure and function. This research area remains ongoing. Despite being composed of the same glucose glycosidic monomer, the primary starch components, amylose and amylopectin, demonstrate distinct characteristics and properties throughout the course of food processing. From an analytical perspective, the coexistence of these two macromolecules may result in interference during physicochemical analysis. As a result, the investigation of amylose and amylopectin is frequently conducted separately (Yu et al. 2018). Consequently, employing a singular technique for the all-encompassing examination of starch granules poses a challenge.

The evaluation of starch source, purity, and integrity (whether natural or modified), as well as its physicochemical, morphological, rheological, and structural properties, is a crucial aspect of starch analysis. Therefore, analytical methods for starch hold significant importance. It is imperative to enhance the precision and comprehensibility of outcomes by improving the processes of sample preparation, elimination of interferents, and utilization of analytical techniques. The physicochemical analysis encompasses the determination of straight and amylopectin composition, chain length distributions (CLDs), and total starch content. Additionally, it is important to consider the presence of fast-digesting starch, slow-digesting starch, and resistant starch (Teng et al. 2021). The categorization of distinct starch granules is predicated upon findings derived from diverse analytical methodologies. Microscopic imaging techniques are necessary for the classification of starch granules based on particle size, surface morphology, and damage. The primary tools utilized for conducting this form of evaluation include optical microscopy, scanning electron microscopy (SEM), transmission electron microscopy (TEM), confocal laser scanning microscopy (CLSM), and atomic force microscopy (AFM). The crystallographic details of starch granules' structure can be determined through various techniques such as X-ray diffraction (XRD), small-angle X-ray diffraction (SAXS), nuclear magnetic resonance (NMR) imaging, and infrared (IR) spectroscopy.

Various techniques such as size exclusion chromatography (SEC), high-performance anion-exchange chromatography coupled with a pulsed amperometric detector (HPAECPAD), capillary electrophoresis (CE), thin-layer chromatography (TLC), and asymmetrical-flow field-flow fractionation (AsFlFFF) can be employed to characterize the CLDs of starch granules.

At present, the examination of starch granules necessitates the utilization of multiple competing methodologies that are capable of scrutinizing intricate carbohydrates. Every technique has the potential to yield valuable insights into the properties of starch, but each method is subject to some degree of information loss. There are presently slow, damaging, and often empirically based chemical and physical methods used in research and by industry to test starch qualities (Liu et al. 2013; Miranda et al. 2015; Kim and Huber 2008; Ma et al. 2022; Arzate-Vázquez et al. 2012). The ability to understand the biophysical and structural characteristics of the starch to be used in foods (such as bread, beer, and whisky) and industrial applications (such as the production of polymers) will allow for an improvement in the starch content in cereals and starchy foods (Barrera et al. 2013). Additionally, this can offer useful details regarding the particular use of starch in the selection or screening of new genotypes or lines in breeding programs (Jadhav et al. 2010; Rodriguez-Garcia et al. 2021).

Instrumental techniques like differential scanning calorimetry (DSC) and the Rapid Visco Analyzer (RVA) are presently utilized in analytical approaches to measure biophysical parameters in cereals and starchy meals (Zhang et al. 2013b, 2021; Miranda et al. 2015; Kim and Huber 2008; Ma et al. 2005, 2022; Barrera et al. 2013; Cameron and Donald 1993). The impacts of rain damage on grain quality at the delivery point are also assessed routinely using tools like the RVA on cereals (such as wheat and barley) (Cameron and Donald 1993; Zhang et al. 2015a; Daniels and Donald 2004).

Rapid approaches to predict and track starch biochemical and biophysical properties have been developed during the past 20 years using methods based on vibrational spectroscopy in conjunction with chemometric techniques (Liu et al. 2013; Miranda et al. 2015; Kim and Huber 2008; Ma et al. 2022). These instruments should be quick and simple to use, need little to no sample preparation, and, in certain cases, prevent sample destruction. Most of these characteristics are typical of near-infrared (NIR) and mid-infrared (MIR) spectroscopy (Liu et al. 2013) [see Fig. 7.1].

For vibrational measurements, Raman technique possesses several merits over infrared. As the incident and scattered radiation belong to either UV or visual spectra, the adaptation of instrumentation becomes less cumbersome with obliteration of atmospheric absorption. The sample size cell becomes also less. With pulsed techniques, the small quantities of transient species can be assessed accordingly. Another recent addition to this line is hyperspectral imaging which is also gradually gaining attention in the characterization of starch (Ma et al. 2005; Zhang et al. 2010, 2013b, 2015a, b; Cameron and Donald 1993; Daniels and Donald 2004; Yang et al. 2019; Kovrlija and Rondeau-Mouro 2017; Choi and Kerr 2003; Drzeżdżon et al. 2019).

Considering the impacts of these vibrational analytical approaches, this chapter aims to summarize the aspects of these spectroscopic tools basically in the vibrational domain in the characterization of starch. The following sections deal systematically with the salient features of these tools with direct context to starch research and characterization.

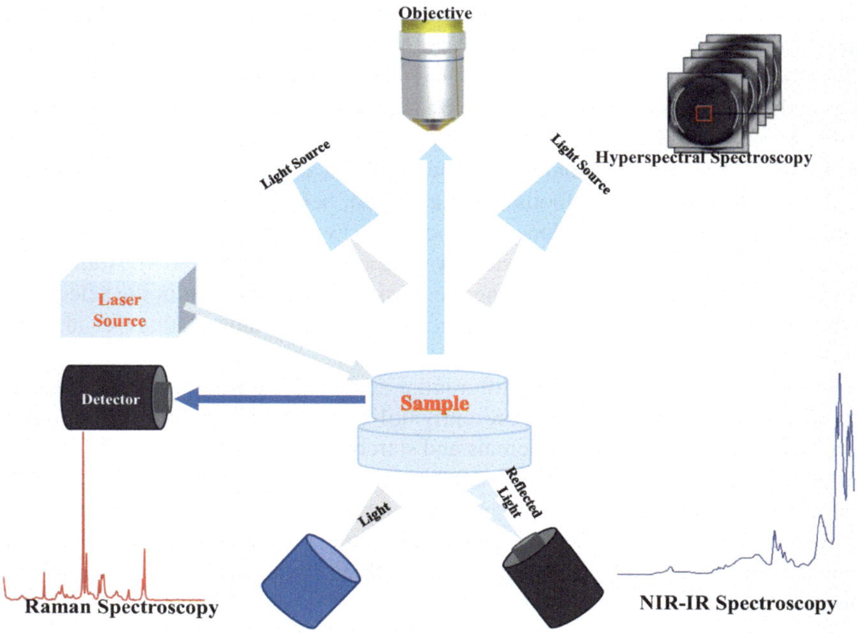

Fig. 7.1 Schematic layout of spectroscopy chiefly in vibrational domain

7.2 NIR Spectroscopy

Near-infrared (NIR) spectroscopy is a valuable analytical technique for characterizing starch. As per name, it is dependent on the absorption of near-infrared light by chemical bonds in the sample, and it can provide wealth of information about the molecular composition, structure, and properties of starch. Below, we outline the aspects in characterizing starch via NIR spectroscopy (Zhang et al. 2010, 2015a, b; Daniels and Donald 2004; Yang et al. 2019; Kovrlija and Rondeau-Mouro 2017; Choi and Kerr 2003; Drzeżdżon et al. 2019).

Quantitative Analysis: It is well known that starch possesses characteristic absorption bands in the NIR region, and by measuring the intensity of these bands, it is possible to develop a calibration curve which can relate absorbance to starch concentration.

Moisture Content: Starch's properties, such as its texture and processing behavior, are influenced by moisture content. NIR spectroscopy can be employed to measure moisture content in starch samples, which is crucial for controlling the quality of starch-based products.

Starch Structure: NIR spectroscopy can provide insights into the structural properties of starch. Different forms of starch (e.g., amylose and amylopectin) have distinct NIR spectra due to variations in their molecular structures. By analyzing the

NIR spectra, one can interrogate the ratio of amylose to amylopectin and the degree of branching in starch molecules.

Purity and Impurity Detection: Likewise, NIR spectroscopy can help identify impurities or contaminants in starch samples. By comparing the NIR spectrum of a starch sample with a reference spectrum of pure starch, it is feasible to sense the presence of foreign substances or contaminants.

Process Monitoring: In industrial starch production, NIR spectroscopy can be used for real-time monitoring of starch properties during the manufacturing process. This enables adjustments to be made in real time to maintain product quality and consistency.

Particle Size and Granule Structure: Starch granules have characteristic sizes and shapes, which can affect their functional properties. NIR spectroscopy can be used to analyze the particle size distribution and granule structure of starch samples, aiding in quality control and process optimization.

Gelatinization and Retrogradation: NIR spectroscopy can track the gelatinization and retrogradation behavior of starch when it is subjected to heat and cooling. This is important in food applications, such as in the production of starchy foods like bread and pasta.

Starch in Different Matrices: NIR spectroscopy can be used to analyze starch in various matrices, including solid foods, gels, pastes, and powders. This versatility makes it a valuable tool in food science and quality control.

NIR spectroscopy is presently used in three main ways in the analysis of cereals and starchy products: (1) as a straightforward analysis and very quick determination of composition, (2) as a screening tool in plant-breeding (for the selection of cross-breeds with the desired qualities), and (3) as an in-line tool to monitor physical and chemical changes during processing. Kaddour and Cuq (Zhang et al. 2015b; Yang et al. 2019) highlighted these three main applications in a recent report.

Cereal grains and other starchy foods primarily consist of starch, and variations in this substance's biophysical and biochemical characteristics (such as amylose, amylopectin, pasting, gelatinization, and viscosity) will directly affect its end-use characteristics (such as those of bread, malt, and polymers). To enhance and evaluate the quality of cereals and other starchy meals, it is important to utilize quick and nondestructive technologies to investigate and track starch qualities such as gelatinization, retrogradation, and water absorption. Recently, it has been investigated how to forecast several quality characteristics using near-infrared reflectance (NIR) and mid-infrared (MIR) spectroscopy (Zhang et al. 2010, 2015b; Yang et al. 2019; Kovrlija and Rondeau-Mouro 2017).

7.3 Vibrational Spectroscopy

The vibrational spectra of a polyatomic molecule are influenced by its chemical structure, interactions with the surrounding environment, and dynamical transformations. The precise determination of various conformations can be achieved using linear absorption and inelastic light-scattering spectroscopy in both gaseous and condensed states. The Fourier-transform infrared absorption and Raman scattering spectroscopy techniques have become indispensable instruments in the fields of chemical analysis, biological assays, and functional materials research. In the last 10 years, there have been significant advancements in the field of coherent nonlinear vibrational spectroscopy, including the emergence of multidimensional techniques such as electronic, IR, THz, IR-Raman, IR-vis, vis-IR, and THz-Raman spectroscopy (Drzeżdżon et al. 2019; Huang et al. 2021). Likewise, there is another coherent multidimensional spectroscopy measurement. These measurements have led to precise estimation of the solvation-induced frequency shifts. It did help in finding the time correlations of the fluctuating frequencies. Meanwhile, vital aspects such as spontaneous fluctuating motions of solvent molecules and dynamical changes in molecular conformations could also be attained from such measurements. Apart from this, vast information of local surroundings involving local electrostatic environments, chemical and biochemical reactions, protein structural dynamics and functions, and characteristic processes of functional materials could be assessable with the proper utilization of coherent vibrational investigations.

To perform a quantitative analysis and interpretation of spectroscopic observations, researchers have employed organic and biochemical techniques to introduce vibrational probes into molecular, biological, and material systems in a site-specific manner (Sow et al. 2018; Sow and Yang 2015; Cozzolino 2012; Agelet and Hurburgh 2010). The addition of vibrational probes to a solvated molecular system through their hydrogen-bonding interactions with surrounding solvent molecules or biomolecular residues may result in invasiveness. Because of their smaller dimension than fluorophores, these probes find application while studying local environments. Entities such as labeled molecules, proteins, nucleic acids, and functional materials are investigated via vibrational spectroscopic investigations which lead to a detailed characterization of local surroundings entailing molecular conformation, protein contacts, energy transfer, and many more. A group of researchers meticulously studied the hitherto experimental findings (Wahr et al. 1996; Pielesz 2012; Cocchi et al. 2004). All these necessitated the imperative interpretative model whose accuracy included atomic-level resolution. However, there is no concrete universal theory engaging the vibrational solvatochromism, electrochromism, and dynamic fluctuation of vibrational frequencies. The absence of such robust theory can be attributed to intrinsically complex intermolecular interactions as well as minuscule vibrational frequency shifts induced by the former.

A complex molecule may possess an infrared spectrum exhibiting a large no. of normal vibrations. The normal mode hitherto may encompass vibrations of all atoms or nearly all. Meanwhile, there is possibility that displacement experienced by a bunch of atoms may be higher as compared to other neighboring bunches. Accordingly, there exist two groups of vibrations—namely, skeletal vibrations and characteristic group vibrations. The former considers similar vibrations exhibited by atom groups, whereas the latter engages a very scant portion.

7.4 Raman Spectroscopy

This is another powerful analytical technique that can be used to characterize starch. Unlike NIR spectroscopy, which measures the absorption of light, Raman spectroscopy measures the inelastic scattering of light. It provides information about the vibrational and rotational energy levels of molecules, allowing for the identification and characterization of chemical compounds. Let us have a look how Raman spectroscopy can be employed to study starch (Agelet and Hurburgh 2010; Wahr et al. 1996; Pielesz 2012; Cocchi et al. 2004; Amir et al. 2013).

Chemical Identification: It is useful to identify the specific chemical constituents present in starch, such as amylose and amylopectin. Each of these components has distinct Raman spectra due to their unique molecular structures. While comparing a reference spectrum with an unknown sample, the composition becomes readily identifiable.

Molecular Structure: Raman spectroscopy provides insights into the molecular structure of starch. It can reveal information about the glycosidic linkages between glucose units, the degree of branching in amylopectin, and the crystallinity of starch granules. For example, the degree of crystallinity affects the peak positions and intensities in the Raman spectrum.

Starch Conformation: Starch molecules can exist in various conformations, including helical structures. Raman spectroscopy can be used to study the conformational changes of starch molecules under different conditions, such as during gelatinization or retrogradation.

Starch Purity and Impurities: Raman spectroscopy can help detect impurities or contaminants in starch samples. Foreign substances will have their own Raman spectrum, and their presence can be identified by comparing the sample's spectrum to reference spectra of pure starch.

Gelatinization and Retrogradation: Raman spectroscopy is sensitive to changes in starch structure that occur during gelatinization (disruption of crystalline structure) and retrogradation (reformation of crystalline structure). Monitoring these changes is important in food processing and product development.

Solid-State Characterization: Raman spectroscopy can be used to analyze starch in its solid-state form. This is particularly useful for studying starch in food products, pharmaceuticals, and other materials where starch may be present in a solid form.

Mapping and Imaging: Raman spectroscopy can be combined with microscopy to create Raman images of starch granules or starch-containing materials. This enables the spatial distribution of starch and its structural features to be visualized.

Dispersive Raman spectroscopy may not be well suited for quantifying amylose content of starch mixtures; however, the method was easily capable of discriminating between waxy and normal starches. This may allow the method to be used for confirming the identity of starch shipments.

The low starch concentration used may have caused the poor Raman signal strength and the unpredictable changes in the X-ray diffraction data. The experiment found that measuring retrogradation in very dilute starch gels could be problematic and that more development is needed in order to apply Raman spectroscopy to a food system like white pan bread.

7.5 Hyperspectral Spectroscopy

Hyperspectral spectroscopy, also known as hyperspectral imaging, is an advanced technique that amalgamates spectroscopy and imaging to collect a vast amount of spectral data at numerous narrow wavelength bands across the electromagnetic spectrum. This technique can be applied to characterize starch and provides valuable insights into its properties and composition. Listed below are the key aspects where hyperspectral spectroscopy can be exploited for starch characterization (Amir et al. 2013; Kang et al. 2015; An et al. 2021; Du et al. 2022; Shea 2007; Syahariza et al. 2013; Cave et al. 2009; Striegel 2008).

Chemical Composition: Hyperspectral spectroscopy can be used to identify the chemical composition of starch samples. Each chemical compound has a unique spectral signature, and starch components like amylose and amylopectin will exhibit distinct spectral patterns. By analyzing the hyperspectral data, you can determine the presence and concentration of different starch constituents.

Molecular Structure: Hyperspectral imaging can reveal information about the molecular structure of starch. It can detect changes in starch structure, such as variations in glycosidic linkages, branching, and crystallinity, which can impact its functional properties.

Contaminant Detection: Hyperspectral imaging can be employed to identify contaminants or impurities in starch samples. Different substances have characteristic spectral signatures, and hyperspectral data can be used to detect foreign materials within starch matrices.

Quality Control: In the food and pharmaceutical industries, hyperspectral spectroscopy can be used for quality control and assurance. By analyzing hyperspectral images of starch-based products, it is possible to assess their composition and ensure they meet desired quality standards.

Spatial Distribution: Hyperspectral imaging provides both spectral and spatial information. This means you can create images that show the distribution of starch and its components within a sample. This is particularly useful when studying starch in heterogeneous materials, such as food products.

Monitoring Changes: Hyperspectral spectroscopy allows for real-time monitoring of starch properties and changes. For instance, it is possible to track alterations in starch structure during processes like gelatinization, retrogradation, or cooking.

Mapping Granule Structure: Hyperspectral imaging can be used to map the structure of starch granules within a sample. This can help characterize the size, shape, and distribution of starch granules, which can affect the properties of starch-containing materials.

7.6 Final Remarks

Several authors have reported the measurement of chemical components of starch, including amylose and amylopectin, as well as physical characteristics of the endosperm such as granule type (A and B) and biophysical properties such as gelatinization, retrogradation, and crystallinity. These measurements were conducted using either near-infrared (NIR) or mid-infrared (MIR) spectroscopic techniques. Nevertheless, divergent outcomes regarding the exactness and correctness of the calibrations achieved were documented. The observed variations were ascribed to the precision of the reference technique employed, such as enzymatic or colorimetric methods, which may be affected by other properties such as lipids and proteins, the diversity in composition, the quantity of samples utilized for constructing the calibration models, and the techniques employed, including NIR, MIR, transmission, ATR, and reflectance.

The efficacy of infrared (IR) spectroscopy in monitoring various stages of wheat processing has been demonstrated through in/on and at-line techniques. This approach is highly suitable for the detection and characterization of alterations in the physical, chemical, and physicochemical properties that take place during the processing of wheat. The literature presents various data indicating that the utilization of IR and specifically NIR spectroscopy can provide significant insights for monitoring wheat processing. This is due to its minimal interference in investigating molecular alterations during the application of mechanical energy. Analysis, control, and comprehension of the origin and composition of starch-based products can be deemed a valuable approach for the industry.

In general, the primary benefits of employing these methodologies are their expeditiousness and simplicity in standard procedures, ascertaining prospective cost savings by diminishing analysis time and expenses. Nevertheless, a significant drawback has arisen from the challenge of measuring and analyzing the data produced by spectroscopic techniques. The extensive range of these applications, whether in regular utilization or in the process of advancement, is evidently not exhibiting any indication of reduction. The integration of these methodologies with multivariate analytical approaches has the potential to serve as a significant instrument for large-scale implementation in research and development or industrial contexts.

Nevertheless, the chemical or biophysical foundation that underlies the cornerstone remains inadequately comprehended. The primary impediments to the broader implementation of these technologies will pertain to the comprehension and analysis of intricate data sets procured through their utilization. The comprehension of models via multivariate techniques, specifically in the context of calibration development, as well as the understanding of the fundamental principles of molecular spectroscopy in rheological analysis through IR spectroscopy, remains insufficiently understood. The forthcoming advancement of these applications will furnish the cereal- and starch-based food sector with expeditious and noninvasive techniques to measure various samples, thereby enabling the industry to swiftly conduct qualitative analysis instead of quantitative analysis.

Acknowledgement Author acknowledges the support received from DST-FIST and UGC SAP DRS II grant in aid to Department of Physics, Tezpur University.

Conflict of InterestThere is no conflict of interest.

References

Agelet LE, Hurburgh CR (2010) A tutorial on near infrared spectroscopy and its calibration. Crit Rev Anal Chem 40:246–260. https://doi.org/10.1080/10408347.2010.515468

Amir RM, Anjum FM, Khan MI, Khan MR, Pasha I, Nadeem M (2013) Application of Fourier transform infrared (FTIR) spectroscopy for the identification of wheat varieties. J Food Sci Technol 50:1018–1023. https://doi.org/10.1007/s13197-011-0424-y

An D, Li Q, Li E, Obadi M, Li C, Li H, Zhang J, Du J, Zhou X, Li N, Xu B (2021) Structural basis of wheat starch determines the adhesiveness of cooked noodles by affecting the fine structure of leached starch. Food Chem 341:128222. https://doi.org/10.1016/j.foodchem.2020.128222

Arzate-Vázquez I, Chanona-Pérez JJ, Calderón-Domínguez G, Terres-Rojas E, Garibay-Febles V, Martínez-Rivas A, Gutiérrez-López GF (2012) Microstructural characterization of chitosan and alginate films by microscopy techniques and texture image analysis. Carbohydr Polym 87:289–299. https://doi.org/10.1016/j.carbpol.2011.07.044

Barrera GN, Calderón-Domínguez G, Chanona-Pérez J, Gutiérrez-López GF, León AE, Ribotta PD (2013) Evaluation of the mechanical damage on wheat starch granules by SEM, ESEM, AFM and texture image analysis. Carbohydr Polym 98:1449–1457. https://doi.org/10.1016/j.carbpol.2013.07.056

Cameron RE, Donald AM (1993) A small-angle X-ray scattering study of the absorption of water into the starch granule. Carbohydr Res 244:225–236. https://doi.org/10.1016/0008-6215(83)85003-4

Cave RA, Seabrook SA, Gidley MJ, Gilbert RG (2009) Characterization of starch by size-exclusion chromatography: the limitations imposed by shear scission. Biomacromolecules 10(8):2245–2253. https://doi.org/10.1021/bm900426n

Choi SG, Kerr WL (2003) 1H NMR studies of molecular mobility in wheat starch. Food Res Int 36:341–348. https://doi.org/10.1016/S0963-9969(02)00225-9

Cocchi M, Foca G, Lucisano M, Marchetti A, Pagani MA, Tassi L, Ulrici A (2004) Classification of cereal flours by chemometric analysis of MIR spectra. J Agric Food Chem 52:1062–1067. https://doi.org/10.1021/jf034441o

Cozzolino D (2012) Recent trends on the use of infrared spectroscopy to trace and authenticate natural and agricultural food products. Appl Spectrosc Rev 47:518–530. https://doi.org/10.1080/05704928.2012.667858

Daniels DR, Donald AM (2004) Soft material characterization of the lamellar properties of starch: smectic side-chain liquid-crystalline polymeric approach. Macromolecules 37:1312–1318. https://doi.org/10.1021/ma030360h

Drzeżdżon J, Jacewicz D, Sielicka A, Chmurzyński L (2019) A review of new approaches to analytical methods to determine the structure and morphology of polymers. TrAC Trends Anal Chem 118:470–476. https://doi.org/10.1016/j.trac.2019.06.013

Du J, Pan R, Obadi M, Li H, Shao F, Sun J, Wang Y, Qi Y, Xu B (2022) In vitro starch digestibility of buckwheat cultivars in comparison to wheat: the key role of starch molecular structure. Food Chem 368:130806. https://doi.org/10.1016/j.foodchem.2021.130806

Huang M, Zhao L, Yang H (2021) Water loss and status in sponge cake: impact of Eucheuma as a flour replacement. J Food Sci 86(3):915–922. https://doi.org/10.1111/1750-3841.15609

Jadhav K, Gupta N, Mujib BA (2010) Maltese cross: starch artifact in oral cytology, divulged through polarized microscopy. J Cytol 27(1):40–41. https://doi.org/10.4103/0970-9371.66698

Kang J, Guo Q, Wang Q, Phillips GO, Cui SW (2015) New studies on gum ghatti (Anogeissus latifolia) part 5: the conformational properties of gum ghatti. Food Hydrocoll 43:25–30. https://doi.org/10.1016/j.foodhyd.2014.04.042

Kannan M (2018) Scanning electron microscopy: principle, components and applications. In: A textbook on fundamentals and applications of nanotechnology. Daya Publishing House, pp 81–92

Kim HS, Huber KC (2008) Channels within soft wheat starch A- and B-type granules. J Cereal Sci 48:159–172. https://doi.org/10.1016/j.jcs.2007.09.002

Kovrlija R, Rondeau-Mouro C (2017) Hydrothermal changes in wheat starch monitored by two-dimensional NMR. Food Chem 214:412–422. https://doi.org/10.1016/j.foodchem.2016.07.051

Kumar R, Mukherjee S, Ayele BT (2018) Molecular aspects of sucrose transport and its metabolism to starch during seed development in wheat: a comprehensive review. Biotechnol Adv 36:954–967. https://doi.org/10.1016/j.biotechadv.2018.02.015

Liu D, Parker ML, Wellner N, Kirby AR, Cross K, Morris VJ, Cheng F (2013) Structural variability between starch granules in wild type and in ae high-amylose mutant maize kernels. Carbohydr Polym 97(2):458–468. https://doi.org/10.1016/j.carbpol.2013.05.013

Ma X, Yu J, Kennedy JF (2005) Studies on the properties of natural fibers-reinforced thermoplastic starch composites. Carbohydr Polym 62:19–24. https://doi.org/10.1016/j.carbpol.2005.07.015

Ma S, Wang Z, Tian X, Sun B, Huang J, Yan J, Bao Q, Wang X (2022) Effect of synergistic fermentation of Lactobacillus plantarum and Saccharomyces cerevisiae on thermal properties of wheat bran dietary fiber-wheat starch system. Food Chem 373:131417. https://doi.org/10.1016/j.foodchem.2021.131417

Miranda K, Girard-Dias W, Attias M, de Souza W, Ramos I (2015) Three-dimensional reconstruction by electron microscopy in the life sciences: an introduction for cell and tissue biologists. Mol Reprod Dev 82(7–8):530–547. https://doi.org/10.1002/mrd.22455

Pielesz A (2012) Vibrational spectroscopy and electrophoresis as a "golden means" in monitoring of polysaccharides in medical plant and gels. Spectrochim Acta A Mol Biomol Spectrosc 93:63–69. https://doi.org/10.1016/j.saa.2012.03.003

Rodriguez-Garcia ME, Hernandez-Landaverde MA, Delgado JM, Ramirez-Gutierrez CF, Ramirez-Cardona M, Millan-Malo BM, Londoño-Restrepo SM (2021) Crystalline structures of the main components of starch. Curr Opin Food Sci 37:107–111. https://doi.org/10.1016/j.cofs.2020.10.002

Shea JJ (2007) Introduction to physical polymer science, 4th edition. IEEE Electr Insul Mag 23:54–55. https://doi.org/10.1109/MEI.2007.288474

Sow LC, Yang H (2015) Effects of salt and sugar addition on the physicochemical properties and nanostructure of fish gelatin. Food Hydrocoll 45:72–82. https://doi.org/10.1016/j.foodhyd.2014.10.021

Sow LC, Chong JMN, Liao QX, Yang H (2018) Effects of κ-carrageenan on the structure and rheological properties of fish gelatin. J Food Eng 239:92–103. https://doi.org/10.1016/j.jfoodeng.2018.05.035

Striegel AM (2008) Observations regarding on-column, flow-induced degradation during SEC analysis. J Liq Chromatogr Relat Technol 31(20):3105–3114. https://doi.org/10.1080/10826070802480024

Syahariza ZA, Sar S, Hasjim J, Tizzotti MJ, Gilbert RG (2013) The importance of amylose and amylopectin fine structures for starch digestibility in cooked rice grains. Food Chem 136:742–749. https://doi.org/10.1016/j.foodchem.2012.08.053

Teng C, Chen D, Wu G, Campanella OH (2021) Non-invasive techniques to study starch structure and starchy products properties. Curr Opin Food Sci 38:196–202. https://doi.org/10.1016/j.cofs.2020.11.008

Wahr JA, Tremper KK, Samra S, Delpy DT (1996) Near-infrared spectroscopy: theory and applications. J Cardiothorac Vasc Anesth 10:406–418. https://doi.org/10.1016/S1053-0770(96)80107-8

Yang X, Chi C, Liu X, Zhang Y, Zhang H, Wang H (2019) Understanding the structural and digestion changes of starch in heat-moisture treated polished rice grains with varying amylose content. Int J Biol Macromol 139:785–792. https://doi.org/10.1016/j.ijbiomac.2019.08.051

Yu W, Tao K, Gilbert RG (2018) Improved methodology for analyzing relations between starch digestion kinetics and molecular structure. Food Chem 264:284–292. https://doi.org/10.1016/j.foodchem.2018.05.049

Zhang J, Chen F, Liu F, Wang Z-W (2010) Study on structural changes of microwave heat-moisture treated resistant Canna edulis ker starch during digestion in vitro. Food Hydrocoll 24:27–34. https://doi.org/10.1016/j.foodhyd.2009.07.005

Zhang H, Zhang W, Xu C, Zhou X (2013a) Morphological features and physicochemical properties of waxy wheat starch. Int J Biol Macromol 62:304–309. https://doi.org/10.1016/j.ijbiomac.2013.09.030

Zhang B, Li X, Liu J, Xie F, Chen L (2013b) Supramolecular structure of A- and B-type granules of wheat starch. Food Hydrocoll 31:68–73. https://doi.org/10.1016/j.foodhyd.2012.10.006

Zhang B, Chen L, Li X, Li L, Zhang H (2015a) Understanding the multi-scale structure and functional properties of starch modulated by glow-plasma: a structure functionality relationship. Food Hydrocoll 50:228–236. https://doi.org/10.1016/j.foodhyd.2015.05.002

Zhang B, Chen L, Xie F, Li X, Truss RW, Halley PJ, Shamshina JL, Rogers RD, McNally T (2015b) Understanding the structural disorganization of starch in water–ionic liquid solutions. Phys Chem Chem Phys 17:13860–13871. https://doi.org/10.1039/C5CP01176K

Zhang B, Qiao D, Zhao S, Lin Q, Wang J, Xie F (2021) Starch-based food matrices containing protein: recent understanding of morphology, structure, and properties. Trends Food Sci Technol 114:212–231. https://doi.org/10.1016/j.tifs.2021.05.033

8 Analytical Tools for Starch Characterization

Gunjana Deka, Gitanjali Gautam, Aditi Duarah, Arnab Roy, and Himjyoti Dutta

Abstract

Starch is a plant-based, granular biological macromolecule with versatile molecular, functional, and nutritional properties. Starch has been extensively researched for morphological, structural, physicochemical, and nutritional properties. Being the most abundantly available edible polysaccharide, the use of starch in foods and biodegradable nonfood systems is inevitable. Extensive industrial usage necessitates starch characterization creating need for further developing and using the appropriate characterization tools. Starch has been synergistically looked from nutritional as well as material science perspectives. Being naturally granular and varied in shape and size, its unique construction pattern had always been a topic of curiosity. Arrangement of amylose and amylopectin within the glycosidic matrix has been greatly debated throughout the last century. Versatility in processed properties, especially gelatinization and retrogradation, and their resulting product characteristics have been widely reported. Digestibility of starch has been a matter of research for the nutrition-aware modern global human population. Concepts of rapidly digestible starch and resistant starch attained consumer popularity at a very rapid rate. In this chapter, different popular tools used for microscopic viewing, molecular and micro-constructional assessment, thermal characterization, rheology measurement, and digestibility analysis have been discussed to support future researchers with the basic knowledge on the principles and working of the analytical tools.

Keywords

Biopolymer · Electron microscopy · Crystallinity · Rheology · Infrared

G. Deka · G. Gautam · A. Duarah · A. Roy · H. Dutta (✉)
Department of Food Technology, Mizoram University, Aizawl, Mizoram, India

N. Mazumder, M. H. Rahman (eds.), *Advanced Research in Starch*,
https://doi.org/10.1007/978-981-99-9527-1_8

8.1 Introduction

The global market trend indicates major increase in plant-based foods across the world (Aschemann-Witzel et al. 2021). This has accelerated the researches on edible plants and nutrition derived from those. Starch is the major form of energy storage in plant cells. In plant cells, starch granules exist in the amyloplasts which are plastid organelles generally having distinct compartments. Besides being used by the plant for energy, consumed starch provides energy to the consumer through the release of glucose and subsequent conversion and release of adenosine triphosphate (ATP) molecules.

Granular starch is a polymer of D-glucose units having intermolecular glycosidic linkages, forming long glycosidic chains. The α-1,4 glycosidic linkages primarily form these chains. The fraction in starch composed of long glycosidic chains with very few periodic α-1,6 branch points is called amylose. The other fraction, amylopectin, has more frequent primary, secondary, and tertiary branch points giving it a brush-like structure. The amylopectin branch chains form intermolecular ordered superhelices, which provide the crystallinity to the glucose biopolymer. Randomly arranged amylose and unbranched amylopectin portions relate to its amorphousness, finally creating the semicrystalline structure. The physicochemical properties imparted by amylose and amylopectin are markedly dissimilar due to the major differences in bonding patterns, molecular sizes, and reactivity with polar and non-polar solvents. The amylose-to-amylopectin ratio varies in starches from different plant species (Pérez and Bertoft 2010). Their ratio in the starch is determinant of the behavior of starch to any processing treatment, primarily cooking.

The behavior of starch, when subjected to cooking or heating in sufficient water, is unique among all natural polymeric materials. The glycosidic granular structure allows starch to undergo gelatinization. At high temperature, the otherwise partially soluble starch attains high hygroscopicity and starts absorbing water by expanding the glycosidic chains and de-coiling of the crystalline superhelices. In the process, the granules incredibly swell. After attaining the state of maximum swelling where it cannot absorb any more water, the granule bursts at a stage called gelatinization and a temperature called gelatinization temperature. The entrapped unbonded water molecules get released, and amylose gets leached into the surrounding water forming a low viscosity hot paste. When cooled down, these released chains attain a tendency to return to the original conformations, recoil and recreate the lost crystallinity, and form a harder gel. This phenomenon, which is partially the reverse of gelatinization, is called retrogradation. The gelatinization and retrogradation properties of starch are primarily governed by amylose and amylopectin ratio. However, several other factors, namely, granule size and shape, porosity, chain length distributions, rate of heating and cooling, and presence of non-starch components like proteins, lipids, and enzymes, among others, also play major roles in these characteristic phase transitions in starch (Wang et al. 2021).

In the food processing industries, starchy plant-based raw materials and ingredients like cereal, legume, and tuber flours are largely used where the starch content and quality are primarily decisive of the product quality. Besides, extracted and purified starch is also a common food processing ingredient used for its plasticizing, thickening, structuring, and nutritive properties. Besides having hundreds of abundant natural sources of raw starch with varied properties, the tremendous industrial demand of the biomaterial has compelled scientists and technologists to work for developing modified and improved starches having engineered properties as desired for specific product purposes. Starches are modified physically and enzymatically without altering the basic glycosidic molecular status and chemically, where new functional groups are introduced to the molecules, by which new functional properties get introduced to the basic glycosidic structure (Iftikhar et al. 2021). In the food industry, apart from use as an edible and functional substance, suitably engineered and modified starches have been also used to create novel food coating and packaging materials. However, such high demand of raw and modified starches is not limited to their usage in the food industry but also extends to other industry. Major utilization of starch in composite materials used for packaging, building material, and other structures highly relates to its biodegradability (Cheng et al. 2021; Yu et al. 2021).

Being a common naturally occurring material with primary nutritional need in humans and herbivores, and extensive industrial applicability, researches on starch have been expansively carried out. The researches have evolved through time. Ideas on nanoscale construction of the granules, the crystalline lamellae, interaction with non-starch components, and functionality are still being generated. Simultaneously, analytical tools for visualizing and hypothesizing the micro- to nano-structural arrangements and analyzing the nutritional and functional properties of starch as powder, slurry, and paste have been developed and improved (Table 8.1). To understand starch, knowledge on these tools and their working principles are of much importance.

Table 8.1 Starch characteristics and analytical tools

Starch characteristics	Tools used
Granule morphology • Granule size and shape • Porous and pitted surface • Growth rings • Blocklet structures	Optical microscopy
Granule morphology • Maltese cross	Polarization microscopy
Spatial distribution Microstructural arrangement Associated proteins and lipids Molecular packing of starch constituents	CSLM
Reaction sites and anionic sites Chemical modification	R-CSLM
Granule morphology ultrastructure • Hilum and concentric lines of growth	SEM and TEM
Surface texture (roughness/smoothness) • Blocklet structures • Protrusions, pores, depressions, and defects	AFM
Molecular weight	HPSEC
Size and shape Molecular weight Conformation	SLS and MALLS

(continued)

Table 8.1 (continued)

Starch characteristics	Tools used
Chain length distribution Separation of poly-, oligo-, and monosaccharides	HPAEC
Crystallinity • Types A, B, C, and V • % Crystallinity • Diffraction angle at 2θ	XRD (WAXS)
Semi crystal structure • Crystalline and amorphous lamellar sizes and shapes embedded in the matrix • Operating range $0.001 < q < 5\ nm^{-1}$	SAXS
Specific bond vibrations of associated functional groups • Vibrational modes of amylose and amylopectin Chemical composition and physical state	FTIR
Structural properties Degree of crystallinity Molecular weight distribution Chemical composition (including presence of impurities) Water relationship with starch • Spin properties of ^{1}H (α-(1 → 4)- and α-(1 → 6)-glycosidic bonds), ^{13}C (C associated with nonreducing ends) and ^{31}P (phosphodiester cross-linking in naturally occurring or chemically modified starches).	NMR
Gelatinization, retrogradation, and retrograded crystalline melting transitions • Onset (T_o), peak (T_p), and conclusion (T_c) temperatures and the enthalpy change (ΔH)	DSC
Thermal transitions in dry starch • Decomposition into smaller compounds followed by ash formation	TGA
Paste viscosity Stability of gelatinizing starch granules Retrogradation • Peak viscosity (PV), pasting temperature (PT), breakdown viscosity (BV), setback viscosity (SB), and final viscosity (FV) or cold paste viscosity (CPV)	RVA
Texture profile • Hardness, fracturability, adhesiveness, gumminess, springiness, chewiness, and resilience	Texture analyzer
Static and dynamic flow behavior of starch • Static parameters: Consistency coefficient (K) and fluid index (n) $n = 1$, Newtonian fluid; $n < 1$, pseudoplastic fluid; $n > 1$, dilatant plastic fluid • Dynamic parameters: Storage modulus (G'), loss modulus (G”), and loss angle tangent ($\tan \delta = G''/G'$) Tan $\delta > 1$, viscous materials; $\delta < 1$, elastic materials	Rheometer

CSLM confocal scanning laser microscopy, *R-CSLM* reflectance confocal laser scanning microscopy, *SEM* scanning electron microscopy, *TEM* transmission electron microscopy, *AFM* atomic force microscopy, *HPSEC* high-performance size exclusion chromatography, *MALLS* multi-angle laser light scattering, *HPAEC* high-performance anion exchange chromatography, *XRD* X-ray diffractometry, *WAXS* wide-angle X-ray scattering, *SAXS* small-angle X-ray scattering, *FTIR* Fourier transform infrared spectroscopy, *NMR* nuclear magnetic resonance, *DSC* differential scanning calorimetry, *TGA* thermogravimetric analysis, *RVA* rapid visco analyzer

8.2 Microscopic Tools

The granules of starch range between 1 and 100 μm in size and variable shapes. The sizes and shapes vary from species to species, maturity stage and may also get altered due to processing. Further, the natural construction of the granules constitutes different glycosidic arrangements forming unique nano-sized blocklets and chains (Sun et al. 2022). Visualizing the granules and their construction necessitates the use of microscopy. For research on starch characterization, use of microscopic tools is inevitable.

8.2.1 Optical Microscopy

Optical microscope magnifies the sample image under visible light using glass lenses, which is the basic imaging technique for morphological analysis of microscopic samples. For starch imaging, compared to advanced microscopy techniques discussed later, optical microscopy is a handy and cost-effective option. However, the comparatively lower magnification (typically 1000×) and resolution are insufficient to create detailed surface imaging and study of finer arrangement of surface and internal structures (Jagadeesan et al. 2022). Starch granule image contrast can be improved by iodine staining and polarization microscope. In polarization microscope, only polarized light (PL) is used to illuminate the sample. Under PL, starch granules exhibit natural birefringence, and the Maltese cross at the granule hilum becomes clearly visible. Maltese cross is considered as the core of starch microstructural growth and is affected by processing and modification, which can be also evidenced by PL microscopy (Chakraborty et al. 2020).

8.2.2 Confocal Scanning Laser Microscopy

The evolution of optical microscopy from light microscopy to polarized microscopy and later through ultraviolet and fluorescent microscopy showed path to a more improved confocal scanning laser microscopy (CSLM). It uses laser beams to enhance contrast and resolution of the image. The unwanted planes of light are blocked in CSLM with the help of an aperture diaphragm (pinhole). The pinhole allows only the emitted light from the focal point to pass through, achieving optical sectioning in the laser path, meaning only light from one focal plane is collected. The focused beam of laser light is emitted directly onto a specific point within the sample. This angle can be changed with addition of a movable mirror on the laser path, and the whole sample can be scanned. The plane of light can be also changed periodically to capture image of differently fluorescent constituents. Suitable dying with fluorescent molecules can be carried out to obtain higher resolution of the captured images. After capturing the images at different angles and different laser lights, a 3D image can be generated using software. The background of the non-observed points would remain black (Chakraborty et al. 2020; Huang et al. 2022).

This microscopy technique allows visualization of spatial distribution of starch granules within a sample, change in microstructural arrangement across the granular matrix over processing and modification, and starch-associated proteins and lipids, among others. Dyes like Nile red and iodine can be used for capturing high-resolution images at appropriate wavelength of the laser lights used. CSLM can be used for obtaining high-resolution 3D starch imaging providing an understanding about the molecular packaging of starch constituents and their associated proteins, which are the indicators of sites of amylose biosynthesis. Reflectance confocal laser scanning microscopy (R-CLSM) is an improvised CLSM approach that provides an easier and faster way of studying starch microstructures, generally determining the reaction sites and anionic sites in starch granules and chemically modified starches (Huang et al. 2022).

8.2.3 Scanning Electron Microscopy

For microscopic imaging of starch granule surfaces, scanning electron microscope (SEM) is the most extensively used tool. Visualizing the exact size and shape of the granules, topography, uniformity and defect, surface roughness, pore size, and their distribution are possible under SEM. Physical interaction of starch with other compounds can be also visualized. It can also provide composition and crystallographic information, element types, relative ratio, arrangement of atoms, and their degree of order (Chakraborty et al. 2020; Akhtar et al. 2018). The instrument is based on the principle that primary electrons released from the tungsten filament of a lamp, when bombarded on the sample, can release secondary electrons (SEs). Starch sample is fixed on the stub using adhesive tape. A light source with a shorter wavelength should be preferably used (Jagadeesan et al. 2022). Heat energy in the range of 1–40 kV is provided to accelerate the primary electrons produced and emitted from the source, which are then focused to a monochromatic beam (dia ≤ 100 nm) using magnetic field lenses and metal slits within the vacuum column. The SEs, along with backscattered primary electrons, are used by a detector to create the surface image of the sample being scanned on a fluorescent screen. Elevated regions receive higher energy and quantity of electrons than lower regions, which is reflected in the relative brightness and darkness of the created image. Electromagnetic objective and condenser lenses, the path of the beam being free of gases (vacuum), and the right voltage ensure maximum and efficient bombardment and image resolution. For the most improved resolution and prevent sample damage at higher magnification, a very thin layer of highly conductive metal like gold, palladium, platinum, or silver is often sputter coated on the sample before analysis. The resolution of SEM ranges between 0.5 and 4 nm, which is typically sufficient to study the surface morphology of starch granules and measure the dimensions. However, one important necessity of drying of the SEM sample limits its usage for wet samples like gelatinized starch. Environmental SEM (ESEM) and liquid phase electron microscopy (LPSEM) are two more recent developments over classical SEM where fresh, wet samples can be also analyzed through a special composite mechanism by which electrons scattered by the surrounding gas-liquid mixture are nullified (de Silveira et al. 2020).

8.2.4 Transmission Electron Microscopy

With a resolution of 0.1–0.2 nm, transmission electron microscopy (TEM) generates better resolution images of nano-sized materials and their finer structure. While SEM uses backscattered and secondary electrons for image generation, TEM uses the transmitted electrons. The positioning of the sample, lenses, and detectors in the electronic beam path is accordingly done. Sufficient electron transmission requires minimum scattering of electrons by the sample. For viewing the internal morphology of samples like starch granules, an ultramicrotome is used to prepare sections of 50–100 nm thickness. Sample thickness is critical for resolution and coverage of the structures to be observed in the generated electron-lucent photograph. Starch granule ultrastructure including hilum and concentric lines of growth can be observed under SEM and TEM if proper microtomy is done (Wulff et al. 2020). Sputter coating of sample is not required in case of TEM. TEM can better reveal miniature changes in arrangement after processing and modification. Like conventional SEM, as discussed above, dehydration during sample preparation and observation could hamper image interpretation, creating necessities for innovations targeted at wet sample analysis (Huang et al. 2022).

8.2.5 Atomic Force Microscopy

AFM has been the most effective tool in starch surface visualization. Unlike light and electron microscopy where 2D microscopic images are formed by projected light and electron beams, 3D atomic force microscopy (AFM) image of sample surface is formed based on the precise movement of a sharp crystal tip over it. Starch sample is simply fixed over a freshly developed mica or silica surface. The crystal tip mounted on a flexible cantilever spring is brought close to the sample. Interatomic forces between the tip and the sample cause the cantilever to deflect. The cantilever is moved laterally across the surface of the sample in a raster pattern, and as it scans, the deflection is measured using an optical laser beam detection system which provides a high-resolution topographical data of the sample surface. In contact mode operation, the tip touches the sample surface, and a repulsive interactive force exists between them. In noncontact mode, the tip does not touch the sample surface, and the interactive force between the two is predominant. With movement of the tip over the surface, the laser detector actually records and draws a force map. Besides the topography measurement, AFM also records frictional force image and phase lag images that indicate the surface textural properties of the sample affecting movement of the tip. AFM can provide high-resolution nanoscale image in atmospheric or aqueous conditions (Huang et al. 2022). With the help of height parameter (R), roughness average (Ra), and root mean square roughness

(Rq), AFM can determine quantitative parameters of surface roughness. Besides surface smoothness and material properties, the blocklet structure of starch construction can be predominantly observed using AFM. Protrusions, pores, and minute depressions and defects on the surface are also precisely recorded. Suitable grinded starch granules can be subjected to AFM for generating images of glycosidic clusters, spherulites, and crystals. This makes AFM the most ideal microscopic tool for microstructural imaging of starch (Zhu 2017). Great care should be taken to maintain machinery parts and operative conditions to obtain the minutest nanoscale detailing of the images created by AFM.

8.3 Constructional Assessment Tools

The glycosidic construction of starch largely encompasses the crystalline and amorphous distributions of the chains. None of the microscopic tools discussed above has sufficiently evolved to an extent that the exact chain size, length, and crystalline structures can be visualized. Chain length distribution based on molecular mass detection is traditionally carried out using gel permeation chromatography (GPC), which was later improvised to high-performance size exclusion chromatography (HPSEC) fitted with appropriate eluent detectors. Starch crystallinity relates to most of its functionalities. X-ray diffraction study for crystallinity and amorphousness detection and measurement has been a common practice. Also, the types of bonds in the polymeric chains, newer bonds formed on chemical modifications, and the corresponding vibrations are detected by infrared (IR) spectroscopic and nuclear magnetic resonance (NMR) techniques (Fig. 8.1).

8.3.1 Chromatography

Popularly abbreviated as HPSEC, high-performance size exclusion chromatography is the modern version of gel permeation chromatography. It is based on fractionation of different molecular weight segments of starch when passed through a gel having appropriate pore sizes for the elution to occur. Initially used for separating amylose and amylopectin, later it was used for separating different molecular weight (more appropriately, hydrodynamic volume, Vh) fractions of amylopectin fine structures obtained after using suitable debranching enzymes like isoamylase or pullulanase enzyme to sub-fractionate the polymer. Molecules with higher Vh elute faster through the containing gel as the stationary phase. Once eluted, their presence is traditionally detected by differential refractive index detector which operates on the principle of the change in refraction of incident light when the eluent passes through the point of contact. Lately, static light scattering (SLS) and multi-angle laser light scattering (MALLS) have been identified as more powerful detectors to determine the size, molecular weight, and shape of the eluted fractions. As the laser beam is directed toward the sample, the scattered light is detected at multiple angles simultaneously. The scattering data collected at different angles are analyzed using

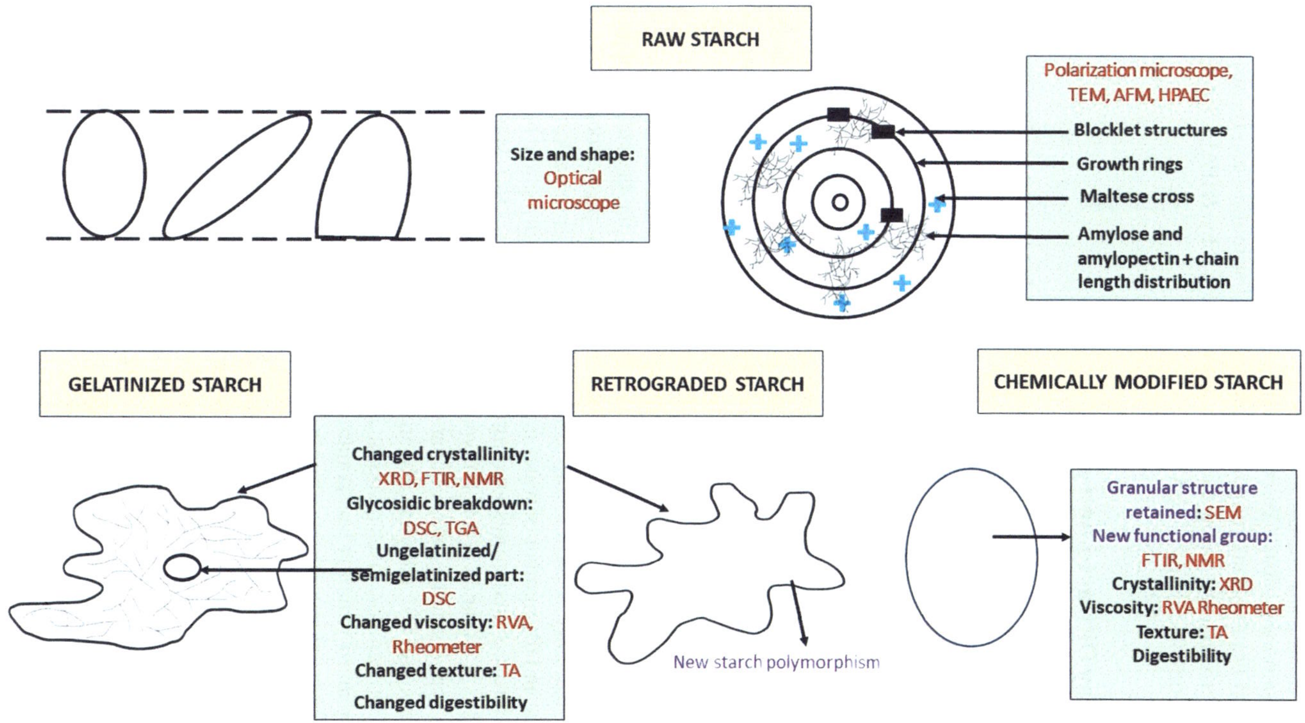

Fig. 8.1 Analytical tools for studying structural properties and their changes in raw and processed starch

mathematical algorithms using Debye plot or Zimm plot. This provides information regarding the molecular weight, size, and shape and conformation of the starch sample (Rübsam et al. 2012; Ye et al. 2019). Dextrins of known molecular weight are used to calibrate the instrument. Chain length distribution can be also assessed using high-performance anion exchange chromatography (HPAEC) coupled with pulsed amperometric detector (PAD).

8.3.2 X-Ray Diffractometry

Starch is a semicrystalline biological macromolecule. The inherent crystallinity is primarily imparted by ordered arrangement of amylopectin side chains forming supercoil helical structures entrapping water molecules in between and secondarily by amylose long chains forming regular arrangements around certain molecules. The degree of crystalline arrangements in starch granules can be measured by X-ray diffraction (XRD) studies. When an X-ray hits an atom, it gets diffracted by the electrons. For orderly arranged crystalline lamellae within starch matrix, where the atoms are arranged in a regular fashion, the diffracted rays undergo "constructive interference" at specific angle, resulting in distinct peak formations in the software generated diffractograph. The diffractographs record different angles (2Θ) at which X-rays are diffracted from the sample. Irregularly arranged atoms in the amorphous regions do not form such peaks. Using a two-phase model, the XRD determines the double helical packing contributed by the long-range crystalline orders of amylopectin side-chains. The model subtracts the amorphous corona from the crystalline domains of X-ray scattering (Epp 2016). The ratio between these two phases depicts the degree of starch crystallinity. Starches from different sources differ in packing of the crystalline lamellae and the number of water molecules per unit turn of the helix. Depending on such differences and corresponding peak positions in XRD, starches have been categorized as A, B, and C (A + B) types. Typical compact A-type pattern, which is common in cereal starches, shows characteristic peaks at 15° (singlet), $2\theta = 17°$ and 18° (doublet), and 23° (singlet). Relatively more open B-type pattern, common in tuber and root starches, gives peaks at $2\theta = 14°$, 15°, 17°, 22°, 24°, and 26°. C-type diffraction pattern which is less common and found in certain legumes, beans, and water chestnut, among others, gives peaks corresponding to mixed A- and B-type starch (Govindaraju et al. 2021). Processing can also alter the native patterns and degrees of crystallinity. A V-type crystalline polymorph of starch representing amylose or long chain amylopectin complexation with fatty acid, iodine, or other molecules can be obtained in native or processed forms of any of these with a characteristic peak at $2\Theta = 20°$ (Jagadeesan et al. 2022; Huang et al. 2022; Pozo et al. 2018; Subroto 2020). Hence, for starch powder XRD, recording of diffraction angle within the range of 2–40° of 2Θ is often considered as adequate (Huang et al. 2022). By measuring the area under the crystalline peaks and finding its proportion to that of the whole diffractograph, relative of percentage crystallinity can be measured. As XRD follows the Bragg's law ($n\lambda = 2d\sin\Theta$), interlamellar spacing (d) corresponding to each diffraction peak can be calculated, and any process-induced change and deformation can be analyzed. Detailed peak analysis also enables to assess reduced or developed crystalline perfection over processing or modification of starch.

For generating a more detailed idea about lamellar sizes and shapes (both crystalline and amorphous) embedded in the matrix, small-angle X-ray scattering (SAXS) is preferably used over XRD, which is also known as wide-angle X-ray scattering (WAXS). Electron density difference ($\Delta\rho$) between the crystalline lamellae (ρ_c) and the amorphous lamellae (ρ_a) ($\Delta\rho= \rho_c-\rho_a$) and the electron density difference ($\Delta\rho_u$) between the background material (ρ_u) and the amorphous lamellae (ρ_a) ($\Delta\rho_u= \rho_u-\rho_a$) along with the volume of the crystalline lamellae in the crystalline lamellae fraction influence the intensity of SAXS (Huang et al. 2022). Unlike pure crystalline materials with ideal lamellar structures, the delicate distributional difference in the domain types in semicrystalline starch has been identified as a major prohibiting factor for inadequate calculations of starch crystal sizes. Use of SAXS in starch characterization is largely based on theoretical correlation models and is still evolving. Newer approaches for appropriate measurement are presently being developed (Lin et al. 2019).

8.3.3 Infrared Spectroscopy

Fourier transform infrared (FTIR) spectroscopy uses light waves in the mid-infrared region (4000–400 cm^{-1}) for detecting and measuring bond vibrations corresponding to IR absorption and therefore identifying functional groups associated with the molecule. Information is provided based on the entire sample's chemical composition and physical state (Govindaraju et al. 2021). Starch powder can be subjected to conventional transmission mode FTIR by pressing into a pellet of potassium bromide and placing in the pellet holder of the machine before running the test. Alternatively, attenuated total reflectance (ATR) can be carried out where the pellet formation is not necessary and the sample can be recovered after the experiment. In the machine, a beam of IR light is projected to a crystal (optical element) at an angle greater than the critical angle for total internal reflection, which influences the light to bounce along the surface of the crystal. As the IR beam travels along the surface of the crystal, it interacts with the sample and gets absorbed by the bonds in the sample resulting in decreased intensity. The extent of intensity drop is proportional to the concentration and chemical composition of the sample. A detector helps in finally collecting and analyzing the obtained data in the form of attenuated light to determine the chemical composition of the sample qualitatively and quantitatively (Huang et al. 2022). For starch, major vibrations are recorded for C-O, C-C, C-O-H, O-H, C-H, and -H bonds; skeletal mode vibrations of the pyranose ring of glucose monomers and bond representing minor proteinous and lipid components associated with the sample are recorded. A list of vibrational wavenumbers can be obtained from (Xu et al. 2021). The same authors have also reported on the absorption band ratios which suggest short-range ordered molecular structures (SMS) that are affected by gelatinization and retrogradation. A more detailed and appropriate idea on SMS can be generated after coupling FTIR with Raman spectroscopy.

8.3.4 Nuclear Magnetic Resonance

Nuclear magnetic resonance (NMR) spectroscopy is majorly used for the structural enumeration of modified starch and to determine the degree of substitution of the chemical substitute. Both solid and liquid states utilize NMR methods such as ^{1}H, ^{13}C, ^{31}P, and ^{17}O NMR. The utilization of these approaches may be characterized by their nondestructive nature, which allows for the acquisition of innovative perspectives on the structural aspects of starch systems. When starch is present with water, it undergoes several transformations during heating and cooling making it an essential structure-forming component in food systems (El Nokab et al. 2022; Riley et al. 2022). When it comes to determining the level of branching in starch polysaccharides in a solution, high-resolution NMR is the preferred method at the molecular level. In the solid state, NMR is the only reliable method for quantifying the number of double helices and single helices, regardless of whether they are part of crystalline structures or not. NMR relaxation techniques can provide information on both local motion and how water interacts with molecular surfaces. It is possible to chemically determine the reducing ends of starch molecules and identify nonreducing ends by comparing the reducing power before and after debranching. However, ^{1}H NMR is better for quantifying branching. This method compares the intensity ratios of anomeric proton signals from α-1,6- and α-1,4-linked residues (Gidley 2014). Recent developments in nonaqueous solvent solubilization and dilute acid alteration of critical peaks have increased the technique's dependability. Other applications of NMR in starch chemical analysis include the measurement of organic substituents and the characterization of phosphate substituent proportions using ^{31}P NMR. Apart from XRD, ^{13}C NMR spectra can be also used to identify crystalline type (A and B in starches) and degree of crystallinity of starch samples (Xu et al. 2021).

8.4 Thermal Properties

8.4.1 Differential Scanning Calorimetry

Differential scanning calorimetry (DSC) is a technique used to study the heat flow associated with phase transitions and chemical reactions in materials. DSC measures the difference between heat flow in the sample and that of a reference material (Wang et al. 2015). Distinct transition temperatures, namely, onset (T_o), peak (T_p), and conclusion (T_c) temperatures, are recorded, and the enthalpy change (ΔH) per unit mass of the sample related to the recorded transition is measured (Jagadeesan et al. 2022). Starch is a semicrystalline biopolymer. As stated previously, gelatinization under raised temperature, when heated in excess water, and retrogradation of gelatinized starch upon cooling are unique thermal transitions in starch. To study these transitions in DSC, miniature starch slurry sample is precisely prepared (at least 1:2, starch/water, w/v) in DSC pan, sufficiently moisture stabilized for uniformity, sealed, and heated at a constant rate of rising temperature against an empty reference pan. Heating is carried out from a temperature of about 20 °C to 150 °C, and all thermal transitions take place within this range. Generally, endothermic transition representing gelatinization appears between 60 °C and 90 °C; the exact temperature values depend on several factors including starch source, whether pre-modified or not, extent of moisturization, and heating rate. For instance, gelatinization temperatures of rice starch, corn starch, potato starch, and wheat starch are approximately 60–80 °C, 60–75 °C, 60–70 °C, and 50–70 °C, respectively. To measure thermal transitions during retrogradation, the gelatinized starch-containing pan is then cooled down to about 20 °C to allow sufficient retrogradation. Besides recording the exothermic thermal transitions during retrogradation, the retrograded starch can be again subjected to heating to record the endothermic melting enthalpy of crystallites formed during retrogradation. For each transition, T_o, T_p, T_c, and ΔH can be recorded. For gelatinization, ΔH represents the energy intake for the dissociation of crystalline double helices in the granules. In retrogradation, ΔH represents recoiling of the glycosidic chains to form retrograded crystallites. In a third transition (if recorded), ΔH quantifies the thermal transition of melting of starch-lipid complexes. Generally, both T_o and ΔH for the later two transitions are lower than the former. Melting of starch-lipid complexes occurs between 90 °C and 130 °C, which gives one or more distinct lower intensity peak(s) and can be distinctly differentiated from gelatinization or melting peaks by the T_o and ΔH (Oyeyinka et al. 2021). Their occurrence can be confirmed by conducting XRD and NMR characterization.

8.4.2 Thermogravimetric Analysis

While DSC of starch slurry generates data on gelatinization, retrogradation, and retrograded crystalline melting transitions, thermal transitions in dry starch can be studied using thermogravimetric analysis (TGA). In material science, TGA is used to study thermal decomposition of materials based on monitoring the changes in weight of the sample upon continuous heating or cooling. In it, sample is accurately weighed in a crucible, and temperature is raised by 5–10 °C/min to a prespecified temperature sufficient to achieve apparent combustion completion under constant rate of airflow. For native starch, distinct drop in material weight as recorded from the graph generated is due to moisture evaporation, starch decomposition into smaller compounds dextrins and glucose, and finally ash formation (Wang et al. 2015). For starch-based composites, changes in TGA curve patterns, newer mass drop, and newer transitions can be recorded and tallied with microscopic, XRD, DSC, and other results.

8.5 Paste Rheology

Upon gelatinization, starch forms a gel. The gel characteristics are principally decisive of the way in which the starch would behave in a processed food system. The change in molecular state of starch during gelatinization and retrogradation is reflected in the change in viscosity of the gel. It can be also affected by extraneous factors such as temperature, moisture content, shearing force, and presence of amylolytic enzymes, among others. Gel viscosity measurement is a key analytical activity in starch industries. For most rapid and automated measurement of paste viscosity, Rapid Visco Analyzer (RVA) is the most evolved equipment used extensively worldwide.

8.5.1 Rapid Visco Analyzer

RVA is exclusively used for starch and starch-rich food powders to measure the rheological changes during cooking and cooling. The basic principle of RVA is that when a sample is subjected to controlled temperature and shear conditions, its viscosity and rheological behavior changes over time. RVA measures this change in viscosity as a function of time and temperature. The slurry with specific dry starch to moisture is prepared in a specifically manufactured cylindrical cup, in which a paddle is rotated at prespecified rates, first slowly for slurry uniformity and then at a much faster constant rate for point-by-point measurement of torque required to maintain constant speed of shearing as recoded by the instrument. Various rheological properties, namely, peak viscosity (PV), pasting temperature (PT), breakdown viscosity (BV), setback viscosity (SB), and final viscosity (FV) or cold paste viscosity (CPV), are determined by analyzing the generated viscosity curve. The stability of gelatinizing starch granules in the thermal environment under shearing is determined by the BV, which is related to and follows the occurrence of PV. Severe leaching of amylose can be inferred due to high granule swelling and subsequent breakage during BD. The pasting of the starch molecules from a disordered structure (high energy) to an ordered structure (low energy) is represented by SB which depicts the short-term retrogradation of starch. The amylose and lipid contents along with the branch chain length distribution of amylopectin govern the pasting properties of starch. Swelling of the starch granules is promoted due to amylopectin, whereas it is inhibited due to amylose and lipids (Roy et al. 2019). The chain length of amylopectin and molecular size of amylose produce combined effect on the starch paste viscosity. In the case of cereal starches, generally the waxy varieties show lower pasting temperatures, rapid viscosity development till PV, and lower SB compared to the higher amylose starches. High amylose starches produce a strong gel after a high SB. Increase in pasting temperature and increased resistance to shear-thinning of starch pastes can be seen in the starch-lipid complexes of native starches of rice and maize (Balet et al. 2019).

8.5.2 Texture Profile Analysis

The cooled gel obtained after the RVA operation can be optionally tested in a texture analyzer for a detailed texture profile analysis (TPA) of the gel. For this, the uniform surface of the gel, adjusted at the appropriate temperature, is subjected to destructive pressing by a suitable probe in a single-bite or two-bite test. Parameters, namely, hardness, cohesiveness, stickiness, and chewiness, among others, can be obtained from the generated graph. Each obtained parameter relates to the molecular arrangements, complexity, and freedom of starch glycosidic fractions occurring inside the gel (Liu et al. 2021).

8.5.3 Rheometry

Starch gel can be also subjected to a rheometric analysis in a rheometer to study the effect of controlled stress and strain on the flow and deformation behavior of the gel. Two main rheological studies are popularly performed for starch flow behavior analysis—static and dynamic (Xie et al. 2012). Consistency coefficient (K) and fluid index (n) are the static parameters (steady-state shear flow). Under certain stress or strain with increasing shear rate, the static measurements rapidly cause an increase in shear stress and further level off and cause an increase in the shear rate. The n is the flow behavior index: $n = 1$, Newtonian fluid; $n < 1$, pseudoplastic fluid; and $n > 1$, dilatant plastic fluid. On the other hand, dynamic rheological measurements measure the viscoelastic modulus of samples to determine the viscous or elastic nature of a fluid. The main parameters of dynamic rheology are storage modulus (G'), loss modulus (G“), and loss angle tangent (tan δ = G”/G'). Solid-like properties of viscoelastic materials can be determined by G' in terms of elasticity and rigidity, while liquid-like properties can be determined by G” as viscosity and fluidity. Viscous materials show tan $\delta > 1$, while elastic materials show $\delta < 1$. For solid and weak gels, the values of tan δ are approximately 0.01 and 0.1, respectively (Xie et al. 2012; Chen et al. 2022).

8.6 Digestibility

Starch is the major nutritional polysaccharide in plant-based food systems. Its fate in the gastrointestinal (GI) tract is important from both nutritional and nutraceutical aspects. While glucose liberated on complete acid and enzymatic hydrolysis of starch is the primary source for energy metabolism, partially digested to non-digested dextrins and enzyme-resistant fractions are used for microbial metabolism in the gut for liberation of health-promoting compounds. Starch digestion in the GI tract involves the activity of salivary and pancreatic α-amylases, which break down starch molecules into smaller oligosaccharides such as maltose, and α-limit dextrin. These α-limit dextrins are branched oligosaccharides. In the small intestine, these oligosaccharides are further converted into glucose through the action of brush border enzymes known as maltase-glucoamylase and sucrose-isomaltase (Noda and Sarker 2014; Yang et al. 2023). Using animal models, this enzymatic hydrolysis pattern for starch samples can be assessed in vivo. However, researchers have tried to mimic the digestion pattern in the laboratory conditions maintaining enzyme strengths, pH, and time of in vitro digestion. As a popular approach, based on the time required for glucose liberation under simulated GI conditions, starch has been classified into three fractions, namely, rapidly digestible starch (RDS) which is the starch amounting to release of glucose after 20 min incubation, slowly digestible starch (SDS) which is the starch amounting to release of glucose at 120th minute of incubation, and resistant starch (RS) which was not digested after prolonged incubation (Singh et al. 2010). The starch-enzyme digestion curves have been described from theoretical mathematical relations such as first-order kinetics[13] and log of slope plots (LOS) (Butterworth et al. 2012). The first-order equation is often used for fitting starch hydrolysis curves. However, only a small number of researchers used the INFOGEST oral-gastrointestinal standardized approach to hydrolyze starch (Minekus et al. 2014). Despite its potential use for measuring the impact of various drugs on digestion, starch gel hydrolysis kinetics are yet to be explored. There is also involvement of extensive manual operations in these analytical processes. While enzymatic kits have been developed and used, automated tools for starch digestibility assessment might be soon developed with the present advent of Industry 4.0 across global plant-based food industries.

8.7 Conclusion

Starch, owing to its abundance and being the major source of dietary glucose, bears the highest economic significance among all polysaccharides. Research on starch and its properties has been abundantly carried out for more than a century. Analytical tools for starch characterization have been accordingly developed, and improvements to each type are being taken up. With such developments, several new discoveries have been made during the recent decade, especially on the basic structural attributes of the granular macromolecule and crystalline polymorphism. These again have created future scopes for newer researches on its material and nutritional properties, widening applicability in both food and nonfood industries.

Acknowledgments The authors would like to acknowledge the financial grant received from the Science and Engineering Research Board, Government of India (SRG/2021/000227).

References

Akhtar K et al (2018) Scanning electron microscopy: principle and applications in nanomaterials characterization. In: Handbook of materials characterization. Springer, Cham, pp 113–145. https://doi.org/10.1007/978-3-319-92955-2_4

Aschemann-Witzel J et al (2021) Plant-based food and protein trend from a business perspective: markets, consumers, and the challenges and opportunities in the future. Crit Rev Food Sci Nutr 61(18):3119–3128. https://doi.org/10.1080/10408398.2020.1793730

Balet S et al (2019) Rapid Visco analyser (RVA) as a tool for measuring starch-related physiochemical properties in cereals: a review. Food Anal Methods 12(10):2344–2360. https://doi.org/10.1007/s12161-019-01581-w

Butterworth PJ et al (2012) Analysis of starch amylolysis using plots for first-order kinetics. Carbohydr Polym 87(3):2189–2197. https://doi.org/10.1016/j.carbpol.2011.10.048

Chakraborty I et al (2020) Advanced microscopy techniques for revealing molecular structure of starch granules. Biophys Rev 12(1):105–122. https://doi.org/10.1007/s12551-020-00614-7

Chen S et al (2022) Modification of starch by polysaccharides in pasting, rheology, texture and in vitro digestion: a review. Int J Biol Macromol 207:81–89. https://doi.org/10.1016/j.ijbiomac.2022.02.170
Cheng H et al (2021) Starch-based biodegradable packaging materials: a review of their preparation, characterization and diverse applications in the food industry. Trends Food Sci Technol 114:70–82. https://doi.org/10.1016/j.tifs.2021.05.017
de Silveira G, Forsberg P, Conners TE (2020) Scanning electron microscopy: a tool for the analysis of wood pulp fibers and paper. In: Surface analysis of paper. CRC, Boca Raton, pp 41–71
El Nokab MEH et al (2022) Solid-state NMR spectroscopy: towards structural insights into starch-based materials in the food industry. Polymers 14(21):4686. https://doi.org/10.3390/polym14214686
Epp J (2016) X-ray diffraction (XRD) techniques for materials characterization. In: Materials characterization using nondestructive evaluation (NDE) methods. Elsevier, Amsterdam, pp 81–124. https://doi.org/10.1016/b978-0-08-100040-3.00004-3
Gidley MJ (2014) Starch NMR. In: Starch polymers. Elsevier, Amsterdam, pp 243–253. https://doi.org/10.1016/b978-0-444-53730-0.00029-4
Govindaraju I et al (2021) Structure and morphological properties of starch macromolecule using biophysical techniques. Die Starke 73(1–2):2000030. https://doi.org/10.1002/star.202000030
Huang J et al (2022) A review of wheat starch analyses: methods, techniques, structure and function. Int J Biol Macromol 203:130–142. https://doi.org/10.1016/j.ijbiomac.2022.01.149
Iftikhar SA, Chakraborty S, Dutta H (2021) Effect of acetylation, hydroxypropylation and dual acetylation-hydroxypropylation on physicochemical and digestive properties of rice starches with different amylose content. Biointerface Res Appl Chem 12:6788–6803. https://doi.org/10.33263/briac125.67886803
Jagadeesan S, Govindaraju I, Mazumder N (2022) An insight into the ultrastructural and physiochemical characterization of potato starch: a review. Am J Potato Res 97(5):464–476. https://doi.org/10.1007/s12230-020-09798-w
Lin L, Chi C, Wu C (2019) How to calculate starch lamellar features with improved accuracy by small angle X-ray scattering. Int J Biol Macromol 141:622–625. https://doi.org/10.1016/j.ijbiomac.2019.09.054
Liu X et al (2021) Mechanistic studies of starch retrogradation and its effects on starch gel properties. Food Hydrocoll 120:106914. https://doi.org/10.1016/j.foodhyd.2021.106914
Minekus M et al (2014) A standardized static in vitro digestion method suitable for food–an international consensus. Food Funct 5(6):1113–1124. https://doi.org/10.1039/c3fo60702j
Noda T, Sarker MZI (2014) Factors influencing starch digestibility. In: Proceedings of the international conference on science, technology and social sciences (ICSTSS) 2012. Springer, Singapore, pp 449–453. https://doi.org/10.1007/978-981-287-077-3_52
Oyeyinka SA, Singh S, Amonsou EO (2021) A review on structural, digestibility and physicochemical properties of legume starch-lipid complexes. Food Chem 349:129165. https://doi.org/10.1016/j.foodchem.2021.129165
Pérez S, Bertoft E (2010) The molecular structures of starch components and their contribution to the architecture of starch granules: a comprehensive review. Die Starke 62(8):389–420. https://doi.org/10.1002/star.201000013
Pozo C et al (2018) Study of the structural order of native starch granules using combined FTIR and XRD analysis. J Polym Res 25(12):1. https://doi.org/10.1007/s10965-018-1651-y
Riley IM et al (2022) The use of time domain 1H NMR to study proton dynamics in starch-rich foods: a review. Compr Rev Food Sci Food Saf 21(6):4738–4775. https://doi.org/10.1111/1541-4337.13029
Roy M et al (2019) Effect of steam parboiling and hot soaking treatments on milling yield, physical, physicochemical, bioactive and digestibility properties of buckwheat (Fagopyrum esculentum L.). J Food Sci Technol 56:3524–3533. https://doi.org/10.1007/s13197-019-03849-9
Rübsam H et al (2012) An overview of separation methods in starch analysis: the importance of size exclusion chromatography and field flow fractionation. Die Starke 64(9):683–695. https://doi.org/10.1002/star.201100188

Singh J, Dartois A, Kaur L (2010) Starch digestibility in food matrix: a review. Trends Food Sci Technol 21(4):168–180. https://doi.org/10.1016/j.tifs.2009.12.001

Subroto E (2020) Review on the analysis methods of starch, amylose, amylopectin in food and agricultural products. Int J Emerg Trends Eng Res 8(7):3519–3524. https://doi.org/10.30534/ijeter/2020/103872020

Sun X et al (2022) Molecular structure and architectural characteristics of outer shells and inner blocklets of normal and waxy wheat A-and B-starch granules. J Cereal Sci 105:103477

Wang S et al (2015) Starch Retrogradation: a comprehensive review: starch retrogradation. Compr Rev Food Sci Food Saf 14(5):568–585. https://doi.org/10.1111/1541-4337.12143

Wang Q, Li L, Zheng X (2021) Recent advances in heat-moisture modified cereal starch: structure, functionality and its applications in starchy food systems. Food Chem 344:128700

Wulff D et al (2020) Characterizing internal cavity modulation of corn starch microcapsules. Heliyon 6(10):e05294. https://doi.org/10.1016/j.heliyon.2020.e05294

Xie F, Halley PJ, Avérous L (2012) Rheology to understand and optimize processibility, structures and properties of starch polymeric materials. Prog Polym Sci 37(4):595–623. https://doi.org/10.1016/j.progpolymsci.2011.07.002

Xu H et al (2021) Methods for characterizing the structure of starch in relation to its applications: a comprehensive review. Crit Rev Food Sci Nutr 63(20):1–18. https://doi.org/10.1080/10408398.2021.2007843

Yang Z et al (2023) Factors influencing the starch digestibility of starchy foods: a review. Food Chem 406:135009. https://doi.org/10.1016/j.foodchem.2022.135009

Ye J et al (2019) Synthesis and characterization of citric acid esterified rice starch by reactive extrusion: a new method of producing resistant starch. Food Hydrocoll 92:135–142. https://doi.org/10.1016/j.foodhyd.2019.01.064

Yu X et al (2021) Research progress of starch-based biodegradable materials: a review. J Mater Sci 56:11187–11208. https://doi.org/10.1007/s10853-021-06063-1

Zhu F (2017) Atomic force microscopy of starch systems. Crit Rev Food Sci Nutr 57(14):3127–3144. https://doi.org/10.1080/10408398.2015.1094650

Modified Starches Used in the Food Industry and Biomedical Applications

9

Tarun Mateti, Malti Kumari, Savita,
Smarak Islam Chaudhury, K. Likhith, and Goutam Thakur

Abstract

Starch is widely used in food products and as a biomaterial due to its attributes, such as compostability, availability, non-toxicity, and low cost. However, starch inherently may not be suitable for specific applications and must be modified to enhance its functionality. Starch modification can be performed using physical, chemical, and enzymatic methods that alter its physicochemical and structural properties. Physical methods involve altering the physical structure of starch without introducing chemical changes to its molecular composition and are simple and inexpensive. Chemical modifications involve exploiting the hydroxyl groups present in starch to introduce covalent bonds between starch molecules, whereas enzymatic modification targets the amylose content, the molecular mass, and the structure of amylopectin chains to

Tarun Mateti, Malti Kumari, Savita, Smarak Islam Chaudhury and K. Likhith contributed equally with all other contributors.

T. Mateti
Centre for BioSystems Science and Engineering, Indian Institute of Science, Bangalore, Karnataka, India

Materials Research Centre, Indian Institute of Science, Bangalore, Karnataka, India

M. Kumari · Savita · S. I. Chaudhury
Materials Research Centre, Indian Institute of Science, Bangalore, Karnataka, India

K. Likhith · G. Thakur (✉)
Department of Biomedical Engineering, Manipal Institute of Technology, Manipal Academy of Higher Education, Udupi, Karnataka, India
e-mail: goutam.thakur@manipal.edu

N. Mazumder, M. H. Rahman (eds.), *Advanced Research in Starch*,
https://doi.org/10.1007/978-981-99-9527-1_9

selectively break down or modify specific chemical bonds within starch molecules. This chapter describes starch modification methods for its use in the food industry and as a biomaterial for biomedical applications such as wound healing and its prospects.

Keywords

Starch · Food industry · Biomedical applications · Physical, Chemical, Enzymatic modifications

9.1 Introduction

Starch is a complex carbohydrate polymer with a long chain of glucose molecules linked together, broken down by enzymes in saliva or pancreatic fluid (Carpenter et al. 2015). The body utilizes starch as a primary energy source, and it is found in various foods, such as grains, cereals, fruits, vegetables, and legumes. It contributes to overall health and well-being by regulating blood sugar levels, promoting healthy digestion, and providing essential vitamins and minerals.

Starch comprises 20–25% amylose and 75–80% amylopectin by weight and serves as short-term energy storage in plants and animals. Starch is formed when plants produce glucose during photosynthesis, which is then transported to various parts, such as leaves, stems, and roots, where it is stored as starch granules. However, starch is only found in animals when they consume starch-containing foods, which are digested and metabolized to produce glucose that is used as a source of energy (Lee et al. 2012).

Starch is widely used in various industrial and biological applications. The food industry uses it as a thickener, stabilizer, and gelling agent in various products as it can improve the texture, consistency, binding, and mouthfeel, making its products—such as sauces, soups, puddings, and baked goods—more appealing to consumers. Starch has also been used as a biomaterial for applications like tissue engineering, drug delivery, and wound healing as it is derived from plant sources, making it a biocompatible, biodegradable, renewable, and sustainable material (Garcia et al. 2020). Despite its widespread use, there are some concerns about the safety of using starch as its properties may not always be suitable for specific applications. For example, consuming a substantial amount of starch is associated with obesity, diabetes, and other health problems (Aller et al. 2011). Hence, modification is needed to enhance its functionality.

In recent years, intensive research efforts have been made to modify starch, including physical, chemical, and enzymatic approaches. These modifications have opened new avenues for enhancing functionality, adaptability, structure, reactivity, and biocompatibility. This chapter explores starch's physical, chemical, and enzymatic modifications for its use in the food industry and biomedical applications.

9.2 Physical Modifications of Starch

Physical modifications of starch play a pivotal role in tailoring its properties as these involve altering characteristics concerning the structure and morphology without involving chemical reactions. These modifications are widely employed in the food industry and biomedical applications to enhance starch's functionality and tailor its characteristics for specific needs. In the food industry, physical modifications of starch play a crucial role in improving the texture, stability, and sensory attributes of various food products, whereas in biomedical applications, they contribute to developing biocompatible and biodegradable materials.

9.2.1 Heat/Hydrothermal Treatment

Heat/hydrothermal treatment causes starch granules to swell, which depends on the temperature, time, and moisture content used in different techniques, such as dry heat treatment, moist heat treatment, and extrusion heating. It is based on utilizing controlled heat to induce structural changes in starch molecules and involves subjecting starch to elevated temperatures, often in the presence of water, to trigger processes such as gelatinization, retrogradation, and partial hydrolysis. When starch is heated, water is absorbed by the amorphous regions, causing them to swell and disrupt the crystalline structure, resulting in a viscous gel characterized by increased solubility, swelling power, and viscosity. The gelatinization process is reversible; the gel structure reforms upon cooling, referred to as retrogradation. Controlled retrogradation can enhance the stability and texture of starch-based products, as it can reduce syneresis (water release from gels) and improve freeze-thaw stability. The heat and moisture can cause the breaking of glycosidic bonds within starch, resulting in the formation of smaller molecular chains, such as dextrins and maltodextrins, which have improved solubility and digestibility. Also, heat/hydrothermal treatment can induce structural changes in starch granules. The amorphous regions of the granules expand and become more accessible, allowing for increased interaction with other molecules, whereas the crystalline regions may change their degree of crystallinity and crystallite size, affecting the functional properties of the modified starch (Majzoobi et al. 2015; Chandla et al. 2017) (Fig. 9.1).

9.2.2 Radiation Treatment

Radiation treatment involves exposing starch to ionizing radiation, such as gamma or electron beams, to induce structural changes. The principle behind radiation treatment lies in the ability of ionizing radiation to break covalent bonds and generate free radicals within the starch matrix, which initiate various reactions that lead to modifications in starch properties. Free radicals formed by radiation can initiate

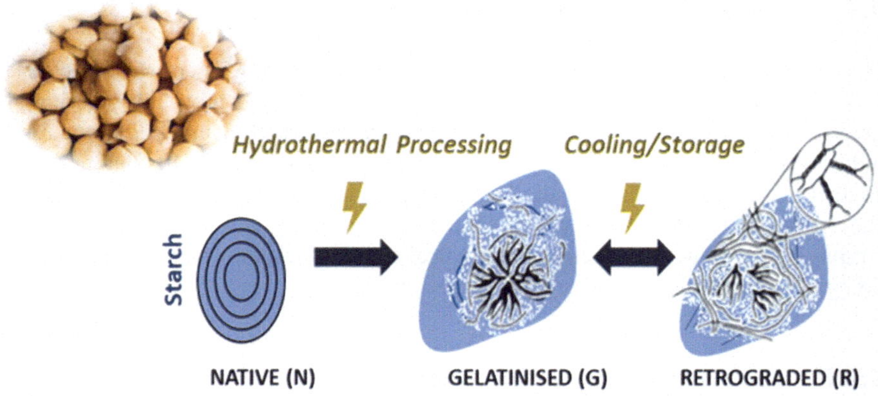

Fig. 9.1 Schematic diagram showing the physical modification of starch during heat/hydrothermal treatment (Edwards et al. 2021) [Open access]

cross-linking reactions between starch molecules, forming new bonds between adjacent starch chains and changing starch properties such as gel strength, water-holding capacity, and viscosity. Ionizing radiation can also induce chain scission, breaking starch molecules into smaller fragments. This process can lead to molecular weight changes, affecting starch viscosity, solubility, and other functional properties. Moreover, the combination of free radicals and water molecules created by radiation can lead to the hydrolysis of starch molecules, breaking down glycosidic bonds and forming smaller molecules, such as dextrins (Sriamornsak et al. 2010; Fiedorowicz et al. 1999).

9.2.3 pH Treatment

pH treatment of starch involves subjecting starch to variations in pH levels, either by adjusting the pH of the starch suspension or incorporating acidic or alkaline solutions. Acidic pH treatment causes starch granules to swell, while alkaline pH treatment leads to depolymerization and solubilization (Wang and Wang 2001). Interestingly, the mechanical properties of such modified starches are similar to their unmodified counterparts (Wicaksono et al. 2016). Changes in pH can affect the ionization and molecular interactions of the starch molecules, leading to alterations in starch solubility, viscosity, swelling, and gelatinization behavior. Furthermore, changes in pH influence the ionic interactions between starch molecules and other components in a formulation, which can lead to the formation of gels with varied textures and stability.

9.2.4 Pressure Treatment

Pressure treatment involves subjecting starch to elevated hydrostatic pressure, typically 100–1000 MPa. This technique leverages the principle that applying high pressure to starch can induce physical and chemical structural changes. High-pressure conditions lead to the compression and expansion of starch granules, which increases the surface area and alters the structural integrity of the granules, enabling better interactions with other components in formulations. Also, treatment at high pressures and longer durations can lead to the partial hydrolysis of starch molecules by the mechanical forces generated, breaking glycosidic bonds and forming smaller molecular fragments like dextrins and maltodextrins (Unal et al. 2023) (Fig. 9.2).

9.2.5 Ultrasound Treatment

Ultrasonic waves can alter the physicochemical properties of starch as it involves subjecting a starch dispersion to high-frequency sound waves, typically in the range of 20–1000 kHz, which create cavitation—the formation, growth, and subsequent collapse of microscopic bubbles within the starch suspension. The collapse of these bubbles generates localized mechanical forces and extreme conditions, such as high temperature and pressure, which promote various physicochemical changes in the starch structure by disrupting starch granules and breaking starch molecules into smaller fragments (Sujka and Jamroz 2013; Hu et al. 2013).

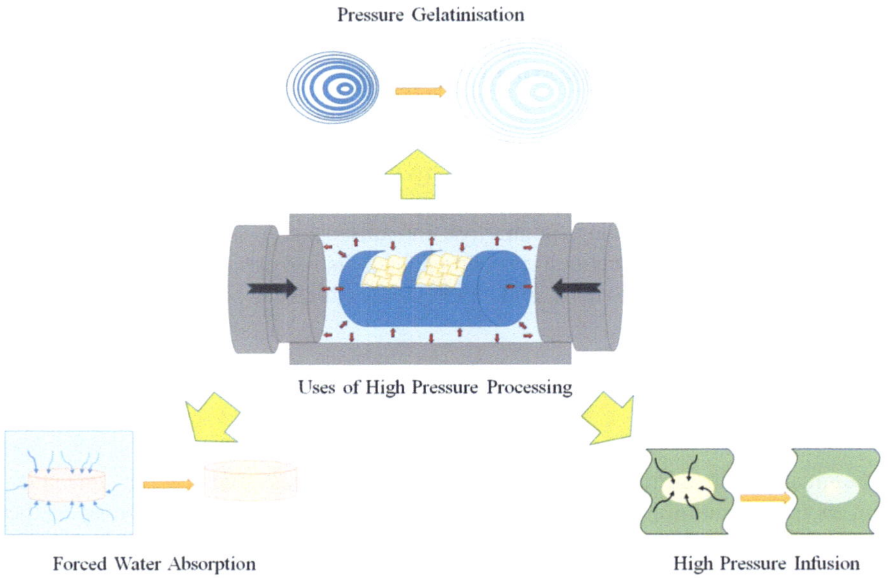

Fig. 9.2 Schematic diagram depicting the physical modifications of starch on pressure treatment (Balakrishna et al. 2020) [Open access]

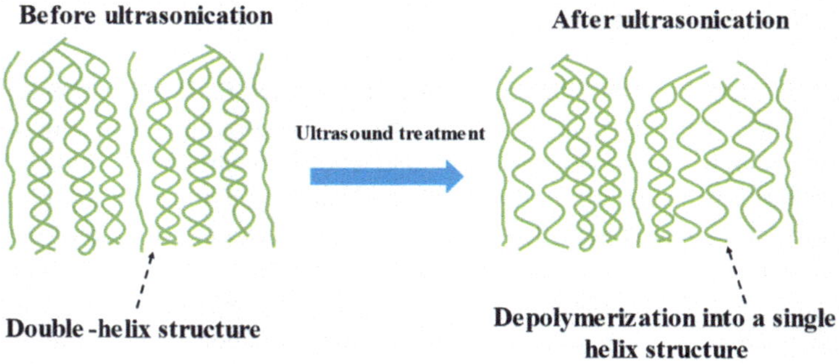

Fig. 9.3 Schematic diagram of starch modification with ultrasound (He et al. 2023) [Open access]

Ultrasound waves can create mechanical vibrations and cavitation bubbles in liquid systems. These vibrations and bubbles generate localized changes in pressure and temperature, leading to molecular interaction and particle size distribution alterations. In areas of low pressure, microscopic bubbles form and rapidly collapse, which generates localized heat and mechanical forces, leading to the disruption of starch granules and molecular structures. The mechanical forces can break down starch aggregates and particles, reducing particle size and leading to improved solubility, dispersion, and interactions with other components. Also, the mechanical energy generated by ultrasound treatment can activate starch molecules, making them more receptive to interactions with other ingredients. This activation can enhance the functional properties of starch, such as thickening and gelling (Fig. 9.3).

9.3 Chemical Modifications of Starch

Chemical modifications of starch involve modifying its physiochemical properties by adding new chemical or functional groups without physically changing the molecule's size or shape. The principal sites for chemical alteration in starch are the three reactive hydroxyl groups in amylose and amylopectin. Chemical modification of starch improves its functional properties, such as increased solubility, stability, viscosity, and resistance to heat, acid, or enzymes, which can be used to create novel products.

9.3.1 Oxidation

Chemical modification of starch by oxidation is a process that involves introducing oxygen-containing functional groups (carboxyl and carbonyl groups) to the starch molecule. This can be achieved by treating the starch with oxidizing agents, such as sodium hypochlorite, hydrogen peroxide, or sodium periodate (Xie et al. 2020). The degree of oxidation can be controlled to achieve specific functional

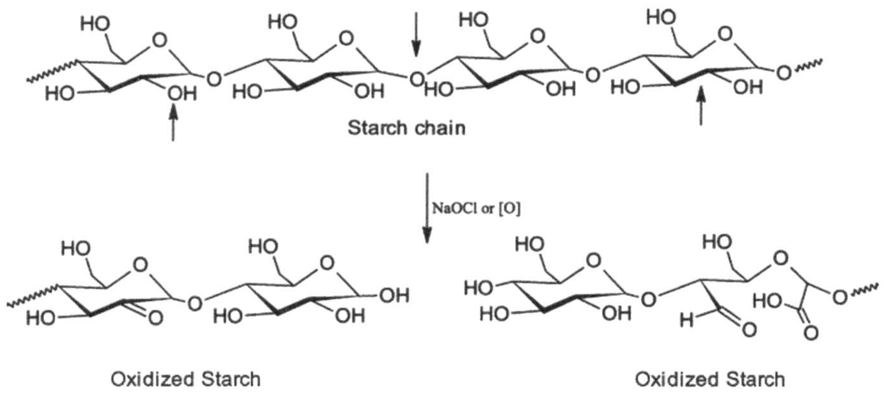

Fig. 9.4 Oxidation reaction of starch to produce oxidized starch (Egharevba 2019) [Open access]

properties; however, excessive oxidation can lead to the degradation of the starch molecules, which can negatively affect their properties (Nakkala et al. 2022; Fonseca et al. 2015).

Oxidative reactions can lead to structural modifications in starch, affecting its solubility, rheological properties, and interactions with other components. Chemical oxidants, such as hydrogen peroxide or sodium hypochlorite, react with starch molecules to introduce oxygen-containing functional groups, including hydroxyl (OH), carbonyl (C=O), and carboxyl (COOH) groups. These groups' presence enhances the reactivity and influences its interactions with other components, such as promoting cross-linking reactions between starch molecules. Also, the oxidative modifications increase the hydrophilicity of starch, enabling it to absorb water more readily, and can lead to the fragmentation of starch molecules and particles, resulting in reduced particle size and improved dispersibility (Fig. 9.4).

9.3.2 Etherification

Chemical modification of starch by etherification involves introducing ether groups to the starch molecule by reacting with etherifying agents, such as ethylene oxide and propylene oxide (Senanayake et al. 2014; Goszczyński et al. 2014). Etherification can alter the molecular structure of starch, enhancing its functional properties and reactivity by modifying its hydroxyl groups. It involves the replacement of hydrogen atoms on hydroxyl groups with ether groups, which include methyl, ethyl, hydroxyethyl, and hydroxypropyl groups. Introducing ether groups disrupts the molecular arrangement, allowing for better hydration and increased viscosity in aqueous solutions. Also, it can improve starch-based gels' stability, emulsions, and suspensions due to altered molecular interactions and the hydrophilic nature of etherified starch (Fig. 9.5).

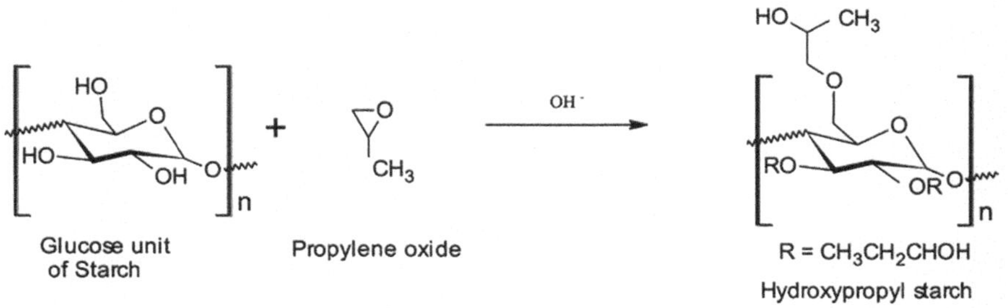

Fig. 9.5 Etherification of starch with propylene oxide (Egharevba 2019) [Open access]

9.3.3 Esterification

Chemical modification of starch by esterification involves introducing ester groups to the starch molecule by reacting with esterifying agents, such as acetic anhydride and succinic anhydride. Depending upon the chemical groups attached to starch, the process can be classified as acetylation (reaction with acetyl group) (Betancur et al. 1997), fatty acylation (reaction with fatty acid) (Bulpin et al. 1982), phosphorylation (reaction with phosphate group) (Nabeshima and Grossmann 2001), and succinylation (reaction with succinic anhydride) (Bao et al. 2003).

Esterification modifies the hydroxyl groups in starch molecules, leading to changes in solubility, viscosity, and interaction capabilities. Esterified starch becomes more hydrophobic due to nonpolar ester groups, which can influence its interactions with other hydrophobic ingredients, such as lipids. The degree of esterification can influence whether the starch is soluble or dispersible in water or other solvents and display modified viscosity and rheological properties compared to native starch. Due to the altered molecular interactions and hydrophobic characteristics of esterified starch, the process can also lead to improved stability of starch-based systems, such as gels, suspensions, and emulsions.

9.3.4 Cationization

Chemical modification of starch by cationization involves introducing positively charged groups to the starch molecule by reacting with cationic reagents, such as epichlorohydrin and 3-chloro-2-hydroxypropyltrimethylammonium chloride (Almonaityte et al. 2020). The degree of cationization can be controlled to achieve specific functional properties and classified as dry, wet, and semidry (Nawaz et al. 2020). Cationized starch becomes positively charged, influencing its behavior in aqueous systems and interactions with negatively charged molecules. Cationic groups enhance the starch's interaction with negatively charged molecules, such as proteins, enhancing its binding, stability, and solubility due to the electrostatic repulsion between cationized starch molecules. This contributes to improved dispersion and an increased ability to bind to surfaces, which can benefit various applications, such as coating and encapsulation (Fig. 9.6).

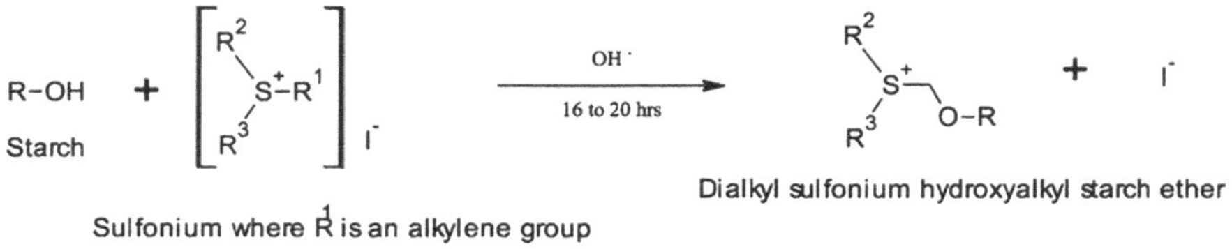

Fig. 9.6 Cationization of starch with sulfonium salt to produce a sulfonium cationic starch (Egharevba 2019) [Open access]

9.3.5 Cross-Linking

Chemical modification of starch by cross-linking involves the formation of covalent bonds between starch molecules or between starch molecules and other molecules, such as proteins or aldehydes. This can be achieved by treating starch with cross-linking agents, such as sodium trimetaphosphate (Gao et al. 2014) and adipic acid dihydrazide (Lv et al. 2020). Cross-linking creates a network of interconnected starch molecules, leading to changes in properties such as viscosity, gelatinization behavior, and mechanical and thermal stress resistance. Cross-linked starch exhibits improved heat and mechanical stress resistance due to the stronger intermolecular interactions within the network structure. This enhanced stability benefits applications requiring processing and storage under challenging conditions. Cross-linked starch may also exhibit modified gelatinization behavior compared to native starch. The cross-linked network structure can impact starch granules' swelling and gelatinization properties, influencing their application in thickening and gelling systems (Ding et al. 2022; Wegrzynowska-Drzymalska et al. 2022).

9.3.6 Acid Treatment

Chemical modification of starch by acid treatment involves the hydrolysis of the starch molecules with acids, such as hydrochloric acid and sulfuric acid (Singh and Ali 2000). However, excessive acid treatment can heavily decrease starch molecular weight and viscosity, negatively affecting its usability. Acid treatment can also produce unwanted by-products, such as 5-hydroxymethylfurfural, which can be toxic and limit its use in food products.

Acid hydrolysis can break down starch molecules into smaller fragments through the cleavage of glycosidic bonds, resulting in shorter polysaccharide chains, dextrins, and even individual glucose molecules. The cleavage of glycosidic bonds reduces the molecular weight of starch molecules and exhibits reduced viscosity which is advantageous for applications requiring thinner consistency. The degradation of starch molecules into smaller fragments increases their solubility in water and other solvents, which can influence the functionality and interactions of acid-treated starch.

9.3.7 Dual Modification

Dual modification combines two or more modification methods (e.g., esterification and cross-linking or oxidation and cationization) to provide unique functional properties to starch. Such can be achieved by combining modification agents or modifying starch sequentially with different agents. This process can be beneficial in specialized applications, such as in producing biodegradable films or coatings. The degree of modification and the selection of modification agents must be carefully controlled to achieve the desired functional properties without negatively affecting starch properties (Abedi et al. 2019).

9.4 Enzymatic Modifications of Starch

Enzymatic modifications of starch can alter its physicochemical properties and create new functional properties for various food and biomedical applications. The enzymatic hydrolysis approach, which employs bio-extracted enzymes such as amylase, pancreatin, pullulanase, and glucoamylases, involves using enzymes to break down starch molecules into smaller units such as glucose, maltose, and dextrins (Diderichsen and Christiansen 1988; Kim et al. 2013; Zhu et al. 2019; Rajan et al. 2008).

Enzymes are highly specific catalysts that target particular bonds in starch molecules, resulting in controlled modifications while minimizing unwanted side reactions. Enzymes can break down glycosidic bonds and, alternatively, create new bonds, forming novel structures within the starch matrix. Enzymatic modification can lead to changes in the molecular structure of starch, including the arrangement of glucose units, branching patterns, and the presence of specific functional groups, which can improve the solubility and reactivity of starch due to the generation of smaller fragments and the introduction of new functional groups (Fig. 9.7).

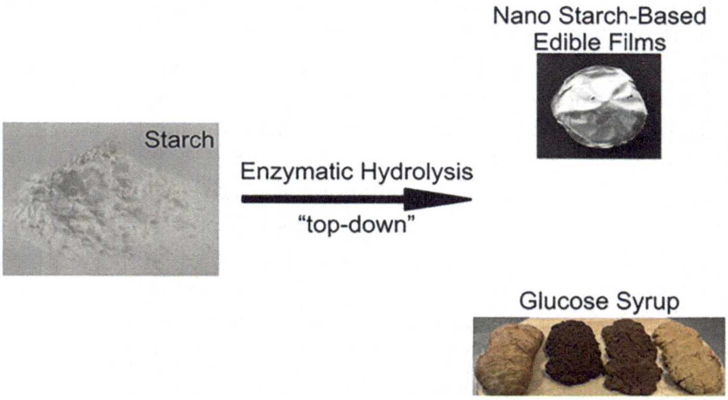

Fig. 9.7 Enzymatic conversion of starch to make films and glucose syrup (Podgorbunskikh et al. 2022) [Open access]

9.5 Modified Starches and Their Applications

Table 9.1 shows the applications of starches modified through physical, chemical, or enzymatic means in the food industry and biomedical sectors. Modified starches produced through heat/hydrothermal treatment can be used in food products, enhancing their texture, stability, gelatinization, and viscosity, whereas those

Table 9.1 Modified starches and their applications

Modification (physical/chemical/enzymatic)	Applications (food/biomedical)	References
Heat/hydrothermal treatment	Food items like noodles having good solubility, color, strength, moisture, texture, and sensory properties	Majzoobi et al. (2015), Chandla et al. (2017)
Radiation treatment	Pharmaceutical tablets with enhanced drug delivery properties and stability	Sriamornsak et al. (2010), Fiedorowicz et al. (1999)
pH treatment	Starch for food items with good gelling, strength, stability, and moisture	Wang and Wang (2001), Wicaksono et al. (2016)
Pressure treatment	Starches with good moisture and gelation for use in food items	Unal et al. (2023)
Ultrasound treatment	Starches having good nutritional value, moisture, texture, and degradation properties, which are suitable for use in food items	Sujka and Jamroz (2013), Hu et al. (2013)
Oxidation	Hydrogels having excellent strength, moisture retention capabilities, stability, and biocompatibility	Xie et al. (2020), Nakkala et al. (2022), Fonseca et al. (2015)
Etherification	Nanoconjugates for cancer therapy	Senanayake et al. (2014), Goszczyński et al. (2014)
Esterification	Starches having excellent physical and functional properties	Betancur et al. (1997), Bulpin et al. (1982), Nabeshima and Grossmann (2001), Bao et al. (2003)
Cationization	Starches having excellent solubility and rheological properties	Almonaityte et al. (2020), Nawaz et al. (2020)
Cross-linking	Designing nanofibers and nanocrystals cross-linked biomaterials	Gao et al. (2014), Lv et al. (2020), Ding et al. (2022), Wegrzynowska-Drzymalska et al. (2022)
Acid treatment	Starches with good gelation and stability	Singh and Ali (2000)
Dual modification	Starches with excellent solubility, stability, moisture retention properties, and texture	Abedi et al. (2019)
Enzymatic modifications	Food items such as rice flour with enhanced physicochemical functionalities and drug delivery systems for cancer therapy	Diderichsen and Christiansen (1988), Kim et al. (2013), Zhu et al. (2019), Rajan et al. (2008)

produced through radiation treatment possess better drug-release properties and can be used to design biomaterials. pH treatment can modify the water-binding capacity, thickening, and gelling properties of starch and can be used in food products, such as dairy, bakery, and confectioneries. In the biomedical field, pH-treated modified starches can be used as drug carriers, wound dressings, and scaffolds for tissue engineering applications, as they offer a natural and biocompatible alternative to synthetic materials. Pressure-treated and ultrasound-modified starches can be used as thickeners, stabilizers, and emulsifiers, as they have altered granules and molecular interactions, leading to solubility, viscosity, gelatinization behavior, texture, mouthfeel, and taste modifications. However, the potential long-term health risks of using such starches in food products are unclear.

Chemically and enzymatically modified starches are widely used in biomedical applications. Oxidization, etherification, and esterification can increase solubility, stability, hydrophilicity, and resistance to high temperatures, making it helpful in manufacturing biomaterials and pharmaceuticals. Cationization can improve starch's water solubility, film-forming ability, and adsorption properties, whereas cross-linking and acid treatment can increase the digestibility of food products and create biomaterials ideal for interacting with human proteins (Wegrzynowska-Drzymalska et al. 2022) by modifying its physical and chemical properties. Finally, enzymatic modifications can improve solubility, sweetness, water-holding capacity, and digestibility and are commonly used to produce sweeteners like glucose syrup and maltodextrin and to create drug carriers for targeted drug delivery.

9.6 Conclusion and Prospects

Modification of starch mitigates the inherent flaws in the native starches, such as loss of viscosity, reversibility, insolubility, and absence of ordered structure, which ultimately provides the feasibility basis for widespread starch applications. The physical, chemical, and enzymatic modifications discussed in this chapter have been the most promising modification methods with abundant newly developed techniques. Understanding the high value of modified starches has prompted its production in the starch industry from various modification reagents and starch sources. Changes in the physicochemical, morphological, thermodynamic, and rheological properties of modified starches may provide a crucial foundation for understanding the efficacy of the industrial starch modification processes. New consumable items with conventional and nonconventional properties can also be sought using dual modification.

To better comprehend starch modification, research should concentrate on novel combinations of techniques to produce starches with diverse and prospective applications and variations in structure, composition, and properties. Moreover, the health of the consumer and the environment should also be considered, although much effort goes into chemically modifying carbohydrates to give them novel nutritional properties.

References

Abedi E, Pourmohammadi K, Abbasi S (2019) Dual-frequency ultrasound for ultrasonic-assisted esterification. Food Sci Nutr 7(8):2613–2624. https://doi.org/10.1002/FSN3.1115

Aller EEJG, Abete I, Astrup A, Alfredo MJ, van Baak MA (2011) Starches, sugars and obesity. Nutrients 3(3):341–369. https://doi.org/10.3390/NU3030341

Almonaityte K, Bendoraitiene J, Babelyte M, Rosliuk D, Rutkaite R (2020) Structure and properties of cationic starches synthesized by using 3-chloro-2-hydroxypropyltrimethylammonium chloride. Int J Biol Macromol 164:2010–2017. https://doi.org/10.1016/J.IJBIOMAC.2020.08.089

Balakrishna AK, Abdul Wazed M, Farid M (2020) A review on the effect of high pressure processing (HPP) on gelatinization and infusion of nutrients. Molecules 25(10):2369. https://doi.org/10.3390/MOLECULES25102369

Bao J, Xing J, Phillips DL, Corke H (2003) Physical properties of Octenyl succinic anhydride modified Rice, wheat, and potato starches. J Agric Food Chem 51(8):2283–2287. https://doi.org/10.1021/JF020371U

Betancur AD, Chel GL, Cañizares HE (1997) Acetylation and characterization of Canavalia ensiformis starch. J Agric Food Chem 45(2):378–382. https://doi.org/10.1021/JF960272E

Bulpin PV, Welsh EJ, Morris ER (1982) Physical characterization of amylose-fatty acid complexes in starch granules and in solution. Starch 34(10):335–339. https://doi.org/10.1002/STAR.19820341004

Carpenter D et al (2015) Obesity, starch digestion and amylase: association between copy number variants at human salivary (AMY1) and pancreatic (AMY2) amylase genes. Hum Mol Genet 24(12):3472–3480. https://doi.org/10.1093/HMG/DDV098

Chandla NK, Saxena DC, Singh S (2017) Processing and evaluation of heat moisture treated (HMT) amaranth starch noodles; an inclusive comparison with corn starch noodles. J Cereal Sci 75:306–313. https://doi.org/10.1016/J.JCS.2017.05.003

Diderichsen B, Christiansen L (1988) Cloning of a maltogenic alpha-amylase from Bacillus stearothermophilus. FEMS Microbiol Lett 56(1):53–59. https://doi.org/10.1111/J.1574-6968.1988.TB03149.X

Ding L, Huang Q, Xiang W, Fu X, Zhang B, Wu JY (2022) Chemical cross-linking reduces in vitro starch digestibility of cooked potato parenchyma cells. Food Hydrocoll 124:107297. https://doi.org/10.1016/J.FOODHYD.2021.107297

Edwards CH, Veerabahu AS, Mason AJ, Butterworth PJ, Ellis PR (2021) α-Amylase action on starch in chickpea flour following hydrothermal processing and different drying, cooling and storage conditions. Carbohydr Polym 259:117738. https://doi.org/10.1016/J.CARBPOL.2021.117738

Egharevba HO (2019) Chemical properties of starch and its application in the food industry. In: Emeje M (ed) Chemical properties of starch. IntechOpen, Rijeka. https://doi.org/10.5772/intechopen.87777

Fiedorowicz M, Tomasik P, You S, Lim S (1999) Molecular distribution and pasting properties of UV-irradiated corn starches. Starch 51(4):126–131

Fonseca LM et al (2015) Oxidation of potato starch with different sodium hypochlorite concentrations and its effect on biodegradable films. LWT Food Sci Technol 60(2):714–720. https://doi.org/10.1016/J.LWT.2014.10.052

Gao F, Li D, Bi CH, Mao ZH, Adhikari B (2014) Preparation and characterization of starch crosslinked with sodium trimetaphosphate and hydrolyzed by enzymes. Carbohydr Polym 103(1):310–318. https://doi.org/10.1016/J.CARBPOL.2013.12.028

Garcia MAVT, Garcia CF, Faraco AAG (2020) Pharmaceutical and biomedical applications of native and modified starch: a review. Starch 72(7–8):1900270. https://doi.org/10.1002/STAR.201900270

Goszczyński TM, Filip-Psurska B, Kempińska K, Wietrzyk J, Boratyński J (2014) Hydroxyethyl starch as an effective methotrexate carrier in anticancer therapy. Pharmacol Res Perspect 2(3):e00047. https://doi.org/10.1002/PRP2.47

He R, Li S, Zhao G, Zhai L, Qin P, Yang L (2023) Starch modification with molecular transformation, physicochemical characteristics, and industrial usability: a state-of-the-art review. Polymers (Basel) 15(13):2935. https://doi.org/10.3390/POLYM15132935

Hu A et al (2013) Ultrasonically aided enzymatical effects on the properties and structure of mung bean starch. Innovative Food Sci Emerg Technol 20:146–151. https://doi.org/10.1016/J.IFSET.2013.08.005

Kim YL, Mun S, Park KH, Shim JY, Kim YR (2013) Physicochemical functionality of 4-α-glucanotransferase-treated rice flour in food application. Int J Biol Macromol 60:422–426. https://doi.org/10.1016/J.IJBIOMAC.2013.04.032

Lee BH et al (2012) Modulation of starch digestion for slow glucose release through 'toggling' of activities of mucosal α-glucosidases. J Biol Chem 287(38):31929–31938. https://doi.org/10.1074/jbc.M112.351858

Lv H et al (2020) Crosslinked starch nanofibers with high mechanical strength and excellent water resistance for biomedical applications. Biomed Mater 15(2):025007. https://doi.org/10.1088/1748-605X/AB509F

Majzoobi M, Pesaran Y, Mesbahi G, Golmakani MT, Farahnaky A (2015) Physical properties of biodegradable films from heat-moisture-treated rice flour and rice starch. Starch 67(11–12):1053–1060. https://doi.org/10.1002/STAR.201500102

Nabeshima EH, Grossmann MVE (2001) Functional properties of pregelatinized and cross-linked cassava starch obtained by extrusion with sodium trimetaphosphate. Carbohydr Polym 45(4):347–353. https://doi.org/10.1016/S0144-8617(00)00273-3

Nakkala K, Godiyal S, Ettaboina SK, Laddha KS (2022) Chemical modifications of turmeric starch by oxidation, phosphorylation, and Succinylation. Starch 74(9–10):2200053. https://doi.org/10.1002/STAR.202200053

Nawaz H et al (2020) Physical and chemical modifications in starch structure and reactivity. In: Chemical properties of starch. IntechOpen, London. https://doi.org/10.5772/INTECHOPEN.88870

Podgorbunskikh E et al (2022) Comprehensive enzymatic conversion of starch for the food industry. Polymers (Basel) 14(21):4575. https://doi.org/10.3390/POLYM14214575

Rajan A, Sudha JD, Abraham TE (2008) Enzymatic modification of cassava starch by fungal lipase. Ind Crop Prod 27(1):50–59. https://doi.org/10.1016/J.INDCROP.2007.07.003

Senanayake S, Gunaratne A, Ranaweera KKDS, Bamunuarachchi A (2014) Effect of hydroxypropylation on functional properties of different cultivars of sweet potato starch in Sri Lanka. Int J Food Sci 2014:1. https://doi.org/10.1155/2014/148982

Singh V, Ali SZ (2000) Acid degradation of starch. The effect of acid and starch type. Carbohydr Polym 41(2):191–195. https://doi.org/10.1016/S0144-8617(99)00086-7

Sriamornsak P, Juttulapa M, Piriyaprasarth S (2010) Microwave-assisted modification of arrowroot starch for pharmaceutical matrix tablets. Adv Mater Res 93–94:358–361. https://doi.org/10.4028/WWW.SCIENTIFIC.NET/AMR.93-94.358

Sujka M, Jamroz J (2013) Ultrasound-treated starch: SEM and TEM imaging, and functional behaviour. Food Hydrocoll 31(2):413–419. https://doi.org/10.1016/J.FOODHYD.2012.11.027

Unal K, Alpas H, Oztop MH (2023) Modification of quinoa starch by high hydrostatic pressure and Ultrasonication and assessment of its physicochemical properties. ACS Food Sci Technol 3(1):31–40. https://doi.org/10.1021/ACSFOODSCITECH.2C00230/ASSET/IMAGES/LARGE/FS2C00230_0005.JPEG

Wang L, Wang Y (2001) Structures and physicochemical properties of acid-thinned corn, potato and rice starches. Starch 53(11):570–576

Wegrzynowska-Drzymalska K et al (2022) Dialdehyde starch nanocrystals as a novel cross-linker for biomaterials able to interact with human serum proteins. Int J Mol Sci 23(14):7652. https://doi.org/10.3390/IJMS23147652/S1

Wicaksono Y, Nuri N, Wisudyaningsih B (2016) Effect of temperature and pH of modification process on the physical-mechanical properties of modified cassava starch. Mol Ther 11(2):248–255. https://doi.org/10.20884/1.JM.2016.11.2.217

Xie X et al (2020) Oxidized starch cross-linked porous collagen-based hydrogel for spontaneous agglomeration growth of adipose-derived stem cells. Mater Sci Eng C 116:111165. https://doi.org/10.1016/J.MSEC.2020.111165

Zhu J et al (2019) Preparation and characterization of pectin/chitosan beads containing porous starch embedded with doxorubicin hydrochloride: a novel and simple colon targeted drug delivery system. Food Hydrocoll 95:562–570. https://doi.org/10.1016/J.FOODHYD.2018.04.042

Andean Grain and Tuber Starch: Characteristics, Modifications, and Industrial Applications

10

Manuel O. Lobo, Rita M. Miranda, Sonia R. Calliope, Cristina N. Segundo, and Norma C. Sammán

Abstract

The Andes region in Latin America is the origin place of many grains. Among them, the best known are native maize (*Zea mays*), quinoa (*Chenopodium quinoa* W.), kiwicha (*Amaranthus caudatus*), and, to a lesser degree, kañiwa (*Chenopodium pallidicaule*). They have been used for centuries for widespread consumption in traditional cooking. In general, they are underutilized grains, and their cultivation is reduced; however, recently, they have been recognized worldwide for their balance of proteins, fats, and starch. For these reasons, the interest in these grains has increased substantially, which has resulted in an intensification of research projects on their properties and applications.

Keywords

Starch · Extrusion · Structural modifications · Techno-functional properties · Applications

10.1 Andean Grains and Tubers

The region also highlights the production of tubers like potato (*Solanum tuberosum* var. andigenum), ocas (*Oxalis tuberosa*), mashua (*Tropaeolum tuberosum*), and papa lisa or Olluco (*Ullucus tuberosus*), each one with specific composition, size, and color characteristics, and roots like yacon (*Smallanthus sonchifolius*). There is relatively little information regarding the characteristics and properties of the starch

M. O. Lobo · R. M. Miranda · S. R. Calliope · C. N. Segundo · N. C. Sammán (✉)
Centro Interdisciplinario de Investigaciones en Tecnologías y Desarrollo Social para el NOA (CIITED, UNJu-CONICET), Facultad de Ingeniería-Universidad Nacional de Jujuy, San Salvador de Jujuy, Argentina

N. Mazumder, M. H. Rahman (eds.), *Advanced Research in Starch*, https://doi.org/10.1007/978-981-99-9527-1_10

of these Andean grains and tubers and the relationships between structural, nutritional, functional, and rheological properties. Knowing this information is essential because the applications and uses in formulating new foods depend on it. The starch granules of these crops have great variations in amylose content; they also have different morphologies and sizes, so it is assumed that these differences will be reflected in their properties. Although some studies have characterized the properties of starches of these grains and tubers, knowledge of the physicochemical, structural, functional, and nutritional properties is still incomplete.

10.2 The Starch of Andean Grains and Tubers

The starch is the most abundant fraction in maize and pseudocereals (quinoa, kiwicha, and kañiwa grains) (Martinez-Lopez et al. 2020; Srichuwong et al. 2017; Repo-Carrasco-Valencia et al. 2009). There are about 260 well-known maize races (Boege 2009; Serratos 2012). The composition of the isolated starch depends on the extraction method used. An important difference between tubers and grains is that the starch is extracted directly from the pulp of the tubers, while for grains grinding and chemical processes are necessary. This is because tubers have very low protein and lipid contents. The purity of the isolated starch is another issue that must be considered since physicochemical, functional, and digestibility characteristics depend on it. An efficient extraction process allows obtaining starches with low fat, protein, fiber, and ash contents. Composition and some properties of isolated starch can be seen in Table 10.1.

The starch content varies in quinoa (*Chenopodium quinoa* W.) between 54% and 67% (db), in kiwicha (*Amaranthus caudatus*) 55–64%, in kañiwa (*Chenopodium pallidicaule*) around 51%, and in different varieties of maize between 66% and 79.2% (db). The starch granules of these three grains, called pseudocereals, show a polygonal/irregular shape of very small diameter (0.46–2.53 μm) for quinoa and kiwicha, whereas kañihua have the smallest size distribution (0.5–1.5 μm). The Andean grain starches display an A-type X-ray diffraction pattern (XRD), which is also exhibited by most cereals (Jan et al. 2017; del Valdez-Arana et al. 2020). In maize, polyhedral and spherical starches are found, most of which have type A crystalline structures, although very few cases have type A and B structures (Salvador-Reyes et al. 2023; Seetharaman et al. 2001; Agama Acevedo et al. 2005). The presence of both types of starches could be related to the climatic conditions and the height above sea level at which they are grown (Salvador-Reyes et al. 2023).

Most of the dry matter of the tubers is starch, sugars, and other polysaccharides. In Andean potatoes, their content is between 65 and 80 g/100 g (db) (Calliope et al. 2018). The starch from Andean potatoes has B-type crystallinity (Velasquez and Bello-Pérez 2021). Martínez et al. (2019) reported that starches isolated from potato cultivars from different parts of the world have a wide granule size distribution, between 5 and 100 μm (mean diameter 20.6 to 30.9 μm). Figure 10.1 shows the morphological differences of starches from Andean grains and tubers. While Bolita raze maize (hard endosperm) presents starch granules with a polyhedral shape,

Table 10.1 Properties of the starch granule extracted from Andean grains and tubers: solubility and swelling power T = 40–90 °C

Parameter/starch	Maize var.	Quinoa	Kiwicha	Kañiwa	References
Composition					
Moisture (%)		10.6–13.2	9.3–10.7	11.1	Velasquez-Barreto et al. (2021b), Fuentes et al. (2019)
Protein (%)		0.6–0.9	0.2–1.0	0.6	
Fat (%)		0.00–1	0.1		
Fiber (%)		0.04–1.6	0.2		
Ash (%)		0.4–2.3	0.3		
Carbohydrate (%)		84.2–86.1	87.7		
Phosphorus (%)		0.05–0.1	0.1		
Physical properties					
Shape	Spherical (soft endosperm), irregular polyhedral (hard endosperm)	Polygonal irregular	Polygonal	Polygonal	Almeida et al. (2022), Salvador-Reyes et al. (2021), Velasquez-Barreto et al. (2021b), Valdez-Arana et al. (2020), Gimenez et al. (2020), Fuentes et al. (2019), Contreras-Gallegos et al. (2015), Steffolani et al. (2013), Kong et al. (2010), Narváez-González et al. (2007)
Diameter (μm)	10–20	0.5–2.5	1–2	0.5–1.5	
Amylose (%)	15.8–29.8	11.3–19.1	3–13	20	
Crystallinity (%)	17.5–21.2	23.8–39.6	24.5–27.9	34.0–35.6	
Hydration properties					
Solubility	2.7–5.7[a]	2.1–11.2	3.3–80.5		Velasquez-Barreto et al. (2021b)
Swelling power	12.8–16.8[a]	3.2–16.9	4.4–39.6		
Thermal properties (gelatinization)					
T_0 (°C)	52.9–66.5	51.9–65.8	59.4–60.6	49.8	Velasquez-Barreto et al. (2021b), Valdez-Arana et al. (2020), Fuentes et al. (2019)
T_p (°C)	67.3–75	59.2–70.9	66.9–67.1	57.5	
T_f (°C)	75–80	66.3–78.4	74.1–76.3	65.5	
ΔH (J/g)	4.5–9.6	7.4–11.4	8.5–12.8	7.1	
Pasting properties					
PT (°C)	74.9–84.7	54.5–69.7	58.7–63.6	50.7	Velasquez-Barreto et al. (2021b), Fuentes et al. (2019), Seetharaman et al. (2001), Narváez-González et al. (2007), Valdez-Arana et al. (2020)
PV (cp)	1200–16,200	746.9–1382	821.4–876	1376	
FV (cp)	1800–2520	935.5–1699	548.5–749	1333	
BDV (cp)	480–927	14.4–131	408.0–385	43	
SBV (cp)	1380–8790	119.7–312.8	135.2		

(continued)

Table 10.1 (continued)

Properties/ starch	Potato	Oca	Mashua	Olluco	References
Composition					
Moisture (%)	9.1–12.3	6.2–14	8.4	7.8–16.2	Martínez et al. (2019, 2022), Velasquez-Barreto et al. (2021a), Calliope et al. (2019), Espín et al. (2004)
Protein (%)	0.0–0.7	0.3–0.6		0.3–0.5	
Fat (%)	0–0.1	0.2–0.5		0.3–0.4	
Fiber (%)	0.0	0.0–1.5	0.0–1.2	1.7	
Ash (%)	0.1–0.6	0.1–0.7	0.3	0.0–0.2	
Phosphorus (mg/100 g)	46.4–119.3	38–41.3			
Physical properties					
Shape	Oval and ellipsoid	Cylindrical, ellipsoid, and oval	Oval, spherical, and truncated granules with small amounts of irregular granules	Ellipsoid, prismatic, conical, and oval	Martínez et al. (2019, 2022), Velasquez-Barreto et al. (2021a, b), Espín et al. (2004)
Diameter (μm)	0.4–120	7–100.2	4.1–62.3	4.5–50.2	
Amylose (%)	18.2–37.5	16.5–33	26.1–27.4	20.4–35.3	
Crystallinity (%)	34.4–36.3	22.9–39.8		18.1–37.1	
Hydration properties					
Solubility (g/100 g)	1.2–9.5	0.5	0.6		Calliope et al. (2019), Espín et al. (2004)
Swelling power (g/100 g)	2.6–4.9	2.1	2		
WAI (g gel/g starch)	2.3–4.9				
Thermal properties (gelatinization)					
T_0 (°C)	56.8–66.5	53.5	57.4	56.1	Martínez et al. (2019, 2022), Velasquez-Barreto et al. (2021a, b), Calliope et al. (2019)
T_p (°C)	58.2–69.5	55.2–64.0	54.9–65.7	52.3–63.1	
T_f (°C)	66.3–79.7	70.4	73	71.3	
ΔH (J/g)	11.5–16.6	9.7–17.7	4.3–12.6	10.2–18	
Pasting properties					
PT (°C)	64.0–64.9	57–65.3	65.2–68.3	58–68.7	Martínez et al. (2019, 2022), Velasquez-Barreto et al. (2021a, b)
PV (Pa s)	8.5–16.2	1.6–13.7	6.2–12.2	2.3–17.8	
FV (Pa s)		0.1	0.2	0.09	
BDV (Pa s)		0.3	0.3	0.2	
BDV (Pa s)		0.3	0.3	0.2	
SBV (Pa s)		0.04	0.08	0.04	

Solubility and swelling power T = 40–90 °C; ([a]) unpublished results; *T0* initial temperature, *Tp* peak temperature, *Tf* final temperature, *PT* pasting temperature, *PV* peak viscosity, *FV* final viscosity, *BDV* breakdown viscosity, *SBV* setback viscosity, *NA* not available

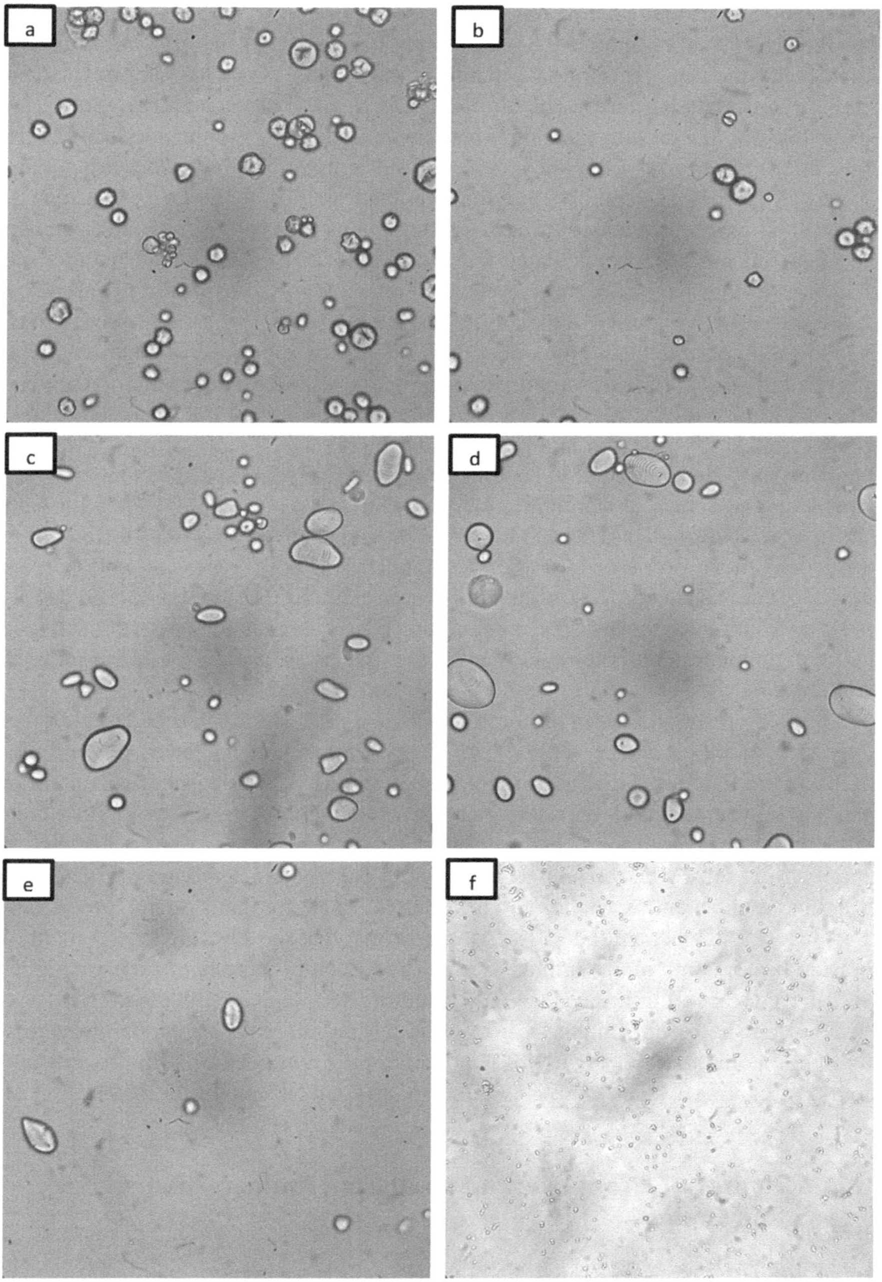

Fig. 10.1 Optical micrographs of starch granules of the Andean grains and tubers (40x amplification); maize races (**a**) *Bolita* and (**b**) *Capia*; potato var. (**c**) *Colorada*, (**d**) *Moradita*, and (**e**) *Santa maría*; (**f**) Kiwicha (*Amaranthus caudatus*)

Capia maize (soft endosperm) has more rounded granules; potato starch granules have the largest sizes among the starches shown here, and the elongation of the granule and the size distribution depend on the variety. Kiwicha has the smallest starch granules among them. The differences in the size of the starch granules can be attributed to the growth time, environmental conditions, and genetic factors (Qu et al. 2018). The type of crystal polymorphism is mainly influenced by the chain length (CL), weight average of amylopectin, and amylose content, i.e., in type A starches, amylopectin molecules have shorter chains (CL between 23 and 29) than type B starches (CL between 30 and 40) (Rojas et al. 2014). Table 10.1 includes the properties of the starches of these grains and tubers reported by different authors.

In maize, the variability in the amylose content of its starches is greater than in pseudocereals, due to the high number of different existing races. The maize can be classified as very soft, soft, intermediate, hard, and very hard according to the constitution of the endosperm, one of the most important parts of the grain. The endosperm is made up of starch and proteins and works as a source of energy for the plant during its development. In general, maize with soft endosperm contains higher percentages of spherical starch granules larger than 12 μm, while those with hard endosperm have smaller sizes and more polyhedral shape. This makes different the physicochemical properties of starch granules in both types of endosperms (Narváez-González et al. 2007; Salvador-Reyes et al. 2021). Other author mentions that the variations in the amylose content may be due to genetic factors of each race and the ambient temperature at which the maize is grown (Agama-Acevedo et al. 2011).

Amylose and amylopectin in starch granules are not evenly distributed; alternate layers of amorphous and crystalline regions exist in them, which have a determining effect on the functional and nutritional properties. Also, the phosphorus content in starch granules influences the pasting and rheological properties. The amylose content and the fraction of long chains of amylopectin can affect the viscoelasticity of starch pastes. However, the higher amylopectin content and small granule size suggest that quinoa and kiwicha starches can be used for nanocrystal production (Sanchez de la Concha et al. 2018). It was shown that the low amylose content of the starches of some maize influences their hydration properties, showing a higher rate of water absorption index (WAI), solubility (S), and swelling power (SP) of their gels. In vitro enzymatic hydrolysis is largely affected by crystallinity, size distribution, and binding of amylose and amylopectin molecules with proteins and lipids (Ai et al. 2013; Lin et al. 2020; Tacer-Caba and Nilufer-Erdil 2019).

10.2.1 Hydration Properties of Starch from Andean Tubers and Grains

Starch solubility (S) and swelling power (SP) provide valuable insights on the interactions between starch chains within crystalline and amorphous zones. The SP may be related to differences in structural and molecular properties (Bello-Perez et al. 2000; Jiang et al. 2020). The quinoa and kiwicha starch solubility increases at temperatures above 50 °C due to the weakening of the hydrogen bonds between the amylose and amylopectin chains. The transport of water into the granules is improved, and the amylose chains are removed by leaching. Kiwicha starches have higher swelling power and solubility at different temperatures than quinoa starches. The relatively low solubility of the starches from both grains is due to the low amylose content and the different lengths of the amylopectin chains, which can restrict starch swelling and strengthen the internal structure. Other authors reported solubility values between 55 °C and 95 °C different from those shown in Table 10.1 for quinoa starch (Jiang et al. 2020; Jan et al. 2017; Li et al. 2016). The differences could be related to the quinoa varieties and harvest growing conditions. Li and Zhu (2017) demonstrated that the amylopectin structure of quinoa starches plays a determining role in swelling power, relative crystallinity, and pasting properties.

In native maize, solubility was found between 2.72 and 5.71 g soluble starch/100 g starch and swelling power between 11.80 and 16.8 g gel/g starch. The variability of these properties is explained by the different amylose contents, sizes, and shapes of the starch granule. Maize starches with harder endosperm swell less than those with softer endosperm (Gonzalez et al. 2005). For example, the Perlita (hard endosperm) and Culli (soft endosperm) races have SP of 13.46 and 16.80 g water/g starch, respectively, at 90 °C (Romero 2023).

Alvis et al. (2008) reported a solubility range between 1.90 and 3.70 g soluble starch/100 g starch and WAI between 2.32 and 5.83 g gel/g starch for starches isolated from ñame, yuca, and Andean potatoes, while Calliope et al. (2019) found solubility values between 1.16 and 9.54 g/100 g starch and WAI between 2.31 and 4.84 g gel/g starch for starches extracted from 16 genotypes of Andean potatoes. The differences in the physicochemical properties between starches of tubers and grains are also due to the different physical characteristics of native starch (granule shape and size) and amylose and amylopectin contents (Table 10.1) (Velasquez and Bello-Pérez 2021).

10.2.2 Thermal Properties of Starch from Andean Tubers and Grains

The starch gelatinization process provides important information about the functionality, energy requirements, and possible uses of the starches and flours that contain it. The starch gelatinization of the Andean grains displays a single thermal transition; the peak temperature of pseudocereals starches is approximately between 57.5 °C and 70.0 °C, slightly lower than those of Andean maize 67.3–75.0 °C (Table 10.1). The low values in the starch gelatinization temperatures of kañiwa and some varieties of quinoa would suggest a less crystalline structure; this would suggest a greater enzymatic susceptibility and therefore greater digestibility (Srichuwong et al. 2017). On the other hand, Velasquez and Bello-Pérez (2021) reported that the starches of Andean tubers mashua, oca, and olluco have low gelatinization temperatures (52.3–65.7 °C) compared to those of maize (57.7–81.1 °C)

Utrilla-Coello et al. (2014) and potatoes (55.9–79.73 °C) (Calliope et al. 2019; Cruz et al. 2016).

Pseudocereals and maize present gelatinization enthalpies between 7.1–12.8 and 4.5–9.5 J/g, respectively, lower than those found in potatoes (11.5–16.6 J/g) and tubers (9–18 J/g) (Table 10.1). Enthalpy increases with increasing amylopectin CL (Cabrera-Ramírez et al. 2021; Carrillo-Navas et al. 2014). Thus, longer branch chains in the amylopectin of the Andean roots and tubers might explain the higher enthalpies compared to that of the Andean grains.

10.2.3 Rheological Properties of Starch from Andean Grains and Tubers

Viscosity profiles describe the pasting behavior of the starches. Figure 10.2 shows a diagram of the viscosity profile of isolate starches from some Andean grains (maize, quinoa, kiwicha, and kañiwa) and maize whole-grain flour with soft endosperm. It can observe that maize forms a more viscous paste than pseudocereals. In Table 10.1, pasting parameters of Andean grain and tuber starches are shown.

Solubility (S) and swelling power (SP) proprieties of starch granules could be related to pasting patterns. Kiwicha starch shows different pasting properties and

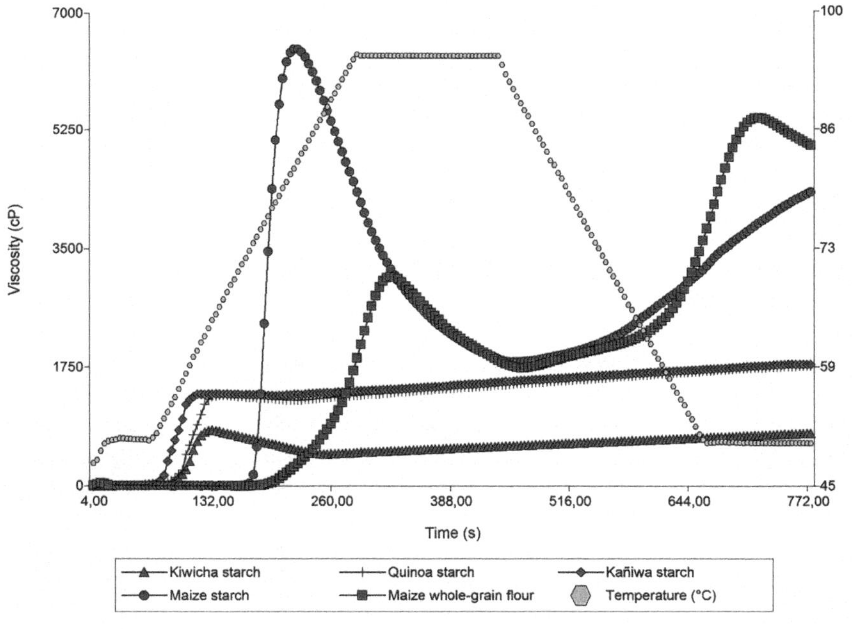

Fig. 10.2 RVA diagram of grain starches from kiwicha, quinoa, kañiwa and maize, and maize whole-grain flour. Showing apparent viscosity (cP) and temperature profile vs. time (s)

gelatinization enthalpies than quinoa starches. Table 10.1 shows that the pasting temperature and the viscosity peak are higher in maize than in pseudocereals. This is possible because both properties depend on the average size of the starch granules; Narváez-González et al. (2006) found a positive correlation between starch granule size and peak viscosity. The larger granules are more crystalline, so they are strong enough to present high swelling. The tubers have much larger values of pasting temperature and viscosity peak than maize, coinciding with their greater granule diameter. Velasquez and Bello-Pérez (2021) observed higher pasting temperatures for olluco starch (67.04 °C) than mashua (65.20 °C) and oca (63.56 °C). Low temperatures of pasting of the starches indicate a lower degree of association in the crystalline and amorphous zones of the starch granules, which allows them to be less resistant to swelling.

The breakdown parameter provides information about the stability during heating. This parameter is lower in quinoa, kañiwa, and amaranth starches compared to maize starches, so they would present greater structural fragility during cooking.

Quinoa starch shows good stability in the freezing and thawing process and high resistance to retrogradation which suggests that it could be used as a thickener for frozen foods and be included in emulsions such as sauces and dressings (Repo-Carrasco-Valencia and Valdez Arana 2017; Lopez-Fernandez et al. 2021).

The setback is related to the tendency to restructuration/retrogradation. High setback viscosities indicate low resistance to retrogradation, primarily attributed to amylose. However, the low differences in the setback values between maize and pseudocereal starch with different amounts of amylose indicate that another phenomenon is involved (Repo-Carrasco-Valencia and Valdez Arana 2017).

Velasquez and Bello-Pérez (2021) observed that the highest pasting temperatures were those of olluco starch (67.04 °C), followed by mashua (65.20 °C) and oca (63.56 °C). The low temperatures of pasting of these starches indicate a lower degree of association in the crystalline and amorphous zones of the granules, which allows them to be less resistant to swelling. Cruz et al. (2016) reported similar pasting temperatures.

10.3 Techno-Functional Properties of Starches in Whole-Grain Flours of Andean Grains

The physicochemical properties of starches from Andean grain are generally affected by the presence of other components of whole-grain flours. Proteins, fibers, lipids, ashes, and micronutrients make up a matrix that interacts with starch and water, modifying its functional properties. Also, in whole-grain flours, the other components of the grain interact physically and/or chemically, forming different supramolecular structures such as complexes of amylose-lipid, protein-amylose, and starch-starch, among others. Thus, the analysis of pasting properties is an effective method to study the behavior of starch within flour. This property relates the functionality of the starch to its structural characteristics, and its parameters allow access to the potential industrial application in products dependent on the viscosity and thickening behavior of starch (Ahmed et al. 2016).

10.3.1 Pasting and Gelatinization Properties

The pasting properties of starches isolated or extracted from flours differ from those in the same flours from which they were obtained. The presence of other components can increase the temperature and pasting time as if they were protecting the starch granules. However, the presence of lipids and fibers decreases all the characteristic viscosities of the starch in the flour (peak viscosity, breakdown, setback, and final viscosity), even eliminating the peak viscosity. This could be mainly because the greater presence of fat in the flour produces a more fluid system and, consequently, a lower viscosity.

Li et al. (2016) determined the behavior of whole-grain flours from seven quinoa varieties, obtaining higher values of pasting temperature and time, with viscosity peak similar to those of starch isolates, up to 5800 cP. Srichuwong et al. (2017) found in amaranth (*Amaranthus caudatus*) that the pasting properties are very similar between whole-grain flours and their isolated starches; the characteristic viscosities also show similar profiles, except for the peak viscosity that decreased from 2000 to 1800 cP due to the presence of components other than starch. del Valdez-Arana et al. (2020), comparing the pasting properties of whole-grain flours from three varieties of quinoa from Peru and its respective isolated starches, found different behaviors; both the temperature (from 66–68 °C to 93–95 °C) and the time (from 4.4–5.3 to 6.8–7.0 min) of pasting increase in whole-grain flours. On the contrary, in whole-grain flour, there is a significant decrease in the characteristic viscosities compared to pure starch, for example, peak viscosity (from 4500–5200 to 500–600 cP), breakdown (from 1000–2000 to 75–110 cP), setback (from 1300–1600 to 380–530 cP), and final viscosity (from 4500–5200 to 800–1000 cP). These results show that the matrix protects the starch granules from heat treatment while increasing plastic behavior due to the presence of lipids and partially preventing the formation of large structures that increase the paste viscosity.

The pasting properties of Andean maize flours are higher than those of their respective isolated starches. In maize with floury endosperm, the PV is higher and the setback lower than those with hard endosperm, due to the greater ease of hydration and less compaction of these granules; in addition, during heating in them, the water would migrate more easily into the granule due to the thinner protein layer that surrounds it (Salvador-Reyes et al. 2021).

In the same way, the thermal properties of starches are modified and/or diluted by the presence of other components. As is known, the study of gelatinization depends fundamentally on the amount of water available for swelling and rupture of the granule. In whole flour, the presence of fiber and protein decreases the availability of moisture for the starch, obtaining gelatinization peaks at higher temperatures and also reducing the enthalpy per gram necessary for the process. On the other hand, there is evidence that lipid-amylose complexes affect gelatinization temperatures in the same way as fiber.

However, no significant correlations are found between To, Tp, Tc, and ΔH and the chemical composition of quinoa flours, except with the starch content. The gelatinization temperatures determined in quinoa flours, initial (52–59 °C), peak

(62–69 °C), and final (73–79 °C), are higher than those of pure starches, while the ΔH of gelatinization is lower than the pure starches, due to dilution of the starch. However, the ΔH of gelatinization corrected for starch concentration is higher per gram of flour, which implies that the protein denaturation process could add to the endothermic gelatinization process (Li et al. 2016; Li and Zhu 2017; Srichuwong et al. 2017). In amaranth (*Amaranthus caudatus*), the same phenomenon was observed, obtaining higher values of To, Tp, and Tc, for the whole-grain flour compared to those of isolated starch (To from 58 °C to 53 °C, Tp from 68 °C to 63.5 °C, and Tc from 80 °C to 76 °C), and, similarly to quinoa, lower values of ΔH for the flours (10.9 to 12.4 J/g) (Srichuwong et al. 2017). On the contrary, del Valdez-Arana et al. (2020) determined a decrease between 2 °C and 4 °C in To, Tp, and Tc, when gelatinizing quinoa flour in relation to quinoa starch in three Peruvian quinoa varieties, while ΔH do not show significant differences for flour and starch.

10.4 Structural Changes of Starches in Extruded Whole-Grain Flour

10.4.1 Impact on Functional Properties

Extrusion is a process that involves several unit operations such as mixing, heating, cutting, shaping, texturing, and drying. In cereals, starch is the component that has the most important role, given that its changes affect the physicochemical, textural, and sensory properties of the extruded product (Marti et al. 2010; Seal et al. 2021). The structure of components from raw materials and their techno-functional properties change due to the effect of extrusion (Paucar Menacho et al. 2022). The magnitude of these changes depends mainly on extrusion process parameters such as residence time, specific mechanical energy, feed rate, screw speed and configuration, geometry, barrel temperature, and nozzle pressure (Yadav et al. 2022). In addition, the composition of the raw material, the humidity, and the granulometry of the feed to the extruder are very important parameters (Riaz 2004). Extrusion induces modifications in the crystalline structure of starch, molecular transition phase changes, mechanical damage expressed as damaged starch, solubility, and modification of water and oil absorption properties (Enríquez-Castro et al. 2022). In addition, some particular conditions of extrusion can modify the digestibility of starch, promoting the formation of rapidly digestible starch (RDS), slowly digestible starch (SDS), and/or resistant starch (RS), generating the formation of complexes amylose-lipid and/or amylose-protein, considered as RS type V (Zhang et al. 2017; Cervantes-Ramírez et al. 2020). There are various investigations on the effect of extrusion on the changes experimented by starch in whole-grain flours, but there are few studies on Andean grain and tuber flours.

In maize, the effect of the extrusion conditions on the starch properties depends on the protein composition and the texture of its endosperm. Hard endosperm maize flour has a more significant expansion, consumes less energy, and cooks more quickly than soft endosperm maize flour (Robutti et al. 2002); due to the hardness of the corn grain due to the microstructural arrangement of the starch granule and the thicker protein matrix that surrounds it. Figure 10.3 (a) shows morphological characteristics of maize starch before extrusion, and 10.3 (b) shows structural changes that occur in the starch of Capia race Andean maize, integral flours, after extrusion; maize starches do not completely disappear, and there are remains of starches, broken and flattened, as a result of the shearing produced during extrusion.

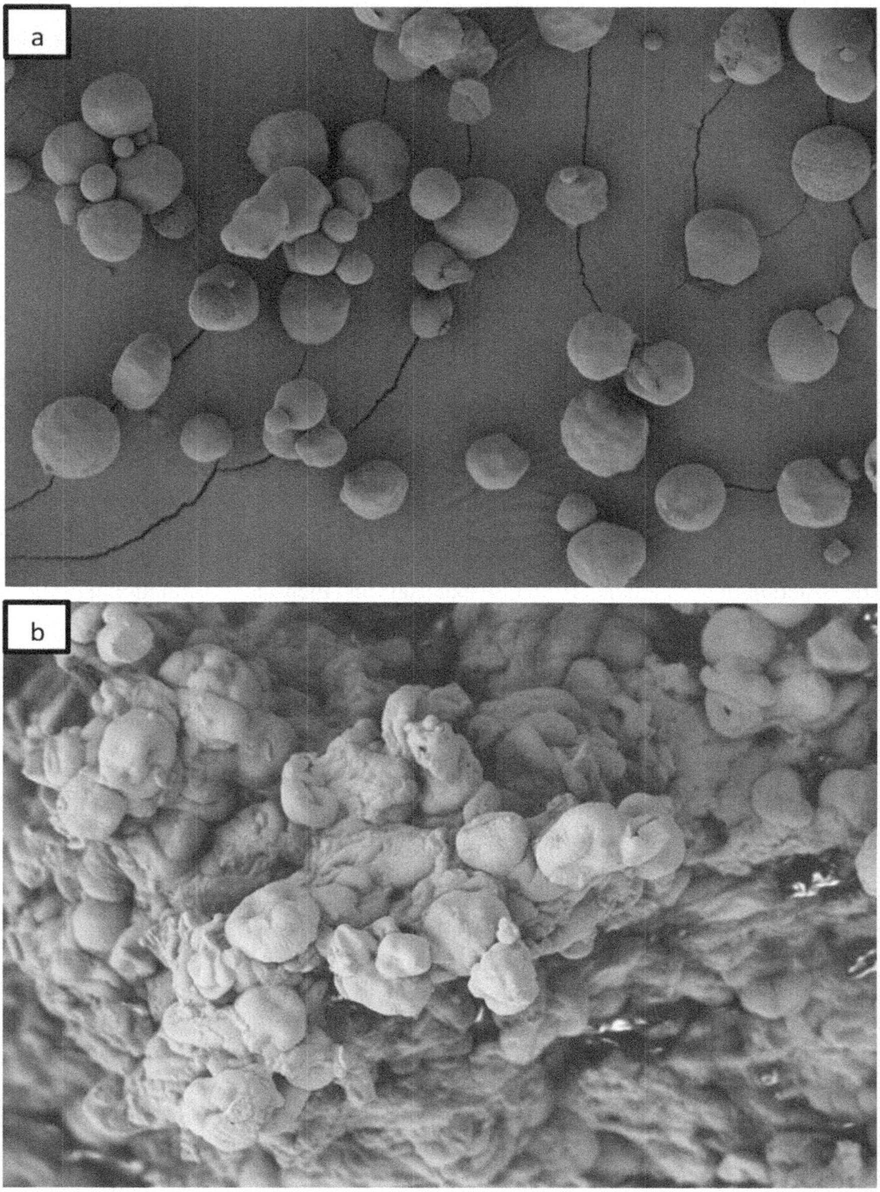

Fig. 10.3 Scanning electron microscopy (SEM) micrographs (1.5 kX amplification) of Andean maize var. Capia race (**a**) native isolated starches and (**b**) extruded (20% moisture and 120 °C) whole-grain flour

There is evidence that the starch present in Andean maize flours with soft endosperm forms more amylose-lipid and amylose-protein complexes by extrusion than those with hard endosperm. This maize type has lower water solubility values because the complexes that involve starch favor the formation of a compact structure that decreases the accessibility of water (Gimenez et al. 2022). Other studies indicate that the damage and gelatinization of starch caused by extrusion depend on the hardness of the maize endosperm; the greatest damage occurs in the hard endosperm maize starch granules, while the highest degree of gelatinization occurs in soft endosperm maize starch granules (Romero 2023; Domínguez et al. 2022). These changes affect the increase in the water absorption index, mainly in maize with soft endosperm, due to its flours being more easily hydrated during heating than those with hard endosperm. In addition, soft endosperm maize starch granules are more brittle and lose their birefringence easily (Salvador-Reyes et al. 2023; Gimenez et al. 2022). Normally the increase in the extrusion temperature process increases the water absorption capacity of the flour. However, the Culli maize race (pigmented, soft endosperm) shows a differentiated behavior when soft extrusion is applied (28% moisture and 80 °C) (Gimenez et al. 2022); its water absorption capacity increases more than drastic conditions of extrusion (18% moisture and 150 °C) because flour fibers limit the starch-water interaction (Salvador-Reyes et al. 2023). There are few investigations in extruded Andean maize with reference to the presence of these structures. Dominguez et al. (2022) found lower content of free lipids in extruded Cuzco maize flour compared to native flour, associated with the formation of the amylose-lipid complex. Salvador-Reyes et al. (2023) found that the formation of amylose-lipid and/or amylose-protein complexes improves milk absorption in extruded breakfast cereals from Culli maize. The extrusion of quinoa, amaranth (kiwicha), and kañiwa flours produces the formation of resistant starch in less quantity than in traditional cereals (Repo-Carrasco-Valencia et al. 2009; Repo-Carrasco-Valencia and Serna (2011)). This is due to the starches of these pseudocereals having smaller granules compared to those of maize starch.

In extruded maize, the loss of type A crystallinity is observed, and a new crystalline order, type V, appears, corresponding to the amylose-lipid complex and retrograded starch; in addition, changes are produced in the partial or total digestibility of the starch (Cueto et al. 2018; Kuktaite et al. 2022). The loss of crystallinity of starches from maize whole-grain flours by extrusion is usually less than that of pseudocereals due to the extensive formation of amylose-lipid complexes.

10.4.2 Rheological Properties of Extruded Whole-Grain Flours

Structural changes of the starch induced by extrusion affect the functional properties of flours; for this reason, the analysis of the pasting properties of flours is important. The increases in cold viscosities, modification of the pasting profile, and the viscoelastic behavior of the extruded material stand out.

Rheological changes in whole-grain flours of any cereal can be observed by the peak viscosity (PV). Alkaline extrusion of maize whole-grain flour produces a decrease in PV (503 cP) compared with native flour (1970 cP). This decrease indicates the presence of non-gelatinized starch remains and an increase in damaged starch (Enríquez-Castro et al. 2022). In general, the PV of extruded maize flours with soft endosperm always is less than those with hard endosperm. This is because in maize of soft endosperm, starch granules hydrate more easily during heating, are more brittle, and lose their birefringence to a greater degree (Salvador-Reyes et al. 2023). The effect of extrusion on the pasting properties of pseudocereals is different from that of maize. In amaranth and quinoa flours, the PV is usually low (<653 cP) or disappears when the extrusion is of high intensity. During the subsequent cooling step, the final viscosity (FV <240 cP) may increase due to the rearrangement of amylose during gel formation and its retrogradation. Under extrusion conditions of moderate intensity, FV may be higher (136–738 cP). The low FV in these grains could be associated with the damage or fragmentation of the amylose chains by extrusion and the reduction of amylose-amylose interactions because they are part of the amylose-lipid complexes (Basilio-Atencio et al. 2020; Espinosa-Ramírez et al. 2021). The low development of viscosity of quinoa and amaranth, characteristic of these grains, is attributed to the low content of starch and amylose compared to other cereals such as maize (Robin et al. 2015).

10.4.3 Use of Extruded Whole-Grain Flours in Food Formulation

Whole-grain flours have a higher amount of nutrients and bio-functional compounds than refined ones (Comettant-Rabanal et al. 2021). The grinding or extrusion of grains affects their technological and nutritional qualities (Román et al. 2018). Whole-grain flour includes parts of grain that are generally eliminated in obtaining refined flour (Milani et al. 2022). These components can generate mechanical problems during extrusion. For example, lipids of whole-grain flours hinder their transport through screw and reduce the expansion capacity of the extrudates (Yadav et al. 2022; Espinosa-Ramírez et al. 2021).

Extruded puffed products must have high degrees of expansion to increase their acceptability by consumers (Kowalski et al. 2016). In these products, starch is essential to form the structure support, which is formed when it passes from the molten to the vitreous state instantaneously due to the effect of the evaporation of water when leaving extruder through the die (Riaz 2004). Andean maize, in general, is classified as soft endosperm grain (Giménez et al. 2020), so it presents a low volume of expansion. However, due to their bioactive properties, they are included in snack formulations (Salvador-Reyes et al. 2023; Neder-Suárez et al. 2021). Low-intensity extrusion conditions make it possible to obtain flours with a low degree of gelatinized starch suitable for making extensible dough (Gimenez et al. 2022; Dominguez et al. 2022). This is due to the endosperm of these floury varieties; the granules are surrounded by a thinner protein matrix that facilitates their gelatinization, obtaining pastes with different viscosities, and also do not retrograde easily. In addition, granules with type A crystalline structure are suitable to be included in baked product formulations because they melt more easily during extrusion and form a structure that would contribute to stabilizing, for example, the alveoli in bread (Roman et al. 2018).

In quinoa, processes such as scarification (elimination of saponins by abrasion) or defatted are important because they modify the composition reducing the content of proteins, lipids, and fiber that negatively affect the expansion of a puffed extruded product (Aluwi et al. 2016). The degrees of expansion of quinoa and kañiwa products are usually relatively lower than those of maize under similar extrusion conditions, due to their higher amount of proteins, lipids, and fiber and lower starch content (Kowalski et al. 2016; Repo-Carrasco-Valencia et al. 2009). Amaranth has a compositional profile similar to quinoa and kañiwa, so good degrees of expansion can be achieved under optimal conditions of extrusion (high temperatures, high shear rates, and low moisture) (Bacilio-Atencio et al. 2020). Due to the difficulties generated by the secondary components in pseudocereals, they are usually mixed with other matrices such as maize or rice to achieve higher degrees of expansion, better textures, and acceptability, adding a nutritional improvement to the final products (Valderrama-Amasifuen et al. 2021).

10.5 Applications of Native and Modified Tuber Starches

Industrially, starch is an important raw material used in a wide range of industries ranging from food and packaging to pharmaceuticals and biofuels. According to the OECD-FAO report (2020), about 55 Mt. of the world's production of roots and tubers is destined for the production of biofuels and starch. A significant volume of international trade in fresh products is not possible due to the perishable nature of these products, so they are marketed processed. In the food industry, the starch of root and tubers are used as ingredients in processed products such as sauces, soups, and bread, with functions involving gelling, stabilizing, thickening, binding, texture modification, preservative, and quality-enhancing and bulking agents.

In the nonfood industry, it is used in pharmaceuticals (tissue and drug carriers), cosmetics, fertilizers, coatings and packaging, textiles, bioplastics, and construction materials, among others (Alassmy et al. 2023).

10.5.1 Applications of Native Tuber Starches in the Food and Nonfood Industry

The Andean potato starch (*Solanum tuberosum andigenum*), accession Imilla negra, is suitable for making paper due to its high phosphate content; in addition, due to the important resistant starch content after cooking, it has great potential application in the food industry as a functional ingredient or in the production of molded objects (Martínez et al. 2019). Also, it is applied in mining and water treatment by its flocculant effects (Xu et al. 2017). Potato starch is preferred to formulate food products because its paste is characterized by high clarity and neutral flavor (Alvani et al. 2011; Martínez et al. 2019). That clarity is attributed to the high content of phosphate esters in the amylopectin chain (Šimková et al. 2013). Potato starch is considered important and unique for its granule size, purity, amylose, and amylopectin chain lengths and ability to exchange certain cations to improve the rheological behavior and to form thick viscoelastic gels on heating and cooling processes (Martínez et al. 2019).

Paulet et al. (2009) studied oca starches from three accessions (Ecuadorian, Bolivian, and Peruvian) with elastic rheological behavior, determined through the G' coefficient, forming soft gels with varied applications in food industries and gastronomy uses. Zhu and Cui (2020) compared the physicochemical characteristics of oca, potato, and maize starches from New Zealand. Obtaining that oca starch has lower gelatinization temperatures, gels with intermediate fluidity and greater viscosity than maize starch gels, and more elastic than potato starch gels. Espín et al. (2004) determined 30% amylose contents in oca, which influences its rheological behavior and digestibility, classifying it as slow digestion together with the Andean potato. In addition, it has greater swelling power, property that can be used in processes such as extrusion. In the same study, the Andean potato and melloco (*Ullucus tuberosus*), known as "*papa lisa*" in Argentina, were characterized as having a low content of digestible starch compared to wheat, suggesting their use in diets to achieve low glycemic responses. In the same study, they found that Andean potatoes and melloco (Ullucus tuberosus), known as "papa lisa" in Argentina, due to their low digestible starch content compared to wheat, could be used in diets that require low glycemic responses. Oca and miso starches (Mirabilis expansa), both with greater digestibility, would be useful in diets for children and the elderly.

Hernández-Medina et al. (2008) reported starch properties of tubers makal (*Xanthosoma yucatanensis*), camote (*Ipomea batata*), yuca (*Manihot esculenta Crantz)*, and sagú (*Maranta arundinacea*) grown in Mexico. Yuca starch has the highest swelling power and solubility; its gels had the highest PV and elasticity (36.2%). Camote starch exhibits the highest gel clarity (51.8%) and makal starch the lowest (10.9%). Based on these characteristics, makal and sagú starches can be used in products that require high processing temperatures. Camote and yuca starches can be included in food systems as thickeners, stabilizers, and gelling agents in refrigerated and frozen foods.

Presently, there is a constant search for new renewable sources to produce natural biopolymers for edible coatings or biodegradable films (Larotonda et al. 2016). Tuber starches are used to apply emerging manufacturing technologies, such as 3D printing because their gels are easier to work with during processing (Jonathan and Karim 2016).

The incorporation of functional components in formulations with high content of starch and bioactive compounds is a growing trend for food and nonfood applications. Torres et al. (2020) studied three varieties of discarded potatoes (var. Agria, Kennebec, and Neiker), of small size and irregular shape, harvested in Galicia, Spain, highlighting the Agria variety as a source of high-quality starch and with great viability to prepare functional hydrogels, with excellent mechanical properties. The use in hydrogel formulation represents added value for discarding potatoes from the economic, social, industrial, and environmental points of view.

10.5.2 Modified Starches

Recent advances in starch processing technologies and genetic manipulation to create new varieties with useful functionalities of starches had a profound impact on the industry. New starch products have been developed, using them as raw materials for food, chemicals, plastics, and fuels. The native starch structure is sometimes inefficient, since certain conditions of technological processes, such as temperature, pH, and pressure, reduce its use in industrial applications by causing low resistance to shear stress, thermal decomposition, high level of retrogradation, and syneresis (Pacheco de Delahaye and Techeira 2009). Modified starches have various applications, such as enzyme-resistant dextrins produced by microwave heating of acidified potato starch with hydrochloric and citric acids, which could be used in the soft drink industry due to their high solubility and low viscosity (Kapusniak and Nebesny 2017). Calliope et al. (2020) characterized native starches of Andean potatoes, varieties Imilla negra, Desiré, Waycha, and Chorcoyeña. Their starches have intermediate and high amylose content and high phosphorus content, conditions related to a greater susceptibility to modification by cross-linking compared to other starches with low amylose content such as rice. The obtained hydrogels have great swelling capacity, a characteristic that can be used in applications such as food packaging systems in trays, placed inside the packaging to absorb the released exudate while controlling the aw of the product.

Hydrophobic-modified starches have an increasing relevance in the application of new biomaterials, especially in the packaging industry, pickering emulsions, and pharmaceuticals.

10.5.3 Applications of Modified Starches in the Industry

Environmental pollution caused by the use of nondegradable plastics has led to the use of degradable natural raw materials such as starch, which has become a promising material due to its good film-forming properties, low cost, and easy processing (Dufresne and Castaño 2017; Dai et al. 2015; Kakaei and Shahbazi 2016).

Dai et al. (2019) evaluated the physicochemical properties of native starches from different sources (waxy corn, yuca, camote, potato, wheat, corn). The film of yuca native starch has the best properties for food preservation. Yuca starches were esterified, cross-linked, and oxidized for edible film applications. Cross-linking provides a more suitable film-forming matrix compared to esterified and oxidized starches, due to better mechanical and water vapor barrier properties.

Some processes used for physical modification include heat treatment, high pressure, microwaves, ultraviolet light, radiation, ohmic heating, wet treatments, and mechanical stress (extrusion). Starches modified by these methods are considered safe and natural materials; however, few studies have been conducted on the effect of physical modification methods on the functional properties of starches from Andean tubers (Bello Perez and Agama-Acevedo 2017).

References

Agama-Acevedo E, Ottenhof MA, Farhat IA, Paredes-López O, Ortíz-Cereceres J, Bello-Pérez LA (2005) Aislamiento y caracterización del almidón de maíces pigmentados. Agrociencia 39(4):419–429. https://www.redalyc.org/articulo.oa?id=30239406

Agama-Acevedo E, Salinas-Moreno Y, Pacheco-Vargas G, Bello-Pérez LA (2011) Características físicas y químicas de dos razas de maíz azul: morfología del almidón. Revista mexicana de ciencias agrícolas 2(3):317–329. http://www.redalyc.org/src/inicio/ArtPdfRed.jsp?iCve=263119714002

Agama-Acevedo E, Bello-Perez LA (2017) Starch as an emulsions stability: the case of octenyl succinic anhydride (OSA) starch. Current Opinion in Food Science 13:78–83. https://doi.org/10.1016/j.cofs.2017.02.014

Ahmed J, Al-Atta H, Arfat YA (2016) Effect of particle size on compositional, functional, pasting and rheological properties of commercial water chestnut flour. Food Hydrocoll 52:888–895. https://doi.org/10.1016/j.foodhyd.2015.08.028

Ai Y, Hasjim J, Jane J (2013) Effects of lipids on enzymatic hydrolysis and physical properties of starch. Carbohydrate Polymers 92(1):120–127. https://doi.org/10.1016/j.carbpol.2012.08.092

Alassmy YA, Abduljawad MM, Al-shamrani KM, Alnafisah MS, El Nokab MEH, Pour ZA, Sebakhy KO (2023) A green/sustainable organocatalytic pathway for the preparation of esterified supercritical CO2-dried potato starch products. J Appl Polym Sci 140(10):e53585. https://doi.org/10.1002/app.53585

Almeida RLJ, Santos NC, Feitoza JVF, da Silva GM, de Sousa MCE, da Silva ER, de Almeida MMM (2022) Effect of heat-moisture treatment on the thermal, structural and morphological properties of Quinoa starch. Carbohydr Polym Technol Appl 3:100192. https://doi.org/10.1016/j.carpta.2022.100192

Aluwi NA, Gu BJ, Dhumal GS, Medina-Meza IG, Murphy KM, Ganjyal GM (2016) Impacts of scarification and germination on the expansion characteristics of select quinoa varieties during extrusion processing. Journal of Food Science 81(12):E2939–E2949. https://doi.org/10.1111/1750-3841.13512

Alvani K, Qi X, Tester RF, Snape CE (2011) Physico-chemical properties of potato starches. Food Chem 125(3):958–965. https://doi.org/10.1016/j.foodchem.2010.09.088

Alvis A, Vélez CA, Villada HS, Rada-Mendoza M (2008) Análisis físico-químico y morfológico de almidones de ñame, yuca y papa y determinación de la viscosidad de las pastas. Inform tecnol 19(1):19–28. https://doi.org/10.4067/S0718-07642008000100004

Basilio-Atencio J, Condezo-Hoyos L, Repo-Carrasco-Valencia R (2020) Effect of extrusion cooking on the physical-chemical properties of whole kiwicha (*Amaranthus caudatus* L) flour variety Centenario: process optimization. LWT 128:109426. https://doi.org/10.1016/j.lwt.2020.109426

Bello-Pérez LA, Agama-Acevedo E, Sáyago-Ayerdi SG, Moreno-Damian E, Figueroa JDC (2000) Some structural, physicochemical and functional studies of banana starches isolated from two varieties growing in Guerrero. México. Starch-Stärke 52(2-3):68–73. https://doi.org/10.1002/(SICI)1521-379X(200004)52:2/3<68::AIDSTAR68>3.0.CO;2-1

Boege E (2009) Centros de origen, pueblos indígenas y diversificación del maíz. Ciencias 92–93:18–28; http://revistas.unam.mx/index.php/cns/article/view/14825

Cabrera-Ramírez AH, Cervantes-Ramírez E, Morales-Sánchez E, Rodriguez-García ME, Reyes-Vega MDLL, Gaytán-Martínez M (2021) Effect of extrusion on the crystalline structure of starch during RS5 formation. Polysaccharides 2(1):187–201. https://doi.org/10.3390/polysaccharides2010013

Calliope S, Lobo M, Sammán N (2018) Biodiversity of Andean potatoes: morphological, nutritional and functional characterization. Food Chem 238:42–50. https://doi.org/10.1016/j.foodchem.2016.12.074

Calliope S, Wagner J, Samman N (2019) Physicochemical and functional characterization of potato starch (*Solanum tuberosum* ssp. Andigenum) from the Quebrada de Humahuaca, Argentina. Starch-Stärke 72(1–2):1900069. https://doi.org/10.1002/star.201900069

Calliope S, Lobo M, Samman N (2020) Caracterización de hidrogeles para uso en envases activos. Una alternativa para el agregado de valor al cultivo de papas andinas. Jornadas de Ciencia y Tecnología de Facultades de Ingeniería del NOA. Publicado en la revista Investigaciones en Facultades de Ingeniería del NOA. ISSN: 1853-6662. Revista No 6. pp 551–558

Carrillo-Navas H, Hernández-Jaimes C, Utrilla-Coello RG, Meraz M, Vernon-Carter EJ, Alvarez-Ramirez J (2014) Viscoelastic relaxation spectra of some native starch gels. Food Hydrocoll 37:25–33. https://doi.org/10.1016/j.foodhyd.2013.10.023

Cervantes-Ramírez JE, Cabrera-Ramirez AH, Morales-Sánchez E, Rodriguez-García ME, de la Luz Reyes-Vega M, Ramírez-Jiménez AK, Gaytán-Martínez M (2020) Amylose-lipid complex formation from extruded maize starch mixed with fatty acids. Carbohydr Polym 246:116555. https://doi.org/10.1016/j.carbpol.2020.116555

Comettant-Rabanal R, Carvalho CWP, Ascheri JLR, Chávez DWH, Germani R (2021) Extruded whole grain flours and sprout millet as functional ingredients for gluten-free bread. LWT 150:112042. https://doi.org/10.1016/j.lwt.2021.112042

Contreras-Gallegos E, Dominguez-Pacheco FA, Hernandez-Aguilar C, Carballo-Carballo A, Cruz-Orea A, Thoen J, Glorieux C (2015) Study of thermal and structural properties of starch granules from different maize genotypes. Food Biophys 10:19–24. https://doi.org/10.1007/s11483-014-9344-5

Cruz G, Ribotta P, Ferrero C, Iturriaga L (2016) Physicochemical and Rheological Characterization of Andean Tuber Starches: Potato (Solanum Tuberosum Ssp. Andigenum), Oca (Oxalis Tuberosa Molina) and Papa lisa (Ullucus Tuberosus Caldas). Starch - Stärke. 68(11–12):1084–1094. https://doi.org/10.1002/star.201600103

Cueto M, Farroni A, Rodríguez SD, Schoenlechner R, Schleining G, del Pilar BM (2018) Assessing changes in enriched maize flour formulations after extrusion by means of FTIR, XRD, and chemometric analysis. Food and bioprocess technology 11:1586–1595. https://doi.org/10.1007/s11947-018-2113-6

Dai L, Qiu C, Xiong L, Sun Q (2015) Characterisation of corn starch-based films reinforced with taro starch nanoparticles. Food Chem 174:82–88. https://doi.org/10.1016/j.foodchem.2014.11.005

Dai L, Zhang J, Cheng F (2019) Effects of starches from different botanical sources and modification methods on physicochemical properties of starch-based edible films. Int J Biol Macromol 132:897–905. https://doi.org/10.1016/j.ijbiomac.2019.03.197

del Valdez-Arana JC, Steffolani ME, Repo-Carrasco-Valencia R, Pérez GT, Condezo-Hoyos L (2020) Physicochemical and functional properties of isolated starch and their correlation with flour from the Andean Peruvian quinoa varieties. Int J Biol Macromol 147:997–1007. https://doi.org/10.1016/j.ijbiomac.2019.10.067

Domínguez NE, Miranda RM, Giménez MA, Lobo MO, Sammán NC (2022) Efecto de la extrusión alcalina en las propiedades fisicoquímicas y funcionales de la harina integral de maíz andino. Innotec 24:10.26461/24.04

Dufresne A, Castaño J (2017) Polysaccharide nanomaterial reinforced starch nanocomposites: a review. Starch-Stärke 69(1–2):1500307. https://doi.org/10.1002/star.201500307

Enríquez-Castro CM, Ramírez-Wong B, Contreras-Jiménez BL, Quintero-Ramos A, de Dios Figueroa-Cárdenas J, Vázquez-Lara F (2022) Effect of extrusion on the crystallinity, viscosity, damage starch, and thermal properties of corn flour, masa, and tortilla. Appl Food Res 2(2):100198. https://doi.org/10.1016/j.afres.2022.100198

Espín S, Villacrés E, Brito Grandes YB (2004) Caracterización físico–química, nutricional y funcional de raíces y tubérculos andinos. In: Barrera VH, Tapia C, Monteros YA (eds) Raíces y Tubérculos Andinos: Alternativas para la conservación y uso sostenible en el Ecuador. INIAP/CIP/COSUDE. (Serie: Conservación y uso de la biodiversidad de raíces y tubérculos andinos: Una década de investigación para el desarrollo (1993–2003). no. 4), Quito, pp 91–116; http://repositorio.iniap.gob.ec/handle/41000/3264

Espinosa-Ramírez J, Rodríguez A, De la Rosa-Millán J, Heredia-Olea E, Pérez-Carrillo E, Serna-Saldívar SO (2021) Shear-induced enhancement of techno-functional properties of whole grain flours through extrusion. Food Hydrocoll 111:106400. https://doi.org/10.1016/j.foodhyd.2020.106400

Fuentes C, Perez-Rea D, Bergenståhl B, Carballo S, Sjööa M, Nilsson L (2019) Physicochemical and structural properties of starch from five Andean crops grown in Bolivia. Int J Biol Macromolecules 125:829–838. https://doi.org/10.1016/j.ijbiomac.2018.12.120

Giménez MA, Segundo CN, Lobo MO, Sammán NC (2020) Physicochemical and techno-functional characterization of native corn reintroduced in the Andean zone of Jujuy, Argentina. In: Proceedings (Vol. 53, no. 1, p. 7). Ed. MDPI. https://doi.org/10.3390/proceedings2020053007

Gimenez MA, Segundo CN, Dominguez NE, Samman NC, Lobo MO (2022) Utilization of hydrothermally treated flours in gluten-free doughs. In: Biology and life sciences forum, p 5. https://doi.org/10.3390/blsf2022017005

Gonzalez RJ, Torres R, De Greef D, Bonaldo A, Robutti J, Borrás F (2005) Efecto de la dureza del endospermo del maíz sobre las propiedades de hidratación y cocción. Arch Latinoam Nutr 55(4):354–360. http://www.alanrevista.org/ediciones/2005/4/art-6/

Hernández-Medina M, Torruco-Uco JG, Chel-Guerrero L, Betancur-Ancona D (2008) Caracterización fisicoquímica de almidones de tubérculos cultivados en Yucatán, México. Food Sci Technol 28:718–726. https://doi.org/10.1590/S0101-20612008000300031

Jan KN, Panesar PS, Rana JC, Singh S (2017) Structural, thermal and rheological properties of starches isolated from Indian quinoa varieties. Int J Biol Macromol 102:315–322. https://doi.org/10.1016/j.ijbiomac.2017.04.02

Jiang F, Du C, Guo Y, Fu J, Jiang W, Du S (2020) Physicochemical and structural properties of starches isolated from quinoa varieties. Food Hydrocoll 101:105515. https://doi.org/10.1016/j.foodhyd.2019.105515

Jonathan G, Karim A (2016) 3D printing in pharmaceutics: a new tool for designing customized drug delivery systems. Int J Pharm 499:376–394. https://doi.org/10.1016/j.ijpharm.2015.12.071

Kakaei S, Shahbazi Y (2016) Effect of chitosan-gelatin film incorporated with ethanolic red grape seed extract and Ziziphora clinopodioides essential oil on survival of listeria monocytogenes and chemical, microbial and sensory properties of minced trout fillet. LWT Food Sci Technol 72:432–438. https://doi.org/10.1016/j.lwt.2016.05.021

Kapusniak JK, Nebesny E (2017) Enzyme-resistant dextrins from potato starch for potential application in the beverage industry. Carbohydr Polym 172:152–158. https://doi.org/10.1016/j.carbpol.2017.05.041

Kong X, Kasapis S, Bertoft E, Corke H (2010) Rheological properties of starches from grain amaranth and their relationship to starch structure. Starch-Stärke 62(6):302–308. https://doi.org/10.1002/star.200900235

Kowalski RJ, Medina-Meza IG, Thapa BB, Murphy KM, Ganjyal GM (2016) Extrusion processing characteristics of quinoa (Chenopodium quinoa Willd.) var. Cherry Vanilla. Revista de ciencia de los cereales 70:91–98. https://doi.org/10.1016/j.jcs.2016.05.024

Kuktaite R, Repo-Carrasco-Valencia R, de Mendoza CC, Plivelic TS, Hall S, Johansson E (2022) Innovatively processed quinoa (Chenopodium quinoa W illd.) food: chemistry, structure and end-use characteristics. J Sci Food Agric 102(12):5065–5076. https://doi.org/10.1002/jsfa.11214

Larotonda FDS, Torres MD, Gonçalves MP, Sereno AM, Hilliou L (2016) Hybrid carrageenan-based formulations for edible film preparation: benchmarking with kappa carrageenan. J Appl Polym Sci 133:42263. https://doi.org/10.1002/app.42263

Li G, Zhu F (2017) Amylopectin molecular structure in relation to physicochemical properties of quinoa starch. Carbohydr Polym 164:396–402. https://doi.org/10.1016/j.carbpol.2017.02.014

Li G, Wang S, Zhu F (2016) Physicochemical properties of quinoa starch. Carbohydr Polym 137:328–338. https://doi.org/10.1016/j.carbpol.2015.10.064

Lin L, Yang H, Chi C, Ma X (2020) Effect of protein types on structure and digestibility of starch-protein-lipids complexes. Lebensmittel-Wissenschaft Technol 134:110175. https://doi.org/10.1016/j.lwt.2020.110175

Lopez-Fernandez MP, Rodríguez SD, Favre LC, Busch VM, Buera MDP (2021) Physicochemical, thermal and rheological properties of isolated Argentina quinoa starch. LWT-Food Sci Technol 135:1–9. https://doi.org/10.1016/j.lwt.2020.110113

Marti A, Seetharaman K, Pagani MA (2010) Rice-based pasta: a comparison between conventional pasta-making and extrusion-cooking. J Cereal Sci 52(3):404–409. https://doi.org/10.1016/j.jcs.2010.07.002

Martínez P, Peña F, Bello-Pérez LA, Núñez-Santiago C, Yee-Madeira H, Velezmoro C (2019) Physicochemical, functional and morphological characterization of starches isolated from three native potatoes of the Andean region. Food Chem X 2:100030. https://doi.org/10.1016/j.fochx.2019.100030

Martínez P, Betalleluz-Pallardel I, Cuba A, Peña F, Cervantes-Uc JM, Uribe-Calderón JA, Velezmoro C (2022) Effects of natural freeze-thaw treatment on structural, functional, and rheological characteristics of starches isolated from three bitter potato cultivars from the Andean region. Food Hydrocoll 132:107860. https://doi.org/10.1016/j.foodhyd.2022.107860

Martinez-Lopez A, Millan-Linares MC, Rodriguez-Martin NM, Millan F, Montserrat-de la Paz S (2020) Nutraceutical value of kiwicha (*Amaranthus caudatus* L.). J Funct Foods 65:103735. https://doi.org/10.1016/j.jff.2019.103735

Milani P, Torres-Aguilar P, Hamaker B, Manary M, Abushamma S, Laar A, Taylor J (2022) The whole grain manifesto: from green revolution to grain evolution. Glob Food Sec 34:100649. https://doi.org/10.1016/j.gfs.2022.100649

Narváez-González ED, de Dios F-CJ, Taba S, Sánchez FR (2006) Kernel microstructure of Latin American races of maize and their thermal and rheological properties. Cereal Chemistry 83(6):605–610. https://doi.org/10.1094/CC-83-0605

Narváez-González ED, de Dios Figueroa-Cárdenas J, Taba S, Castaño-Tostado E, Martínez-Peniche RÁ (2007) Efecto del tamaño del gránulo de almidón de maíz en sus propiedades térmicas y de pastificado. Rev Fitotec Mex 30(3):269–269. https://doi.org/10.35196/rfm.2007.3.269

Neder-Suárez D, Lardizabal-Gutiérrez D, Zazueta-Morales JDJ, Meléndez-Pizarro CO, Delgado-Nieblas CI, Ramírez WB, Quintero-Ramos A (2021) Anthocyanins and functional compounds change in a third-generation snacks prepared using extruded blue maize, black bean, and chard: an optimization. Antioxidants 10(9):1368. https://doi.org/10.3390/antiox10091368

OECD/FAO (2020) OCDE-FAO Perspectivas Agrícolas 2020-2029. OECD Publishing, Paris. https://doi.org/10.1787/a0848ac0-es

Pacheco de Delahaye E, Techeira N (2009) Propiedades químicas y funcionales del almidón nativo y modificado de ñame (*Dioscorea alata*). Interciencia 34(4):280–285; Recuperado en 17 de julio de 2023, de. http://ve.scielo.org/scielo.php?script=sci_arttext&pid=S0378-18442009000400012&lng=es&tlng=es

Paucar Menacho LM, Schmiele M, Lavado-Cruz AA, Verona-Ruiz AL, Mollá C, Peñas E, Martínez-Villaluenga C (2022) Andean sprouted Pseudocereals to produce healthier Extrudates: impact in nutritional and physicochemical properties. Food Secur 11(20):3259. https://doi.org/10.3390/foods11203259

Paulet G, Bello-Pérez L, Salas F, Buleje E (2009) Características viscoelásticas y estimaciones de masas moleculares en almidón de oca (*Oxalis tuberosum*). Revista de la Sociedad Química del Perú 75(2):266–276. [citado 2023-07-17]. Disponible en: <http://www.scielo.org.pe/scielo.php?script=sci_arttext&pid=S1810-634X2009000200014&lng=es&nrm=iso>. ISSN 1810-634X

Qu J, Xu S, Zhang Z, Chen G, Zhong Y, Liu L, Zhang R, Xue J, Guo D (2018) Evolutionary, Structural and Expression Analysis of Core Genes Involved in Starch Synthesis. Sci. Rep. 8(1):1–16. https://doi.org/10.1038/s41598-018-30411-y

Repo-Carrasco-Valencia R, Peña J, Kallio H, Salminen S (2009) Dietary fiber and other functional components in two varieties of crude and extruded Kiwicha (*Amaranthus caudatus*). J Cereal Sci 49(2):219–224. https://doi.org/10.1016/j.jcs.2008.10.003

Repo-Carrasco-Valencia RAM, Serna LA (2011) Quinoa (Chenopodium quinoa, Willd.) as a source of dietary fiber and other functional components. Food Science and Technology 31:225–230. ISSN 0101-2061. https://www.scielo.br/j/cta/a/64zr4jJ7KCB8kZgZyQ3Hs6Q/

Repo-Carrasco-Valencia R, Arana JV (2017) Carbohydrates of kernels. Pseudocereals: chemistry and technology:49–70. https://doi.org/10.1002/9781118938256.ch3

Riaz MN (2004) Introducción a los extrusores y sus principios. In: Extrusores en las aplicaciones de alimentos. Acribia, España

Robin F, Théoduloz C, Srichuwong S (2015) Properties of extruded whole grain cereals and pseudocereals flours. Int J Food Sci Technol 50(10):2152–2159. https://doi.org/10.1111/ijfs.12893

Robutti J, Borras F, González R, Torres R, De Greef D (2002) Endosperm properties and extrusion cooking behavior of maize cultivars. LWT 35(8):663–669. https://doi.org/10.1006/fstl.2002.0926

Rojas W, Pinto M, Alanoca C, Gómez L, Leónlobos P, Alercia A, Diulgheroff S, Padulosi S, Bazile D (2014) Chapter 1.5—Estado de la conservación ex situ de los recursos genéticos de quinua. In: Bazile DEAE (ed) Estado del arte de la quinua en el mundo en 2013. FAO (Santiago de Chile) y CIRAD, Montpellier, pp 65–94; ISBN 978-92-5-308558-3. http://www.fao.org/quinoa2013/publications/detail/es/item/279313/icode/?nomobile=1

Roman L, de la Cal E, Gomez M, Martinez MM (2018) Specific ratio of A-to B-type wheat starch granules improves the quality of gluten-free breads: optimizing dough viscosity and pickering stabilization. Food Hydrocoll 82:510–518. https://doi.org/10.1016/j.foodhyd.2018.04.034

Román L, Reguilón MP, Gómez M (2018) Physicochemical characteristics of sauce model systems: influence of particle size and extruded flour source. J Food Eng 219:93–100. https://doi.org/10.1016/j.jfoodeng.2017.09.024

Romero MJ (2023) Características tecnológicas de harinas integrales de maíces andinos modificadas térmicamente. Tesis de grado. Facultad de Ingeniería-Universidad Nacional de Jujuy, Jujuy

Salvador-Reyes R, Rebellato AP, Pallone JAL, Ferrari RA, Clerici MTPS (2021) Kernel characterization and starch morphology in five varieties of Peruvian Andean maize. Food Res Int 140:110044. https://doi.org/10.1016/j.foodres.2020.110044

Salvador-Reyes R, Sampaio UM, de Menezes Alves Moro T, Brito ADCD, Behrens J, Campelo PH, Pedrosa Silva Clerici MT (2023) Andean purple maize to produce extruded breakfast cereals: impact on techno-functional properties and sensory acceptance. J Sci Food Agric 103(2):548–559. https://doi.org/10.1002/jsfa.12165

Sanchez de la Concha BB, Agama-Acevedo E, Nuñez-Santiago MC, Bello- PLA, Garcia HS, Alvarez-Ramirez J (2018) Acid hydrolysis of waxy starches with different granule size for nanocrystal production. J Cereal Sci 79:193–200. https://doi.org/10.1016/j.jcs.2017.10.018

Seal CJ, Courtin CM, Venema K, de Vries J (2021) Health benefits of whole grain: effects on dietary carbohydrate quality, the gut microbiome, and consequences of processing. Compr Rev Food Sci Food Saf 20(3):2742–2768. https://doi.org/10.1111/1541-4337.12728

Seetharaman K, Tziotis A, Borras F, White PJ, Ferrer M, Robutti J (2001) Thermal and functional characterization of starch from Argentinean corn. Cereal Chem 78(4):379–386. https://doi.org/10.1094/CCHEM.2001.78.4.379

Serratos JA (2012) El origen y la diversidad del maíz en el continente americano, 2nd edn. Universidad Autónoma de la Ciudad de México. https://docplayer.es/2523922-El-origen-y-la-diversidad-del-maiz-en-el-continente-americano.html. Accessed 18 Jul 2023

Šimková D, Lachman J, Hamouz K, Vokál B (2013) Effect of cultivar, location and year on total starch, amylose, phosphorus content and starch grain size of high starch potato cultivars for food and industrial processing. Food Chem 141(4):3872–3880. https://doi.org/10.1016/j.foodchem.2013.06.080

Srichuwong S, Curti D, Austin S, King R, Lamothe L, Gloria-Hernandez H (2017) Physicochemical properties and starch digestibility of whole grain sorghums, millet, quinoa and amaranth flours, as affected by starch and non-starch constituents. Food Chem 233:1–10. https://doi.org/10.1016/jfoodchem.2017.04.019

Steffolani ME, León AE, Pérez GT (2013) Study of the physicochemical and functional characterization of quinoa and kañiwa starches. Starch-Stärke 65(11–12):976–983. https://doi.org/10.1002/star.201200286

Tacer-Caba Z, Nilufer-Erdil D (2019). Resistant starch. In encyclopedia of food chemistry (pp. 571–575). Elsevier. 10.1016/B978-0-08-100596- 5.22407-4

Torres AV, Zamudio-Flores PB, Salgado-Delgado R, Bello-Pérez LA (2007) Morphological, thermal, and mechanical studies of film elaborated with the blend low-density polyethylene and chemical-modified banana starch. J Appl Polym Sci 106(6):3994–3999. https://doi.org/10.1002/app.27051

Torres MD, Fradinho P, Rodríguez P, Falqué E, Santos V, Domínguez H (2020) Biorefinery concept for discarded potatoes: recovery of starch and bioactive compounds. J Food Eng 275:109886. https://doi.org/10.1016/j.jfoodeng.2019.109886

Utrilla-Coello RG, Hernández-Jaimes C, Carrillo-Navas H, González F, Rodríguez E, Bello-Perez LA, Alvarez-Ramirez J (2014) Acid hydrolysis of native corn starch: morphology, crystallinity, rheological and thermal properties. Carbohydr Polym 103:596–602. https://doi.org/10.1016/j.carbpol.2014.01.046

Valderrama-Amasifuen F., Arteaga H., Flores Perez AE.Obregón J. and Barraza-Jáuregui G (2021). Snacks a base de maíz morado, quinua y kiwicha. Características físicas y sensoriales. https://www.laccei.org/LACCEI2021-VirtualEdition/full_papers/FP569.pdf

Velasquez Barreto FF, Bello-Pérez LA (2021) Chemical, structural, technological properties and applications of Andean tuber starches: a review. Food Rev Intl 39(3):1293–1308. https://doi.org/10.1080/87559129.2021.1933022

Velasquez-Barreto FF, Arteaga MH, Alvarez-Ramirez J, Bello-Pérez LA (2021a) Structural, functional, and chemical properties of small starch granules: Andean quinoa and kiwicha. Food Hydrocoll 120:106883. https://doi.org/10.1016/j.foodhyd.2021.106883

Velásquez-Barreto FF, Bello-Pérez LA, Nuñez-Santiago C, Yee-Madeira H, Sánchez CEV (2021b) Relationships among molecular, physicochemical and digestibility characteristics of Andean tuber starches. Int J Biol Macromol 182:472–481. https://doi.org/10.1016/j.ijbiomac.2021.04.039

Xu X, Huang X-F, Visser RGF, Trindade LM (2017) Engineering potato starch with a higher phosphate content. PLoS One 12:e0169610. https://doi.org/10.1371/journal.pone.0169610

Yadav GP, Dalbhagat CG, Mishra HN (2022) Effects of extrusion process parameters on cooking characteristics and physicochemical, textural, thermal, pasting, microstructure, and nutritional properties of millet-based extruded products: a review. J Food Process Eng 45(9):e14106. https://doi.org/10.1111/jfpe.14106

Zhang B, Selway N, Shelat KJ, Dhital S, Stokes JR, Gidley MJ (2017) Tribology of swollen starch granule suspensions from maize and potato. Carbohydr Polym 155:128–135. https://doi.org/10.1016/j.carbpol.2016.08.064

Zhu F, Cui R (2020) Comparison of physicochemical properties of oca (*Oxalis tuberosa*), potato, and maize starches. Int J Biol Macromol 148:601–607. https://doi.org/10.1016/j.ijbiomac.2020.01.028

Unlocking the Potential of Resistant Starch: A Genetic Perspective on Enhancing Human Health

11

Bhaswati Sarmah, Vishwa Jyoti Baruah, and Nirmal Mazumder

Abstract

The chapter delves into the genetic aspects of resistant starch and its potential impact on human health. Resistant starch is a type of dietary fiber that resists digestion, reaching the colon intact, and has shown promise in improving various aspects of health. The genetic underpinnings of how individuals metabolize and respond to resistant starch have been explored. It discusses the role of specific genes in influencing an individual's ability to derive health benefits from this type of starch. Understanding these genetic factors can help tailor dietary recommendations and interventions to optimize the health benefits of resistant starch for different individuals. Emphasis has also been given to the importance of personalized nutrition and how genetic insights into resistant starch utilization can be harnessed to enhance overall well-being. By shedding light on the genetic nuances of resistant starch, this chapter opens up exciting possibilities for more targeted and effective dietary strategies, potentially improving health outcomes for individuals.

Keywords

Cancer · Chronic diseases · Diabetes · Metabolic disorder · Resistant starch

B. Sarmah (✉)
Section of Plant Breeding and Genetics, Cornell University, Ithaca, NY, USA

V. J. Baruah
School of Integrative Plant Science, Cornell University, Ithaca, NY, USA

N. Mazumder
Department of Biophysics, Manipal School of Life Sciences, Manipal Academy of Higher Education, Manipal, Karnataka, India

N. Mazumder, M. H. Rahman (eds.), *Advanced Research in Starch*, https://doi.org/10.1007/978-981-99-9527-1_11

11.1 Introduction

There is a mounting body of evidence indicating that dietary modifications can play a crucial role in preventing or mitigating chronic health conditions in developed nations. A significant aspect of the American diet comprises starchy foods like white bread, cakes, and noodles, which predominantly consist of easily digestible starch. This heightened digestibility has raised concerns that the rapid breakdown of such starches might contribute to chronic diseases in both humans and animals. To address this concern, researchers have increasingly turned their attention to starches that resist digestion by enzymes, commonly referred to as "resistant starches" (Englyst et al. 1992).

Starch, a prominent carbohydrate used for energy storage in living organisms, is a complex polysaccharide composed of sugar units connected via a-1,4- and a-1,6-glycosidic bonds (Hizukuri et al. 1989). Its digestibility can be assessed in vitro, leading to its classification into three primary categories: rapidly digestible starch (RDS), slowly digestible starch (SDS), and resistant starch (RS), as defined by Englyst and Hudson in Englyst and Hudson 1996. Rapidly digestible starch (RDS) comprises the portion of starch that undergoes complete digestion and absorption within the small intestine. This fraction rapidly undergoes hydrolysis and is converted into individual glucose molecules in a mere 20 min through enzymatic digestion (Englyst et al. 1992). Slowly digestible starch (SDS), on the other hand, requires a more extended period for digestion. SDS is not fully broken down until approximately 120 min of enzymatic digestion (Englyst et al. 1992). While it takes longer to digest, it can ultimately be absorbed in the small intestine.

Resistant starch (RS) is the starch fraction that poses a challenge to enzymatic digestion. Specifically, RS is defined as the portion of starch that resists digestion in the small intestine and remains undigested by enzymes even after a 120-min period of enzymatic digestion (Englyst et al. 1992). Instead of being absorbed in the small intestine like RDS and SDS, RS proceeds to the colon, where it serves as a substrate for fermentation by gut microbiota. This classification system allows for a more nuanced understanding of starch behavior during digestion. It distinguishes between starches that are rapidly broken down, those that require more time for digestion, and those that resist enzymatic hydrolysis altogether. Such distinctions are invaluable in the fields of nutrition and health, as they have significant implications for the impact of starch-containing foods on blood sugar levels and overall dietary health.

RS is a type of starch that behaves more like a dietary fiber. It has gained attention for its potential health benefits, including improved glycemic control and digestive health. Resistant starches have been the subject of extensive research and review, encompassing various aspects. These include general examinations (Sajilata et al. 2006), assessments of their health-related properties (Nugent 2005), efforts to enhance their presence in food products (Thompson 2000), evaluations of their functional attributes as food ingredients (Fuentes-Zaragoza et al. 2011), and investigations into their role in promoting gut health, potentially through the production of beneficial compounds like butyrate (Brouns et al. 2002). Despite the growing awareness of resistant starches and their potential health benefits, it is crucial to note that the estimated daily intake of resistant starch among Americans is rather modest, hovering around approximately 5 g per day. This intake falls significantly short of the recommended minimum of 6 g of resistant starch per meal, which is suggested to confer health advantages (Murphy et al. 2008). In essence, the prevailing research underscores the pivotal role that dietary choices, particularly the type of starches consumed, can play in the development and management of chronic health conditions in industrialized nations. It highlights the need for a shift in dietary patterns, with a focus on incorporating more resistant starches into daily meals to potentially mitigate the risk of chronic diseases. Increasing the consumption of these starches, which are less susceptible to rapid digestion, holds promise for promoting better health outcomes and overall well-being.

11.2 Classification of Resistant Starches

Resistant starch is a specific fraction of starch that withstands digestion by amylases in the small intestine. Instead, it proceeds to the colon, where it becomes a substrate for fermentation by gut microbiota (Englyst and Cummings 1985). A classification system was introduced to categorize starch based on its digestive rate, distinguishing between rapidly digestible starches, slowly digestible starches, and resistant starches (Englyst et al. 1992). These distinctions are determined through in vitro digestion tests. Presently, there are five types of resistant starch (RS) identified, each with unique characteristics and properties.

Type I RS (RSI): This type of resistant starch is found in the endosperm of cereal grains or seeds. Starch granules are encased in a matrix of proteins and cell wall material, which hinders starch digestion and reduces glycemic responses (O'Dea et al. 1980). When these grains or seeds are cooked whole or coarsely ground, their cell walls and protein matrices prevent water penetration, inhibiting starch gelatinization and enzymatic hydrolysis. As a result, the starch remains resistant and passes to the colon without digestion. Examples include bread made from whole grains and pasta made from durum wheat by extrusion (Granfeldt et al. 1991).

Type II RS (RSII): Uncooked starch from sources like potatoes, green bananas, gingko, and high-amylose maize, which have a B- or C-type crystalline structure, is highly resistant to enzymatic hydrolysis (Jane et al. 2003). However, most of these starches become highly digestible after cooking due to starch gelatinization and the loss of B- and C-type crystalline structures. High-amylose starch produced by specific genetic mutations retains its resistance even after cooking, as it has longer branch chains and a high gelatinization temperature (Regina et al. 2006; Li et al. 2008; Jiang et al. 2010).

Type III RS (RSIII): RSIII is retrograded amylose and starch, formed when amylose molecules, with their linear structures, tend to form double helices, especially at low temperatures with sufficient moisture content (4–5 °C). Retrograded amylose has high gelatinization temperatures, up to 170 °C, and cannot be dissociated by cooking. However, with storage, particularly in a refrigerator, retrograded amylose loses its water-binding capacity, rendering it resistant to enzymatic hydrolysis (Jane and Robyt 1984; Witt et al. 2010; Sievert and Pomeranz 1990).

Type IV RS (RSIV): This type is chemically modified starch, created either through cross-linking or the addition of chemical derivatives. Highly cross-linked starch loses its ability to swell during cooking and remains granular after cooking, making it resistant to enzymatic susceptibility (Woo and Seib 2002; Han and BeMiller 2007). Starch modified with chemical derivatives, such as octenyl succinic groups or acyl groups, undergoes structural changes that restrict enzymatic hydrolysis, resulting in resistant starch. A portion of this starch can be fermented by bacteria amylases to produce short-chain fatty acids (He et al. 2008).

Type V RS (RSV): When starch interacts with lipids, amylose and long branches of amylopectin form single-helical complexes with fatty acids and fatty alcohols (Jane and Robyt 1984; Ai et al. 2013). This complex formation prevents starch binding and cleavage by amylase enzymes. Additionally, the amylose-lipid complex restricts the swelling of starch granules and enzyme hydrolysis (Hasjim et al. 2010; Seneviratne and Biliaderis 1991). Notably, this type of resistant starch is thermally stable, as the complex can reform after cooking.

The starch digestibility is influenced by multiple factors, including non-starch components in the digestive process, starch structure, and processing methods before digestion. It highlights that the influence of these factors on digestibility is complex and that the classification systems used serve to categorize starch based on the most significant extrinsic factor affecting digestibility. Furthermore, it is important to acknowledge the botanical role of starch in plants, emphasizing the importance of glucose reserves stored in structures that are readily available to the plant yet capable of long-term survival in storage organs like seeds or tubers. The complexity of starch structure, including crystallinity and chain length, significantly influences its digestibility, with high-amylose starch being more resistant than low-amylose starch.

11.3 Genetic Factors That Potentially Influence RS Content

The quantity of resistant starch (RS) is under genetic regulation, with various genes primarily associated with starch synthesis playing a significant role. Mutations in genes responsible for granule-bound starch synthase (GBSS), soluble starch synthase (SS), and starch-branching enzyme (SBE) have been documented as having an impact on RS levels in rice, maize, and wheat (as observed in studies by Zhou et al. 2016; Baysal et al. 2020; Li et al. 2021; Liu et al. 2021). Table 1 provides a detailed list of these genetic factors that influence RS formation in rice. Additionally, the expression of starch-synthesis-related genes (SSRGs) can be regulated by several transcription factors, mRNA splicing factors, and epigenome modifiers, potentially exerting an influence on RS content.

The enzyme granule-bound starch synthase (GBSS) plays a pivotal role in the biosynthesis of amylose, facilitating the conversion of individual ADP-glucose molecules into linear chains of glucose residues connected by a-1,4-glycosidic bonds

(Tetlow 2011). Specifically, GBSSI, encoded by the Waxy (Wx) gene, stands out as a crucial enzyme for amylose biosynthesis within the endosperm (Shure et al. 1983; Wang et al. 1990). Multiple studies have consistently pointed to a positive correlation between the amylose content (AC) and RS levels in rice (Berry 1986; Sievert and Pomeranz 1989; Chen et al. 2017b). The substantial diversity in AC in rice primarily arises from allelic variation at the Wx locus (Tian et al. 2009; Zhang et al. 2019). Cultivated rice features two major functional Wx alleles, Wx_a and Wx_b, predominantly distributed in indica and japonica cultivars, respectively (Sano 1984; Sano et al. 1986; Hirano and Sano 1998). This correlation is exemplified by rice varieties possessing the Wx_a allele, characterized by high AC, which exhibit significantly greater RS content compared to rice varieties carrying the wx allele (Itoh et al. 2003; You et al. 2022). This pattern, where a higher amylose level corresponds to elevated RS content, extends beyond rice and is observed in other cereals like wheat and maize (Li et al. 2020; Lin et al. 2016). The Wx_b allele carries a G-to-T single base mutation at the 50 splice site of the first intron compared to the Wxa allele. This mutation results in reduced mRNA and protein levels of Wx (Sano et al. 1986; Cai et al. 1998; Isshiki et al. 1998). In general, rice cultivars with the Wxa allele produce higher amylose quantities, whereas those with the Wxb allele generate lower amylose levels (Sano et al. 1986; Hirano and Sano 1998; Isshiki et al. 1998; Larkin and Park 2003). This pattern aligns with indica cultivars (Wxa) having higher RS levels than japonica cultivars (Wxb) (Bao et al. 2017). The increased RS observed in the Wxa background can be primarily attributed to higher AC, promoting the formation of RS3 and RS5. Furthermore, additional natural alleles of the Wx gene- and genome-edited Wx mutants have been documented to influence the AC of rice (Liu et al. 2009; Zhang et al. 2019; Huang et al. 2020; Zeng et al. 2020; Zhou et al. 2021). However, their specific effects on RS content in rice have yet to be reported. Given the strong association between AC and RS, there is a compelling possibility that these naturally occurring Wx alleles and genome-edited mutants may also impact RS content in rice. Consequently, future research exploring the effects of these genetic variations on RS content promises to yield valuable insights into the regulation of resistant starch levels in rice.

Various starch synthases (SSs) have been found to be responsible for modifying amylopectin chains and have significant roles in the formation of resistant starch (RS) in rice. There are four primary SS members (SSI, SSII, SSIII, and SSIV) in rice, each with distinct substrate preferences (Ohdan et al. 2005). Rice contains eight SS isoforms, and most of them are active during the endosperm filling stages (Ohdan et al. 2005; Tian et al. 2009). However, only a few of these SSs have been reported to impact RS formation (Zhou et al. 2016; Raja et al. 2017; You et al. 2022). Notably, SSI has emerged as a crucial contributor to RS formation in rice, with rice accessions carrying deleterious SSI variants displaying high RS content (Raja et al. 2017). Suppressing SSI expression in both japonica and indica rice cultivars significantly elevates amylose content (AC) and compromises eating and cooking quality. Additionally, reducing SSI expression influences amylopectin chain distribution and increases granule-bound starch synthase (GBSS) activity and AC, suggesting that SSI's influence on RS formation may involve AC and amylopectin chain distribution (Zhao et al. 2019). Another SS, SSIIa, encoded by the alkali degeneration (ALK) gene, has been implicated in RS formation in rice. ALK has different alleles, including ALKa and ALKb (SSIIaj) mainly found in japonica cultivars and ALKc (SSIIai) mainly present in indica cultivars. The SSIIai allele, associated with active SSIIa, has been linked to higher RS content in indica cultivars than the SSIIaj allele in japonica cultivars (You et al. 2022). Functionally, SSIIa elongates short A and B1 chains of amylopectin, resulting in the accumulation of B1 chains in indica rice (Nakamura et al. 2005). This leads to high-crystallinity starch that is more resistant to enzymatic digestion, thereby increasing RS content (Bao et al. 2017). Additionally, SSIIIa and SSIIIb are important contributors to RS formation in rice. SSIIIa is primarily expressed in developing rice endosperm, and mutations in this gene, as seen in the b10 mutant, increase RS content by promoting amylose and lipid biosynthesis. This, in turn, forms the amylose-lipid complex RS5, contributing to higher RS levels (Zhou et al. 2016). Similar trends are observed in wheat and maize, where mutations in SSII or SSIII result in increased AC and alterations in amylopectin chain length distribution, potentially impacting RS content (Shimbata et al. 2012; Zhu et al. 2013).

Starch-branching enzymes (SBEs) are another class of enzymes responsible for creating a-1,6-glycosidic bonds, which introduce branches into amylopectin (Han et al. 2007; Tian et al. 2009). In rice, SBEI and SBEIIb are primarily expressed in the developing endosperm, while SBEIIa shows consistent expression in various rice tissues (Ohdan et al. 2005). SBEIIb has emerged as a crucial factor affecting RS content in rice, as deficiencies in SBEIIb activity significantly increase RS content in the rice endosperm (Butardo et al. 2011; Baysal et al. 2020; Guo et al. 2020; Miura et al. 2021). Mutations in the SBEIIb gene, as seen in amylose extender (ae) mutants, result in a marked increase in RS content, accompanied by elevated amylose levels and gelatinization temperature (GT) (Mizuno et al. 1993; Nishi et al. 2001; Asai et al. 2014; Miura et al. 2021). Moreover, novel ae mutants have been developed with even higher RS content (Wada et al. 2018). In addition to natural mutations, CRISPR-Cas9 technology has been used to create sbeIIb mutants, further confirming the impact of SBEIIb on RS levels (Sun et al. 2017). These mutants exhibited increased amylose content, long and intermediate amylopectin chains, and RS content. Downregulating SBEIIb expression through RNA interference (RNAi) also yielded rice with high RS content (Butardo et al. 2011). Interestingly, the increased amylose content in SBEIIb mutants is not due to a higher proportion of amylose chains but rather arises from an increase in long and intermediate debranched amylopectin chains. This change leads to the formation of B-type starch crystals, contributing to the higher RS content (Butardo et al. 2011). SBEIIb deficiency also results in elevated sugar, lipid, and protein levels in the endosperm, collectively contributing to the high RS content observed in these mutants. This pattern of increased RS content linked to SBEIIb deficiency is observed not only in rice but also in wheat, where CRISPR-Cas9-induced knockout of SBEIIa significantly raised RS content and amylose content (Li et al. 2021).

Several starch-synthesis-related genes (SSRGs), including GBSSI, SSIIa, SSIIIa, and SBEIIb, have been identified as influential factors in the formation of resistant starch (RS). Combinations of these SSRGs have also been found to have a cumulative effect on RS content in rice and other crops. Apart from these well-established SSRGs, there are indications that other SSRGs may also play roles in RS formation in crops like rice. One example is SSIIb, which, when downregulated, leads to notable reductions in amylose content (AC) and mid-length amylopectin chains while increasing short amylopectin chains and starch granule size in the rice endosperm (Huang et al. 2021; Xu et al. 2020). Suppression of SSIIb expression in both Wxa and Wxb backgrounds results in decreased AC and changes in the distribution of amylopectin chain lengths, ultimately improving eating and cooking quality (Huang et al. 2021). These changes in AC, amylopectin chain distribution, and starch granule size in the ss2b mutant suggest a potential role for SSIIb in RS formation. Another SSRG, SSIV, is primarily responsible for initiating starch granules in plants. While its direct involvement in RS formation has not been reported, studies have shown that the RS content is higher in the ss3a ss4b double mutant compared to the wild type and ss3a. Additionally, the starch granules in the ss3a ss4b double mutant are smaller and more spherical. These observations raise the possibility that SSIVb may have an influence on RS formation, although further research is needed to confirm its specific role. These findings suggest that multiple SSRGs beyond the well-known ones may contribute to the complex process of RS formation in crops like rice (Toyosawa et al. 2016; Tsuiki et al. 2016).

11.4 Transcription Factors That Potentially Influence RS Content

Several transcription factors have been identified as potential regulators of SSRGs that directly impact RS formation in crops like rice. These transcription factors include members of various families, such as bZIP, NF-Y, MYC, and MADS-box proteins (Zhu et al. 2003; Bello et al. 2019; Wang et al. 2020a; Niu et al. 2020; Feng et al. 2022). In rice, the bZIP family member OsbZIP33 has been found to interact with the ACGT motif in the promoter regions of Wx and SBEI genes, promoting their expression (Yang et al. 2001; Cai et al. 2002). Another bZIP transcription factor, OsbZIP58, interacts with the Dof family transcription factor RPBF, leading to the upregulation of several genes involved in starch synthesis, including OsAGPL3, Wx, SSIIa, SBEI, SBEIIb, and ISA2. The osbzip58 mutant exhibits lower total starch, amylose content (AC), and lipid content, accompanied by changes in amylopectin chain length (Kawakatsu et al. 2009; Wang et al. 2013; Xu et al. 2020; Yamamoto et al. 2006).

The DREB-type transcription factor SALT-RESPONSIVE ERF1 (SERF1) represses the expression of both Wx and RPBF genes, and the serf1 mutant displays increased total starch and protein content. The MYC family member OsBP-5 activates the Wx gene's transcription by binding to its promoter region, and mutations in OsBP-5 lead to reduced AC in rice. NF-YB1 forms a complex with NF-YC12 and bHLH144, activating Wx gene expression (Schmidt et al. 2014). Mutations in NF-YB1, NF-YC12, or bHLH144 result in decreased total starch and AC. Another transcription factor, OsbZIP76, interacts with the OsNF-YB family members OsNF-YB1 and OsNF-YB9, influencing amylose accumulation in rice (Zhu et al. 2003). The MADS-box transcription factor OsMADS14 interacts with NF-YB1 and binds to the promoters of OsAGPL2 and Wx, affecting the shape and size of starch granules and leading to changes in total starch, AC, and soluble sugar content. NAC family transcription factors OsNAC20 and OsNAC26 form heterodimers to regulate starch biosynthesis by directly binding to SSRG promoters (Wang et al. 2020a). Mutations in these factors result in decreased total starch and protein content, along with increased AC and soluble sugar content. These transcription factors collectively form a complex regulatory network that influences the expression of Wx and other key genes involved in starch synthesis, ultimately affecting amylose synthesis and starch structure, which may, in turn, impact RS content. Further research is needed to explore the precise roles of these transcription factors in RS formation (Feng et al. 2022).

11.5 Splicing Factors That Potentially Influence RS Content

In rice, there are several dull endosperm (Du) genes, namely, Du1, Du2, and Du3, which have been identified as regulators of amylose synthesis by specifically influencing the splicing process of Wxb pre-mRNA (Isshiki et al. 2000, 2008; Zeng et al. 2007). In du mutants, the efficiency of splicing the Wxb pre-mRNA is reduced, leading to a decrease in the production of the Wx protein. This, in turn, results in a dull endosperm phenotype characterized by low levels of amylose (Zeng et al. 2007; Isshiki et al. 2008). Du1 is responsible for encoding a member of the pre-mRNA processing (Prp1) family (Zeng et al. 2007), while Du3 encodes a rice counterpart of the cap-binding protein 20 kD subunit (CBP20), which is part of the nuclear cap-binding complex (CBC). The CBC is involved in pre-mRNA splicing (Isshiki et al. 2008). Considering that Wx is a pivotal gene in the formation of resistant starch (RS), it is plausible that the Du genes have the potential to influence the RS content in rice.

11.6 Epigenetic Modifications That Potentially Influence RS Content

Epigenetic modifications are known to play a critical role in regulating various aspects of plant growth and development. In the context of starch synthesis in rice, previous research has revealed that DNA methylation levels in major genes related to starch synthesis, known as SSRGs, are often lower in the rice endosperm compared to other tissues. This suggests that DNA methylation is a crucial factor in the regulation of starch synthesis (Zemach et al. 2010). Furthermore, it has been observed that the promoter region of the Wx gene, a key player in starch synthesis, contains two neighboring CpG islands that may influence its DNA methylation status. Notably, a low level of DNA methylation in the Wx gene's promoter region is closely linked to high amylose content (AC) in the rice endosperm (Anacleto et al. 2019). These findings indicate that DNA methylation has the potential to impact the formation of resistant starch (RS) in rice. More recently, DNA methylation of SSRGs has also been detected in maize endosperm. Interestingly, the coding regions of SSRGs with low expression levels exhibited high levels of DNA methylation in maize. This suggests that epigenetic modifications, particularly DNA methylation, which are associated with starch synthesis, could offer a promising approach for precisely adjusting RS content in rice and potentially other cereal crops.

11.7 Posttranslational Modifications That Potentially Influence RS Content

Phosphorylation, a significant posttranslational modification of proteins, holds the capacity to alter the catalytic activities of enzymes involved in starch synthesis, known as starch-synthesis-related enzymes (SSREs), in cereal crops. This, in turn, has an impact on starch production within the endosperm (Chen et al. 2016; Pang et al. 2018). In the context of rice, the oligomerization of GBSSI/Wx is a critical factor influencing its enzymatic activity, thereby facilitating amylose synthesis in the endosperm. However, phosphorylation exerts a suppressive effect on GBSSI oligomerization (Liu et al. 2013). Notably, researchers have identified nine phosphorylation sites in GBSSI, with Ser415 being a pivotal site significantly associated with GBSSI enzyme activity. The substitution of Ser415 with Pro415 in Wxlv leads to heightened GBSSI activity compared to Wx_a, resulting in elevated amylose content (AC) within the rice endosperm (Zhang et al. 2019). This observation hints at the potential of phosphorylation to impact resistant starch (RS) formation by altering the AC of the rice endosperm. Moreover, research has shown that phosphorylation is a widespread occurrence among SSREs in cereal crops (Tetlow et al. 2004; Walley et al. 2013; Makhmoudova et al. 2014; Cao et al. 2015; Chen et al. 2016, 2017a). This collective evidence underscores the significant role of phosphorylation in the process of starch biosynthesis. Consequently, it is conceivable that the phosphorylation of SSREs could potentially influence RS formation, although further investigation is required to confirm this possibility in future studies.

11.8 Other Genetic Factors That Potentially Influence RS Content

Starch has been suggested to engage with various constituents, including proteins, lipids, dietary fibers, ions, and sugars, potentially impacting the formation of resistant starch (RS). Essentially, genes responsible for protein, lipid, and sugar synthesis and metabolism could potentially play a role in determining RS levels. However, the effects of these genes on RS content in rice have not been quantified. Protein content, as the second major component of rice endosperm, exhibits an inverse relationship with starch content. Generally, an increase in protein content coincides with a reduction in starch content, implying that genes associated with protein synthesis may have the potential to influence RS formation. For example, GPA3 (glutelin precursor accumulation 3) encodes a regulator involved in post-Golgi vesicular trafficking. This regulator forms a complex with Rab5a-guanine exchange factor VPS9 and Rab5a, governing dense vesicle-mediated post-Golgi trafficking in rice. In the gpa3 mutant, amylose content (AC) significantly decreased, while protein and lipid contents increased (Ren et al. 2014). Another instance is OsAAP10, an amino acid permease responsible for loading amino acids into the rice endosperm. Mutations in OsAAP10 resulted in reductions in both protein content and AC (Wang et al. 2020b). Recent research has indicated a positive correlation between lipid content and RS content in rice (Shen et al. 2021). This further supports the notion that genetic alterations in lipid-related genes could potentially influence RS content. Additionally, several genes involved in carbon metabolism exert an influence on both starch and lipid contents.

11.9 Health Benefits of Resistant Starch

11.9.1 Potential Impacts of Resistant Starch on Gut Microbiota

The mammalian gastrointestinal (GI) tract harbors a greater number of bacterial cells than the host itself, with a well-established role of GI microbiota in influencing host health (Sekirov et al. 2010). Dysbiosis, characterized by altered microbiota, has been associated with various human diseases such as diabetes, obesity, inflammatory bowel diseases, and colorectal cancer. Recent research has also revealed connections between GI microbiota and neurological diseases as well as their influence on host behavior. Consequently, there is a growing interest in how dietary choices can modulate GI microbiota to enhance host health or contribute to diseases (Louis et al. 2007; Bouhnik et al. 2004; Higgins 2013).

Diet has long been recognized as a factor influencing the microbial communities in the GI tract. Changes in these microbiota induced by diet can benefit the host through the breakdown of dietary fibers and the production of short-chain fatty acids (SCFAs), which serve as an essential energy source for the host and play roles in immune modulation. While limited studies have specifically examined the effects of different types of resistant starch on microbiota, it is evident that high-fiber diets profoundly impact mammalian microbiota composition. Alterations in microbiota composition can occur rapidly following dietary changes, driven both directly by bacteria capable of digesting resistant starch and indirectly through factors like decreased pH due to SCFA accumulation. Additionally, metabolic cross-feeding among different bacteria can lead to shifts in community composition (Phillips et al. 1995; Tomlin and Read 1990).

Advancements in culture-independent bacterial community analysis and next-generation sequencing technologies have significantly accelerated our understanding of GI microbiota composition. These methods, including 16S rRNA gene sequencing and whole-genome sequencing, have been used in large-scale studies to characterize the microbiota in various body sites, including the GI tract. Such studies have identified distinct "enterotypes" in human populations characterized by differences in the abundance of specific genera like *Bacteroides*, *Prevotella*, and *Ruminococcus*, which can be influenced by dietary fiber content. Research by Martínez et al. used pyro-sequencing to investigate how resistant starches impact fecal microbiota in humans. The study participants consumed two different forms of resistant starch, RSII and RSIV, which led to changes in microbiota composition. However, responses to dietary starch varied among individuals, reflecting the known microbial variation in human populations. Strain differences in starch degradation capabilities and host-specific factors also contributed to this variation (Bouhnik et al. 2004; Higgins 2013).

Deep-sequencing studies focusing on the effects of resistant starches on colonic microbiota are in their early stages but hold promise for elucidating how specific bacterial taxa interact with different starch forms. Understanding individual variations in response to dietary resistant starch will be crucial for potential therapeutic and preventive applications to improve human health. Another area of intense interest is the use of prebiotics, including resistant starch, to modify colonic microbiota for the benefit of host health (McOrist et al. 2011; Cummings et al. 1996). Prebiotics, a category that includes various nondigestible carbohydrates, have the potential to correct or prevent human diseases such as obesity, diabetes, inflammatory bowel diseases, and cancer by modulating GI microbiota in interactions with the host (DuPont and DuPont 2011; Holmes et al. 2012).

11.9.2 Potential Impacts of Resistant Starch on Chronic Diseases

The consumption of diets rich in fiber has traditionally been associated with protection against colorectal cancer (Zhang and Hamaker 2010). More recently, resistant starch has gained attention for its potential role in preventing colon cancer and inflammatory bowel diseases. While research on the effects of resistant starch on human colonic health is limited, a couple of human intervention studies have provided insights. One study involved a 4-week intervention with red meat, which increased certain markers associated with colorectal cancer in the colons of participants (Le Leu et al. 2012). However, when the intervention included red meat along with butyrylated resistant starch, these markers were not elevated. This suggests that resistant starch might protect the human colon from potential harmful effects of dietary red meat. In another study with individuals at high risk for colon polyps and cancer due to Lynch syndrome, a diet containing maize starch was compared to a placebo diet over 29 months. This study found no impact on polyp or colon cancer development at a 4-year follow-up (Burn et al. 2008; Le Leu et al. 2002).

In contrast to the limited number of human interventions, several studies have explored the relationship between resistant starch and colon cancer prevention in laboratory animals. These studies often involved rats treated with a colon carcinogen and/or fed diets high in protein to damage the colonic epithelium (Le Leu et al. 2007). Feeding high-fiber or resistant starch diets resulted in increased fecal bulk, higher fecal pH, elevated butyrate levels, reduced cell proliferation markers, increased apoptosis, and decreased colon carcinogenesis. The addition of high-amylose resistant starch to high-protein diets reduced protein fermentation products and paralleled a decrease in colorectal carcinoma development (Paturi et al. 2012). However, studies using potato fiber or potato resistant starch in rats on control or high-protein diets did not observe an impact on DNA damage in the colon.

The mechanism by which dietary resistant starch may alter colon carcinogenesis is still a subject of investigation. Proposed mechanisms include changes in fecal water-holding capacity, modification of the microbiota, and increased production of short-chain fatty acids (SCFAs) (Le Leu et al. 2012; Burn et al. 2008). SCFAs, especially butyrate, have gained significant attention due to their roles in colonic epithelial metabolism, differentiation, and modulation of signaling pathways related to mucosal physiology. In cell culture, butyrate has demonstrated anti-tumorigenic properties, such as reducing cell proliferation and inducing apoptosis in colorectal tumor cell lines. However, it remains unclear whether the concentrations of butyrate achieved in the colon of animals and humans fed resistant starches are optimal for these suppressive effects (Zhang and Hamaker 2010).

Colon cancer is thought to result from dysregulation of molecular pathways governing epithelial proliferation, maturation, and apoptosis, involving both genetic and epigenetic components. Resistant starch is believed to influence the development or progression of colon cancer through multiple mechanisms, including alterations in the microbiota and protection against unrepaired DNA damage (Le Leu et al. 2012; Burn et al. 2008). Butyrate, which is produced during the microbial fermentation of resistant starch, plays a pivotal role in many of these mechanisms. Additionally, butyrate is known to inhibit histone deacetylase, modulating gene transcription, and has been associated with the suppression of preneoplastic and neoplastic phenotypes in vitro. Furthermore, butyrate exerts protective effects against intestinal mucosal inflammation, which is linked to inflammation-mediated colorectal cancer (Fung et al. 2012).

Research into the potential modification of epigenetic mechanisms by products of resistant starch metabolism, like butyrate, offers promise for dietary strategies in colorectal cancer prevention. These mechanisms are potentially reversible and could provide valuable insights into the role of resistant starch in maintaining colonic homeostasis. The complex interactions between microbiota, dietary components, colonic epithelium, the immune system, and the nervous and endocrine systems are ongoing areas of investigation, and future research may uncover integrative mechanisms (Bordonaro et al. 1999; Bordonaro 2013).

11.9.3 Potential Impacts and Mechanisms of Action of Resistant Starch in Prevention or Therapy of Metabolic Diseases

Diabetes is a prevalent health issue, with 8% of the US population affected, particularly those over the age of 60. Both type 1 and type 2 diabetes are characterized by elevated blood sugar levels, leading to systemic tissue damage. Lifestyle changes, such as dietary modifications, can play a crucial role in managing diabetes. One such change is replacing ordinary starch in foods with resistant starch due to its low glycemic index (Knowler et al. 2002). Research has shown that consuming foods rich in resistant starch, such as corn porridges and crackers, can lead to lower post-meal glucose levels and reduced insulin responses compared to foods containing regular starch (Alexander 2012). Additionally, consuming less digestible starches may reduce the glycemic response to subsequent meals, a phenomenon known as the "second meal effect." This dietary shift contributes to diabetes management and may also aid in weight management and body composition (Brighenti et al. 2006).

Studies in animals, such as Goto-Kakizaki rats (a nonobese model of type 2 diabetes), have demonstrated improved glycemic control with dietary resistant starch (Mutlu et al. 2011; McGill et al. 2008). Low-glycemic-index carbohydrates like resistant starch might also help reduce diabetic complications by protecting kidney function and maintaining adequate nutritional status, especially regarding vitamin D. Some research has indicated that diabetes can lead to vitamin D insufficiency, potentially due to increased excretion of vitamin D metabolites in urine (Hurskainen et al. 2012; Husemoen et al. 2012; Alam et al. 2012). However, replacing cornstarch with high-amylose maize, rich in resistant starch, has shown promise in preventing such vitamin D losses in diabetic rats (Lee et al. 2012; Nykjaer et al. 1999).

Obesity and weight management are significant concerns linked to excessive energy intake. Increasing dietary fiber consumption has been associated with satiety and lower body mass index (BMI) (Wanders et al. 2011; Du et al. 2010). Resistant starches, considered a type of dietary fiber, may offer similar benefits for weight management. Studies in rodents have suggested that replacing rapidly digestible starch with resistant starch can reduce body weight (Slavin 2005). However, long-term studies are needed to confirm these effects in humans. There are several reasons to believe that consuming resistant starch could support weight management in humans. First, it reduces the energy density of the diet, potentially increasing feelings of fullness and promoting weight loss. Second, it may enhance satiety, although results in humans have been mixed. Lastly, resistant starch might influence energy expenditure or fat oxidation, although there is presently limited evidence to support this hypothesis (Aziz et al. 2009). Further research is necessary to fully understand the impact of resistant starch on human body weight and its potential as a dietary strategy for weight management.

11.10 Future Prospects

There are a lot of challenges in studying resistant starch, and more research is required to better understand its mechanisms of action and its full range of health benefits. It highlights the importance of conducting clinical trials and population studies to establish the link between RS intake and various health outcomes more definitively. There are significant opportunities for improving human health through an understanding of the genetic aspects of resistant starch. However, it also presents various challenges and suggests potential future research directions in this emerging field. Understanding the genetic basis of how individuals metabolize and respond to resistant starch is a complex endeavor. Future research should aim to comprehensively identify the key genetic factors involved and assess the extent of genetic variability in the human population. One of the primary challenges is translating genetic insights into personalized nutrition recommendations. Research should focus on developing practical tools and guidelines for tailoring resistant starch consumption to an individual's genetic profile. It is crucial to investigate the long-term health outcomes associated with genetic-based dietary interventions. Research could delve into the impacts on chronic diseases, gut microbiota, and metabolic health.

Genetic information in the context of nutrition and health raises ethical concerns. Future research should address issues related to privacy, consent, and potential discrimination, ensuring that genetic data is handled responsibly. Moreover, the accessibility of genetic testing and personalized dietary advice is a significant challenge. Future research should explore how to make these services more affordable and accessible to diverse populations to avoid exacerbating health disparities. Another critical challenge is disseminating knowledge about the potential benefits of resistant starch and its genetic implications to healthcare professionals and the general public. Future research should develop effective strategies for nutritional education and awareness. More clinical trials are needed to establish the efficacy and safety of genetic-based dietary interventions using resistant starch. This requires collaboration between geneticists, nutritionists, and clinicians. As the field progresses, regulatory frameworks need to be established to ensure the responsible and ethical use of genetic information in nutrition. This includes guidelines for genetic testing, data storage, and consent.

In conclusion, while the genetic perspective on resistant starch offers exciting prospects for enhancing human well-being, addressing the challenges and pursuing the suggested research directions is essential to unlock its full potential and make personalized nutrition a reality for individuals seeking improved health outcomes.

References

Ai Y, Hasjim J, Jane J-l (2013) Effects of lipids on enzymatic hydrolysis and physical properties of starch. Carbohydr Polym 92:120–127

Alam U, Najam O, Al-Himdani S, Benoliel S, Jinadev P, Berry JL, Kew M, Asghar O, Petropoulos IN, Malik RA (2012) Marked vitamin D deficiency in patients with diabetes in the UK: ethnic and seasonal differences and an association with dyslipidaemia. Diabet Med 29:1343–1345

Alexander D (2012) Postprandial effects of resistant starch corn porridges on blood glucose and satiety responses in non-overweight and overweight adults. Iowa State University, Ames

Anacleto R, Badoni S, Parween S, Butardo VM, Misra G, Cuevas RP, Kuhlmann M, Trinidad TP, Mallillin AC, Acuin C et al (2019) Integrating a genome-wide association study with a large-scale transcriptome analysis to predict genetic regions influencing the glycaemic index and texture in rice. Plant Biotechnol J 17:1261–1275. https://doi.org/10.1111/pbi.13051

Asai H, Abe N, Matsushima R, Crofts N, Oitome NF, Nakamura Y, Fujita N (2014) Deficiencies in both starch synthase IIIa and branching enzyme IIb lead to a significant increase in amylose in SSIIainactive japonica rice seeds. J Exp Bot 65:5497–5507. https://doi.org/10.1093/jxb/eru310

Aziz AA, Kenney LS, Goulet B, Abdel-Aal ES (2009) Dietary starch type affects body weight and glycemic control in freely fed but not energy restricted obese rats. J Nutr 139:1881–1889
Bao JS, Zhou X, Xu FF, He Q, Park YJ (2017) Genome-wide association study of the resistant starch content in rice grains. Starch-Starke 69:1600343. https://doi.org/10.1002/star.201600343
Baysal C, He WS, Drapal M, Villorbina G, Medina V, Capell T, Khush GS, Zhu CF, Fraser PD, Christou P (2020) Inactivation of rice starch branching enzyme IIb triggers broad and unexpected changes in metabolism by transcriptional reprogramming. Proc Natl Acad Sci U S A 117:26503–26512. https://doi.org/10.1073/pnas.2014860117
Bello BK, Hou Y, Zhao J, Jiao G, Wu Y, Li Z, Wang Y, Tong X, Wang W, Yuan W et al (2019) NF-YB1-YC12-bHLH144 complex directly activates Wx to regulate grain quality in rice (Oryza sativa L.). Plant Biotechnol J 17:1222–1235. https://doi.org/10.1111/pbi.13048
Berry CS (1986) Resistant starch: formation and measurement of starch that survives exhaustive digestion with amylolytic enzymes during the determination of dietary fibre. J Cereal Sci 4:301–314. https://doi.org/10.1016/S0733-5210(86)80034-0
Bordonaro M (2013) Crosstalk between Wnt signaling and RNA processing in colorectal cancer. J Cancer 4:96–103
Bordonaro M, Mariadason JM, Aslam F, Heerdt BG, Augenlicht LH (1999) Butyrate-induced apoptotic cascade in colonic carcinoma cells: modulation of the beta-catenin-Tcf pathway and concordance with effects of sulindac and trichostatin a but not curcumin. Cell Growth Differ 10:713–720
Bouhnik Y, Raskine L, Simoneau G, Vicaut E, Neut C, Flourié B, Brouns F, Bornet FR (2004) The capacity of nondigestible carbohydrates to stimulate fecal bifidobacteria in healthy humans: a double-blind, randomized, placebo-controlled, parallel-group, dose-response relation study. Am J Clin Nutr 80:1658–1664
Brighenti F, Benini L, Del Rio D, Casiraghi C, Pellegrini N, Scazzina F, Jenkins DJA, Vantini I (2006) Colonic fermentation of indigestible carbohydrates contributes to the second-meal effect. Am J Clin Nutr 83:817–822
Brouns F, Kettlitz B, Arrigoni E (2002) Resistant starch and "the butyrate revolution". Trends Food Sci Technol 13:251–261
Burn J, Bishop DT, Mecklin J, Macrae F, Moeslein G, Olschwang S, Bisgaard M, Ramesar R, Elliott F, Mathers J (2008) Results of the CAPP-2- trial (aspirin and resistant starch) in HNPCC gene carriers. EJC Suppl 6:25
Butardo VM, Fitzgerald MA, Bird AR, Gidley MJ, Flanagan BM, Larroque O, Resurreccion AP, Laidlaw HKC, Jobling SA, Morell MK et al (2011) Impact of down-regulation of starch branching enzyme IIb in rice by artificial microRNA- and hairpin RNAmediated RNA silencing. J Exp Bot 62:4927–4941. https://doi.org/10.1093/jxb/err188
Cai XL, Wang ZY, Xing YY, Zhang JL, Hong MM (1998) Aberrant splicing of intron 1 leads to the heterogeneous 5' UTR and decreased expression of waxy gene in rice cultivars of intermediate amylose content. Plant J 14:459–465. https://doi.org/10.1046/j.1365-313X.1998.00126.x
Cai Y, Xie DL, Wang ZY, Hong MM (2002) Interaction of rice bZIP protein REB with the 5′-upstream region of both rice sbe1 gene and waxy gene. Chin Sci Bull 47:310–314. https://doi.org/10.1360/02tb9074
Cao H, Yan X, Chen GX, Zhou JW, Li XH, Ma WJ, Yan YM (2015) Comparative proteome analysis of A- and B-type starch granule-associated proteins in bread wheat (Triticum aestivum L.) and Aegilops crassa. J Proteome 112:95–112. https://doi.org/10.1016/j.jprot.2014.08.002
Chen GX, Zhou JW, Liu YL, Lu XB, Han CX, Zhang WY, Xu YH, Yan YM (2016) Biosynthesis and regulation of wheat amylose and amylopectin from proteomic and phosphoproteomic characterization of granule-binding proteins. Sci Rep 6:33111. https://doi.org/10.1038/srep33111
Chen GX, Zhen SM, Liu YL, Yan X, Zhang M, Yan YM (2017a) In vivo phosphoproteome characterization reveals key starch granule-binding phosphoproteins involved in wheat water deficit response. BMC Plant Biol 17:168. https://doi.org/10.1186/s12870-017-1118-z
Chen MH, Bergman CJ, McClung AM, Everette JD, Tabien RE (2017b) Resistant starch: variation among high amylose rice varieties and its relationship with apparent amylose content,

pasting properties and cooking methods. Food Chem 234:180–189. https://doi.org/10.1016/j.foodchem.2017.04.170
Cummings JH, Beatty E, Kingman S, Bingham S, Englyst H (1996) Digestion and physiological properties of resistant starch in the human large bowel. Br J Nutr 75:733–747
Du H, Boshuizen HC, Forouhi NG, Wareham NJ, Halkjaer J, Tjonneland A, Overvad K, Jakobsen MU, Boeing H et al (2010) Dietary fiber and subsequent changes in body weight and waist circumference in European men and women. Am J Clin Nutr 91:329–336
DuPont AW, DuPont HL (2011) The intestinal microbiota and chronic disorders of the gut. Nat Rev Gastroenterol Hepatol 8:523–531
Englyst HN, Cummings JH (1985) Digestion of the polysaccharides of some cereal foods in the human small-intestine. Am J Clin Nutr 42:778–787
Englyst HN, Hudson GJ (1996) The classification and measurement of dietary carbohydrates. Food Chem 57:15–21. https://doi.org/10.1016/0308-8146(96)00056-8
Englyst HN, Kingman SM, Cummings JH (1992) Classification and measurement of nutritionally important starch fractions. Eur J Clin Nutr 46(Supp 2):S33–S50
Feng T, Wang L, Li L, Liu Y, Chong K, Theißen G, Meng Z (2022) OsMADS14 and NF-YB1 cooperate in the direct activation of OsAGPL2 and waxy during starch synthesis in rice endosperm. New Phytol 234:77–92. https://doi.org/10.1111/nph.17990
Fuentes-Zaragoza E, Sanchez-Zapata E, Sendra E, Sayas E, Navarro C, Fernandez-Lopez J, Perez-Alvarez JA (2011) Resistant starch as prebiotic: a review. Starch-Stärke 63:406–415
Fung KY, Cosgrove L, Lockett T, Head R, Topping DL (2012) A review of the potential mechanisms for the lowering of colorectal oncogenesis by butyrate. Br J Nutr 108:820–831
Granfeldt Y, Bjorck I, Hagander B (1991) On the importance of processing conditions, product thickness and egg addition for the glycemic and hormonal responses to pasta—a comparison with bread made from pasta ingredients. Eur J Clin Nutr 45:489–499
Guo D, Ling X, Zhou X, Li X, Wang J, Qiu S, Yang Y, Zhang B (2020) Evaluation of the quality of a high-resistant starch and low-glutelin rice (Oryza sativa L.) generated through CRISPR/Cas9-mediated targeted mutagenesis. J Agric Food Chem 68:9733–9742. https://doi.org/10.1021/acs.jafc.0c02995
Han J-A, BeMiller JN (2007) Preparation and physical characteristics of slowly digesting modified food starches. Carbohydr Polym 67:366–374
Han Y, Sun FJ, Rosales-Mendoza S, Korban SS (2007) Three orthologs in rice, Arabidopsis, and Populus encoding starch branching enzymes (SBEs) are different from other SBE gene families in plants. Gene 401:123–130. https://doi.org/10.1016/j.gene.2007.06.026
Hasjim J, Lee S-O, Hendrich S, Setiawan S, Ai Y, Jane J-l (2010) Characterization of a novel resistant-starch and its effects on postprandial plasma-glucose and insulin responses. Cereal Chem 87:257–262
He J, Liu J, Zhang G (2008) Slowly digestible waxy maize starch prepared by octenyl succinic anhydride esterification and heat-moisture treatment: glycemic response and mechanism. Biomacromolecules 9:175
Higgins JA (2013) Resistant starch: a promising dietary agent for the prevention/treatment of inflammatory bowel disease and bowel cancer. Curr Opin Gastroenterol 29:190–194
Hirano HY, Sano Y (1998) Enhancement of Wx gene expression and the accumulation of amylose in response to cool temperatures during seed development in rice. Plant Cell Physiol 39:807–812. https://doi.org/10.1093/oxfordjournals.pcp.a029438
Hizukuri S, Takeda Y, Maruta N, Juliano BO (1989) Molecular- structures of rice starch. Carbohydr Res 189:227–235. https://doi.org/10.1016/0008-6215(89)84099-6
Holmes E, Kinross J, Gibson GR, Burcelin R, Jia W, Pettersson S, Nicholson JK (2012) Therapeutic modulation of microbiota-host metabolic interactions. Sci Transl Med 6:137rv6
Huang L, Sreenivasulu N, Liu Q (2020) Waxy editing: old meets new. Trends Plant Sci 25:963–966. https://doi.org/10.1016/j.tplants.2020.07.009
Huang L, Gu Z, Chen Z, Yu J, Chu R, Tan H, Zhao D, Fan X, Zhang C, Li Q et al (2021) Improving rice eating and cooking quality by coordinated expression of the major starch synthesis related

genes, SSII and Wx, in endosperm. Plant Mol Biol 106:419–432. https://doi.org/10.1007/s11103-021-01162-8

Hurskainen AR, Virtanen JK, Tuomainen TP, Nurmi T, Voutilainen S (2012) Association of serum 25-hydroxyvitamin D with type 2 diabetes and markers of insulin resistance in a general older population in Finland. Diabetes Metab Res Rev 28:418–423

Husemoen LL, Skaaby T, Thuesen BH, Jorgensen T, Fenger RV, Linneberg A (2012) Serum 25(OH)D and incident type 2 diabetes: a cohort study. Eur J Clin Nutr 66:1309–1314

Isshiki M, Morino K, Nakajima M, Okagaki RJ, Wessler SR, Izawa T, Shimamoto K (1998) A naturally occurring functional allele of the rice waxy locus has a GT to TT mutation at the 5′ splice site of the first intron. Plant J 15:133–138. https://doi.org/10.1046/j.1365-313X.1998.00189.x

Isshiki M, Nakajima M, Satoh H, Shimamoto K (2000) Dull: rice mutants with tissue-specific effects on the splicing of the waxy pre-mRNA. Plant J 23:451–460. https://doi.org/10.1046/j.1365-313x.2000.00803.x

Isshiki M, Matsuda Y, Takasaki A, Wong HL, Satoh H, Shimamoto K (2008) Du3, a mRNA cap-binding protein gene, regulates amylose content in japonica rice seeds. Plant Biotechnol 25:483–487. https://doi.org/10.5511/plantbiotechnology.25.483

Itoh K, Ozaki H, Okada K, Hori H, Takeda Y, Mitsui T (2003) Introduction of Wx transgene into rice wx mutants leads to both high and low-amylose rice. Plant Cell Physiol 44:473–480. https://doi.org/10.1093/pcp/pcg068

Jane J, Robyt JF (1984) Structure studies of amylose-v complexes and retrograded amylose by action of alpha-amylases, and a new method for preparing amylodextrins. Carbohydr Res 132:105–118

Jane JL, Ao Z, Duvick SA, Wiklund M, Yoo S-H, Wong K-S, Gardner C (2003) Structures of amylopectin and starch granules: how are they synthesized? J Appl Glycosci 50:167–172

Jiang H, Campbell M, Blanco M, Jane J-L (2010) Characterization of maize amylose-extender (ae) mutant starches: part II. Structures and properties of starch residues remaining after enzymatic hydrolysis at boiling- water temperature. Carbohydr Polym 80:1–12

Kawakatsu T, Yamamoto MP, Touno SM, Yasuda H, Takaiwa F (2009) Compensation and interaction between RISBZ1 and RPBF during grain filling in rice. Plant J 59:908–920. https://doi.org/10.1111/j.1365-313X.2009.03925.x

Knowler WC, Barrett-Connor E, Fowler SE, Hamman RF, Lachin JM, Walker EA, Nathan DM (2002) Diabetes prevention program res G. Reduction in the incidence of type 2 diabetes with lifestyle intervention or metformin. N Engl J Med 346:393–403

Larkin PD, Park WD (2003) Association of waxy gene single nucleotide polymorphisms with starch characteristics in rice (Oryza sativa L.). Mol Breed 12:335–339. https://doi.org/10.1023/B:Molb.0000006797.51786.92

Le Leu RK, Hu Y, Young GP (2002) Effects of resistant starch and nonstarch polysaccharides on colonic luminal environment and genotoxin-induced apoptosis in the rat. Carcinogenesis 23:713–719

Le Leu RKB, Brown IL, Hu Y, Esterman A, Young GP (2007) Suppression of azoxymethane-induced colon cancer development in rats by dietary resistant starch. Cancer Biol Ther 6:1621–1626

Le Leu R, Conlon M, Winter J, Humphreys K, Michael M, Hu Y, Bird A, Topping D, Young G (2012) Effect of high red meat intake and resistant starch in humans on risk factors for colorectal cancer. J Gastroenterol Hepatol 27:24–25

Lee JI, Oh SJ, Ha WC, Kwon HS, Sohn TS, Son HS, Cha BY (2012) Serum 25-hydroxyvitamin D concentration and arterial stiffness among type 2 diabetes. Diabetes Res Clin Pract 95:42–47

Li L, Jiang HX, Campbell M, Blanco M, Jane JL (2008) Characterization of maize amylose-extender (ae) mutant starches. Part I: relationship between resistant starch contents and molecular structures. Carbohydr Polym 74:396–404

Li HT, Sartika RS, Kerr ED, Schulz BL, Gidley MJ, Dhital S (2020) Starch granular protein of highamylose wheat gives innate resistance to amylolysis. Food Chem. 330:127328. https://doi.org/10.1016/j.foodchem.2020.127328

Li J, Jiao G, Sun Y, Chen J, Zhong Y, Yan L, Jiang D, Ma Y, Xia L (2021) Modification of starch composition, structure and properties through editing of TaSBEIIa in both winter and spring wheat varieties by CRISPR/Cas9. Plant Biotechnol J 19:937–951. https://doi.org/10.1111/pbi.13519

Lin LS, Guo DW, Huang J, Zhang XD, Zhang L, Wei CX (2016) Molecular structure and enzymatic hydrolysis properties of starches from high-amylose maize inbred lines and their hybrids. Food Hydrocolloids 58:246–254. https://doi.org/10.1016/j.foodhyd.2016.03.001

Liu L, Ma X, Liu S, Zhu C, Jiang L, Wang Y, Shen Y, Ren Y, Dong H, Chen L et al (2009) Identification and characterization of a novel waxy allele from a Yunnan rice landrace. Plant Mol Biol 71:609–626. https://doi.org/10.1007/s11103-009-9544-4

Liu DR, Huang WX, Cai XL (2013) Oligomerization of rice granule-bound starch synthase 1 modulates its activity regulation. Plant Sci 210:141–150. https://doi.org/10.1016/j.plantsci.2013.05.019

Liu Q, Hu YP, Hu MY, Sun LJ, Chen XY, Li QY, Wang PN, Wang LA, Zhang YJ, Li H (2021) Identification and molecular characterization of mutant line deficiency in three waxy proteins of common wheat (Triticum aestivum L.). Sci Rep 11:3510. https://doi.org/10.1038/s41598-021-82865-2

Louis P, Scott KP, Duncan SH, Flint HJ (2007) Understanding the effects of diet on bacterial metabolism in the large intestine. J Appl Microbiol 102:1197–1208

Makhmoudova A, Williams D, Brewer D, Massey S, Patterson J, Silva A, Vassall KA, Liu F, Subedi S, Harauz G et al (2014) Identification of multiple phosphorylation sites on maize endosperm starch branching enzyme IIb, a key enzyme in amylopectin biosynthesis. J Biol Chem 289:9233–9246. https://doi.org/10.1074/jbc.M114.551093

McGill AT, Stewart JM, Lithander FE, Strik CM, Poppitt SD (2008) Relationships of low serum vitamin D3 with anthropometry and markers of the metabolic syndrome and diabetes in overweight and obesity. Nutr J 7:4

McOrist AL, Miller R, Bird A, Keogh J, Noakes M, Topping D, Conlon M (2011) Fecal butyrate levels vary widely among individuals but are usually increased by a diet high in resistant starch. J Nutr 141:883–889

Miura S, Koyama N, Crofts N, Hosaka Y, Abe M, Fujita N (2021) Generation and starch characterization of non-transgenic BEI and BEIIb double mutant rice (Oryza sativa) with ultra-high level of resistant starch. Rice 14:3. https://doi.org/10.1186/s12284-020-00441-0

Mizuno K, Kawasaki T, Shimada H, Satoh H, Kobayashi E, Okumura S, Arai Y, Baba T (1993) Alteration of the structural properties of starch components by the lack of an isoform of starch branching enzyme in rice seeds. J Biol Chem 268:19084–19091. https://doi.org/10.1016/S0021-9258(17)46738-X

Murphy MM, Douglass JS, Birkett A (2008) Resistant starch intakes in the United States. J Am Diet Assoc 108:67–78

Mutlu A, Mutlu GY, Ozsu E, Cizmecioglu FM, Hatun S (2011) Vitamin D deficiency in children and adolescents with type 1 diabetes. J Clin Res Pediatr Endocrinol 3:179–183

Nakamura Y, Francisco PB, Hosaka Y, Sato A, Sawada T, Kubo A, Fujita N (2005) Essential amino acids of starch synthase IIa differentiate amylopectin structure and starch quality between japonica and indica rice varieties. Plant Mol Biol 58:213–227. https://doi.org/10.1007/s11103-005-6507-2

Nishi A, Nakamura Y, Tanaka N, Satoh H (2001) Biochemical and genetic analysis of the effects of amylose-extender mutation in rice endosperm. Plant Physiol 127:459–472. https://doi.org/10.1104/pp.010127

Niu B, Deng H, Li T, Sharma S, Yun Q, Li Q, Chen C (2020) OsbZIP76 interacts with OsNF-YBs and regulates endosperm cellularization in rice (Oryza sativa). J Integr Plant Biol 62:1983–1996. https://doi.org/10.1111/jipb.12989

Nugent AP (2005) Health properties of resistant starch. Nutr Bull 30:27–54. https://doi.org/10.1111/j.1467-3010.2005.00481.x

Nykjaer A, Dragun D, Walther D, Vorum H, Jacobsen C, Herz J, Melsen F, Christensen EI, Willnow TE (1999) An endocytic pathway essential for renal uptake and activation of the steroid 25-(OH) vitamin D3. Cell 96:507–515

O'Dea K, Nestel PJ, Antonoff L (1980) Physical factors influencing postprandial glucose and insulin responses to starch. Am J Clin Nutr 33:760–765

Ohdan T, Francisco PB Jr, Sawada T, Hirose T, Terao T, Satoh H, Nakamura Y (2005) Expression profiling of genes involved in starch synthesis in sink and source organs of rice. J Exp Bot 56:3229–3244. https://doi.org/10.1093/jxb/eri292

Pang YH, Zhou X, Chen YL, Bao JS (2018) Comparative phosphoproteomic analysis of the developing seeds in two Indica rice (Oryza sativa L.) cultivars with different starch quality. J Agr Food Chem 66:3030–3037. https://doi.org/10.1021/acs.jafc.8b00074

Paturi G, Nyanhanda T, Butts CA, Herath TD, Monro JA, Ansell J (2012) Effects of potato fiber and potato-resistant starch on biomarkers of colonic health in rats fed diets containing red meat. J Food Sci 77:H216–H223

Phillips J, Muir JG, Birkett A, Lu ZX, Jones GP, O'Dea K, Young GP (1995) Effect of resistant starch on fecal bulk and fermentation-dependent events in humans. Am J Clin Nutr 62:121–130

Raja RB, Agasimani S, Jaiswal S, Thiruvengadam V, Sabariappan R, Chibbar RN, Ram SG (2017) EcoTILLING by sequencing reveals polymorphisms in genes encoding starch synthases that are associated with low glycemic response in rice. BMC Plant Biol 17:13. https://doi.org/10.1186/s12870-016-0968-0

Regina A, Bird A, Topping D, Bowden S, Freeman J, Barsby T, Kosar-Hashemi B, Li ZY, Rahman S, Morell M (2006) High-amylose wheat generated by RNA interference improves indices of large-bowel health in rats. Proc Natl Acad Sci U S A 103:3546–3551. https://doi.org/10.1073/pnas.0510737103

Ren Y, Wang Y, Liu F, Zhou K, Ding Y, Zhou F, Wang Y, Liu K, Gan L, Ma W et al (2014) GLUTELIN PRECURSOR ACCUMULATION3 encodes a regulator of post-Golgi vesicular traffic essential for vacuolar protein sorting in rice endosperm. Plant Cell 26:410–425. https://doi.org/10.1105/tpc.113.121376

Sajilata MG, Singhal RS, Kulkarni PR (2006) Resistant starch—a review. Compr Rev Food Sci 5:1–17. https://doi.org/10.1111/j.1541-4337.2006.tb00076.x

Sano Y (1984) Differential regulation of waxy gene-expression in rice endosperm. Theor Appl Genet 68:467–473. https://doi.org/10.1007/Bf00254822

Sano Y, Katsumata M, Okuno K (1986) Genetic studies of speciation in cultivated rice. 5. Inter- and intraspecific differentiation in the waxy gene expression of rice. Euphytica 35:1–9. https://doi.org/10.1007/Bf00028534

Schmidt R, Schippers JH, Mieulet D, Watanabe M, Hoefgen R, Guiderdoni E, Mueller-Roeber B (2014) SALT-RESPONSIVE ERF1 is a negative regulator of grain filling and gibberellin-mediated seedling establishment in rice. Mol Plant 7:404–421. https://doi.org/10.1093/mp/sst131

Sekirov I, Russell SL, Antunes CM, Finlay BB (2010) Gut microbiota in health and disease. Physiol Rev 90:859–904

Seneviratne HD, Biliaderis CG (1991) Action of a-amylases on amylose-lipid complex superstructures. J Cereal Sci 13:129–143

Shen Y, Wu D, Fogliano V, Pellegrini N (2021) Rice varieties with a high endosperm lipid content have reduced starch digestibility and increased g-oryzanol bioaccessibility. Food Funct 12:11547–11556. https://doi.org/10.1039/d1fo03039f

Shimbata T, Ai YF, Fujita M, Inokuma T, Vrinten P, Sunohara A, Saito M, Takiya T, Jane JL, Nakamura T (2012) Effects of homoeologous wheat starch synthase IIa genes on starch properties. J Agr Food Chem 60:12004–12010. https://doi.org/10.1021/jf303623e

Shure M, Wessler S, Fedoroff N (1983) Molecular-identification and isolation of the waxy locus in maize. Cell 35:225–233. https://doi.org/10.1016/0092-8674(83)90225-8

Sievert D, Pomeranz Y (1989) Enzyme-resistant starch.1. Characterization and evaluation by enzymatic, thermoanalytical, and microscopic methods. Cereal Chem 66:342–347

Sievert D, Pomeranz Y (1990) Enzyme-resistant starch. 2. Differential scanning calorimetry studies on heat-treated starches and enzyme-resistant starch residues. Cereal Chem 67:217–221

Slavin JL (2005) Dietary fiber and body weight. Nutrition 21:411–418

Sun Y, Jiao G, Liu Z, Zhang X, Li J, Guo X, Du W, Du J, Francis F, Zhao Y et al (2017) Generation of high-amylose rice through CRISPR/Cas9-mediated targeted mutagenesis of starch branching enzymes. Front Plant Sci 8:298. https://doi.org/10.3389/fpls.2017.00298

Tetlow IJ (2011) Starch biosynthesis in developing seeds. Seed Sci Res 21:5–32. https://doi.org/10.1017/S0960258510000292

Tetlow IJ, Wait R, Lu ZX, Akkasaeng R, Bowsher CG, Esposito S, Kosar-Hashemi B, Morell MK, Emes MJ (2004) Protein phosphorylation in amyloplasts regulates starch branching enzyme activity and protein-protein interactions. Plant Cell 16:694–708. https://doi.org/10.1105/tpc.017400

Thompson DB (2000) Strategies for the manufacture of resistant starch. Trends Food Sci Technol 11:245–253

Tian Z, Qian Q, Liu Q, Yan M, Liu X, Yan C, Liu G, Gao Z, Tang S, Zeng D et al (2009) Allelic diversities in rice starch biosynthesis lead to a diverse array of rice eating and cooking qualities. Proc Natl Acad Sci U S A 106:21760–21765. https://doi.org/10.1073/pnas.0912396106

Tomlin J, Read NW (1990) The effect of resistant starch on colon function in humans. Br J Nutr 64:589–595

Toyosawa Y, Kawagoe Y, Matsushima R, Crofts N, Ogawa M, Fukuda M, Kumamaru T, Okazaki Y, Kusano M, Saito K et al (2016) Deficiency of starch synthase IIIa and IVb alters starch granule morphology from polyhedral to spherical in rice endosperm. Plant Physiol 170:1255–1270. https://doi.org/10.1104/pp.15.01232

Tsuiki K, Fujisawa H, Itoh A, Sato M, Fujita N (2016) Alterations of starch structure lead to increased resistant starch of steamed rice: identification of high resistant starch rice lines. J Cereal Sci 68:88–92. https://doi.org/10.1016/j.jcs.2016.01.002

Wada T, Yamaguchi O, Miyazaki M, Miyahara K, Ishibashi M, Aihara T, Shibuta T, Inoue T, Tsubone M, Toyosawa Y et al (2018) Development and characterization of a new rice cultivar, 'Chikushi-kona 85', derived from a starch-branching enzyme IIb-deficient mutant line. Breed Sci 68:278–283. https://doi.org/10.1270/jsbbs.17069

Walley JW, Shen ZX, Sartor R, Wu KJ, Osborn J, Smith LG, Briggs SP (2013) Reconstruction of protein networks from an atlas of maize seed proteotypes. P Natl Acad Sci U S A 110:E4808–E4817. https://doi.org/10.1073/pnas.1319113110

Wanders AJ, van den Borne J, de Graaf C, Hulshof T, Jonathan MC, Kristensen M, Mars M, Schols HA, Feskens EJM (2011) Effects of dietary fibre on subjective appetite, energy intake and body weight: a systematic review of randomized controlled trials. Obes Rev 12:724–739

Wang ZY, Wu ZL, Xing YY, Zheng FG, Guo XL, Zhang WG, Hong MM (1990) Nucleotide-sequence of rice waxy gene. Nucleic Acids Res 18:5898. https://doi.org/10.1093/nar/18.19.5898

Wang JC, Xu H, Zhu Y, Liu QQ, Cai XL (2013) OsbZIP58, a basic leucine zipper transcription factor, regulates starch biosynthesis in rice endosperm. J Exp Bot 64:3453–3466. https://doi.org/10.1093/jxb/ert187

Wang J, Chen Z, Zhang Q, Meng S, Wei C (2020a) The NAC transcription factors OsNAC20 and OsNAC26 regulate starch and storage protein synthesis. Plant Physiol 184:1775–1791. https://doi.org/10.1104/pp.20.00984

Wang S, Yang Y, Guo M, Zhong C, Yan C, Sun S (2020b) Targeted mutagenesis of amino acid transporter genes for rice quality improvement using the CRISPR/Cas9 system. Crop J 8:457–464. https://doi.org/10.1016/j.cj.2020.02.005

Witt T, Gidley MJ, Gilbert RG (2010) Starch digestion mechanistic information from the time evolution of molecular size distributions. J Agric Food Chem 58:8444–8452

Woo KS, Seib PA (2002) Cross-linked resistant starch: preparation and properties. Cereal Chem 79:819–825

Xu H, Li XF, Zhang H, Wang LC, Zhu ZG, Gao JP, Li CS, Zhu Y (2020) High temperature inhibits the accumulation of storage materials by inducing alternative splicing of OsbZIP58 during filling stage in rice. Plant Cell Environ 43:1879–1896. https://doi.org/10.1111/pce.13779

Yamamoto MP, Onodera Y, Touno SM, Takaiwa F (2006) Synergism between RPBF Dof and RISBZ1 bZIP activators in the regulation of rice seed expression genes. Plant Physiol 141:1694–1707. https://doi.org/10.1104/pp.106.082826

Yang DC, Wu LY, Hwang YS, Chen LF, Huang N (2001) Expression of the REB transcriptional activator in rice grains improves the yield of recombinant proteins whose genes are controlled by a reb-responsive promoter. Proc Natl Acad Sci U S A 98(20):11438–11443

You H, Liang C, Zhang O, Xu H, Xu L, Chen Y, Xiang X (2022) Variation of resistant starch content in different processing types and their starch granules properties in rice. Carbohyd Polym 276:118742. https://doi.org/10.1016/j.carbpol.2021.118742

Zemach A, Kim MY, Silva P, Rodrigues JA, Dotson B, Brooks MD, Zilberman D (2010) Local DNA hypomethylation activates genes in rice endosperm. P Natl Acad Sci U S A 107:18729–18734. https://doi.org/10.1073/pnas.1009695107

Zeng D, Yan M, Wang Y, Liu X, Qian Q, Li J (2007) Du1, encoding a novel Prp1 protein, regulates starch biosynthesis through affecting the splicing of Wxb pre-mRNAs in rice (Oryza sativa L.). Plant Mol Biol 65:501–509. https://doi.org/10.1007/s11103-007-9186-3

Zeng DC, Liu TL, Ma XL, Wang B, Zheng ZY, Zhang YL, Xie XR, Yang BW, Zhao Z, Zhu QL et al (2020) Quantitative regulation of waxy expression by CRISPR/Cas9-based promoter and 5'UTR-intron editing improves grain quality in rice. Plant Biotechnol J 18:2385–2387. https://doi.org/10.1111/pbi.13427

Zhang G, Hamaker BR (2010) Cereal carbohydrates and colon health. Cereal Chem 87:331–341

Zhang C, Zhu J, Chen S, Fan X, Li Q, Lu Y, Wang M, Yu H, Yi C, Tang S et al (2019) Wx, the ancestral allele of rice waxy gene. Mol Plant 12:1157–1166. https://doi.org/10.1016/j.molp.2019.05.011

Zhao Q, Du XX, Han ZY, Ye Y, Pan G, Asad MAU, Zhou QF, Cheng FM (2019) Suppression of starch synthase I (SSI) by RNA interference alters starch biosynthesis and amylopectin chain distribution in rice plants subjected to high temperature. Crop J 7:573–586. https://doi.org/10.1016/j.cj.2019.03.009

Zhou HJ, Wang LJ, Liu GF, Meng XB, Jing YH, Shu XL, Kong XL, Sun J, Yu H, Smith SM et al (2016) Critical roles of soluble starch synthase SSIIIa and granule-bound starch synthase waxy in synthesizing resistant starch in rice. Proc Natl Acad Sci U S A 113:12844–12849. https://doi.org/10.1073/pnas.1615104113

Zhou H, Xia D, Zhao D, Li YH, Li PB, Wu B, Gao GJ, Zhang QL, Wang GW, Xiao JH et al (2021) The origin of Wx la provides new insights into the improvement of grain quality in rice. J Integr Plant Biol 63:878–888. https://doi.org/10.1111/jipb.13011

Zhu Y, Cai XL, Wang ZY, Hong MM (2003) An interaction between a MYC protein and an EREBP protein is involved in transcriptional regulation of the rice Wx gene. J Biol Chem 278:47803–47811. https://doi.org/10.1074/jbc.M302806200

Zhu F, Bertoft E, Kallman A, Myers AM, Seetharaman K (2013) Molecular structure of starches from maize mutants deficient in starch synthase III. J Agric Food Chem 61:9899–9907. https://doi.org/10.1021/jf402090f

The manufacturer's authorised representative in the EU is Springer Nature Customer Service Centre GmbH, Europaplatz 3, 69115 Heidelberg, Germany. If you have any concerns regarding our products, please contact ProductSafety@springernature.com

Printed and bound by CPI Group (UK) Ltd, Croydon, CR0 4YY
01/10/2026
02225906-0007